T0344648

THE MISSING GENE

THE MISSING GENE

Psychiatry, Heredity, and the
Fruitless Search for Genes

Jay Joseph, Psy.D.

Algora Publishing
New York

ISBN: 0-87586-410-4 (softcover)
ISBN: 0-87586-411-2 (hardcover)
ISBN: 0-87586-412-0 (ebook)

Library of Congress Cataloging-in-Publication Data —

Joseph, Jay.
The missing gene: psychiatry, heredity, and the fruitless search for genes /
by Jay Joseph.
    p. cm.
Includes bibliographical references and index.
ISBN 0-87586-410-4 (trade paper: alk. paper) — ISBN 0-87586-411-2 (hard
    cover: alk. paper) — ISBN 0-87586-412-0 (ebook) 1. Genetic psychology.
    2. Mental illness—Genetic aspects. I. Title.
[DNLM: 1. Biological Psychiatry—trends. 2. Mental Disorders —genetics.
    3. Diseases in Twins—genetics. 4. Genetic Research. WM 102 J83m 2005]
RC455.4.G4J675 2005
616.89'042—dc22

                                                        2005025161

Front Cover:

Printed in the United States

TABLE OF CONTENTS

FOREWORD

David Cohen, Ph.D.

Jay Joseph's book represents a rare achievement. Chapter after chapter of impeccable scholarship lead readers through the most detailed critical examination to date of the assumptions, methods, and conclusions of the field of psychiatric genetics. This is no small feat, given that belief in the genetic origins of mental disorders is one of the most enduring and fervently held among mental health professionals as well as laypersons; and given that a prodigious quantity of studies purporting to demonstrate genetic influences on schizophrenia, depression, bipolar disorder, and ADHD have appeared over the last half century and more.

The book also serves as an exemplar for critical analysis: studies and methods are scrutinized logically, and the ethical, ideological, and political undercurrents of the whole field are also described. And all of it in plain English! Throughout its pages, part critical analysis and part cultural history, Joseph's scholarship never ceases to impress, and his unearthing of inconsistencies, contradictions, and fabrications in the writings of some of the field's leaders makes for fascinating, though at times embarrassing, reading. I felt that all graduate students in the health sciences should read this book if only to understand the topics of conceiving and evaluating scientific studies, especially the fundamental topic of what is needed to reject a null hypothesis.

But there is of course much more to this book. After reading it, I am convinced that progress in psychiatric genetics can only occur if researchers in this field are compelled, either by their sponsors or by their own scientific integrity, to read and try to refute Jay Joseph's charge that such profound errors and biases — in design, measurement, interpretation, and reporting — pervade all the foundational twin and adoption studies that we must take seriously his conclusion that genes for psychiatric disorders are unlikely to exist. And, until researchers and apologists publicly tackle all the elements of Joseph's critique, the reasonable conclusion must be that psychiatric genetics is, through and through, a scientistic, not a scientific enterprise. How else to explain that, with zero direct evidence of genetic influences despite decades of intensive search and with countless failed predictions, the field of psychiatric genetics appears unable to consider that its primary hypotheses may be wrong, or that it simply repeats, mantra style, that the disorders in question are more "complex" than originally anticipated and that future discoveries will confirm this?

Joseph's relentless deconstruction obviously stands in stark contrast to authoritative textbook accounts of a field that has absorbed the energies of thousands of researchers and spent hundreds of millions of dollars, that is commonly seen as standing on the cutting edge of bioscientific research. And therein lies precisely the value of close and painstaking examinations of evidence that this book offers and that are, sadly, rarely undertaken nowadays. Researchers, clinicians, and students increasingly rely on biased, incomplete secondary sources or renowned scientific personalities to inform themselves on the "science" of psychiatric genetics. And what do these sources tell them? As Joseph documents, quote after authoritative quote repeats or even augments the same errors, all the while citing original studies which have obviously not been consulted. And when an unusual textbook author appears to have been curious enough to read the original reports, even gross biases or errors in these reports are almost never discussed. This way, for example, the identical twin concordance rate for "schizophrenia" continues, religiously, to be reported as 50% or higher when from the less biased studies one should report a 20-22% range.

Joseph's book illustrates that the conduct of science rests on fragile pillars, continually swayed by strong forces encouraging the preservation of error, bias, and prejudice, rather than the critical examination of received wisdom. Philosopher of science Karl Popper taught that scientific knowledge progresses when scientists devise tests to refute their cherished assumptions, not when they seek confirmatory evidence of their views. From that insightful perspective, critics, often accused of wasting precious resources or "confusing the public," in contrast to the selfless scientists who "seek to alleviate human suffering," represent the guarantors of progress when scientific inquiry has stultified understanding. Indeed, Joseph's illuminating historical chapters on pellagra and polio, and up-to-the-minute chapter on autism, illustrate how the "genetic disorder" concept is likely delaying discovery of the true causes of a condition at the cost of unnecessary suffering. Autism may be the most "biological" of all disorders diagnosed by psychiatrists. Yet, the sway of flawed twin studies — with their discredited assumption that identical and fraternal twins experience similar environments — remains so powerful that, rather than search actively for likely biological environmental factors that would render *any possible genetic influence* virtually irrelevant, researchers persist on calling autism a "highly heritable genetic disorder."

As Joseph demonstrates, "Never in the recent history of psychiatry have so many definitive claims been made in support of genetics, in the face of so little evidence, as in the case of autism." Indeed, after reading this book it is difficult not to substitute a large list of disorders for "autism" in the preceding sentence and still retain its accuracy.

In 1979, in a small classic entitled *From Genesis to Genocide*, Stephan Chorover argued that theories of human nature and human behavior are very much linked to the control of human behavior. This theme also runs through Jay Joseph's book. He shows that, hard as some leading psychiatric genetic researchers today try to ignore, distort, and misrepresent the origins and implications of their discipline — eugenics and "racial hygiene" — the removal of undesirable persons, not the alleviation of suffering, has always been a goal of the psychiatric genetics field. Cutting through the hype of hyperbolic statements and promises of the "genomic era" for the "understanding and treatment of mental disorders," Joseph's sober account compels us to reconsider afresh just what possible use would be the knowledge on "genetic influence" that researchers are trying so hard, but still failing, to generate? On balance, it is impossible to deny that the impact of the field of psychiatric genetics has been overwhelmingly negative. What precise future does this field promise to us and our descendants?

In our age, when the boundaries between science and marketing have been blurred almost beyond recognition, Jay Joseph has produced a first-rate scientific work and performed an invaluable public service.

David Cohen, PhD
July 17, 2005

Author's Preface

The authors of a 2005 twin study published in a major American political science journal argued that people's inherited tendency toward particular political viewpoints had an important impact on the 2004 United States presidential election.[1] This study, and the subsequent media attention it attracted, is a good indicator of the bizarre lengths that genetic theories, based primarily on twin research, have taken us over the past few decades.

Around the same time, Ronald Kessler and his colleagues published a study in the prestigious *Archives of General Psychiatry* in which they concluded that "about half of Americans will meet the criteria for a *DSM-IV* disorder [Fourth Edition of the American Psychiatric Association's *Diagnostic and Statistical Manual of Mental Disorders*] sometime in their life, with first onset usually in childhood or early adolescence."[2] Critics, of course, could explain this finding on the basis of the DSM's tendency to label a wide range of subjective states and deviant behavior as "mental disorders." The authors of the DSM-IV seem to be saying to us (with apologies to comedian George Carlin), "You are all diseased."

But suppose that Kessler, et al., despite whatever flaws their study may have contained, did show that a sizable portion of Americans will experience some level of chronic or acute psychological dysfunction or distress during their lifetimes. Having established this, the question remains open whether the causes are mainly biological, or whether they reflect the impact of a wide range of psychologically harmful environmental influences that people experience in American society.

An important pillar of the biological argument is genetic theory, which holds that most people diagnosed with psychiatric disorders are genetically predisposed to manifest these disorders. However, if 50% of Americans will develop

---

1. Alford et al., 2005.
2. Kessler et al., 2005, p. 593.

a genetically-based mental disorder, according to genetic theory a sizable percentage of the 50% who *do not* develop a mental disorder nevertheless carry pathological genes. Thus, according to the logic of the American Psychiatric Association's *Diagnostic and Statistical Manual of Mental Disorders*, and currently ascendant theories of genetic causation, a sizable majority of Americans carry pathological genes predisposing them to mental disorders.

Or perhaps not. In this book, I assess the surprisingly shaky foundations of genetic theories in psychiatry in the context of the current crisis in psychiatric molecular genetic research. It was expected that genes for the major psychiatric disorders would have been found by now. However, they have not been found. Indeed, a prominent genetic researcher could write in the July, 2005 edition of the *American Journal of Psychiatry* as follows: "The strong, clear, and direct causal relationship implied by the concept of 'a gene for ...' does not exist for psychiatric disorders. Although we may wish it to be true, we do not have and are not likely to ever discover 'genes for' psychiatric illness."[1] It remains to be seen whether statements such as this will lead to the abandonment of gene searches for psychiatric disorders in the near future.

A belief in the hereditary basis of mental disorders is a very old one. In this book I will show that, following a critical reading of the scientific literature, this age-old belief has little evidence in its favor.

Jay Joseph, Psy.D.
Berkeley, California

---

1. Kendler, 2005a, p. 1250.

LIST OF ABBREVIATIONS

ADD	Attention-Deficit Disorder
ADHD	Attention-Deficit Hyperactivity Disorder
AD-HKD	ADHD/Hyperactivity Disorder
AH	Adoptive Hyperactive
AN	Adoptive Normal
APA	American Psychiatric Association
B1	Chronic Schizophrenia
B2	Acute Schizophrenia
B3	Borderline Schizophrenia
BH	Biological Hyperactive
BN	Biological Normal
BPD	Bipolar Disorder
C	Schizoid or Inadequate Personality
CBCL	Child Behavior Checklist
D1	Uncertain Chronic Schizophrenia
D2	Uncertain Acute Schizophrenia
D3	Uncertain Borderline Schizophrenia
DC	Dichorionic (Twins)
DSM	Diagnostic and Statistical Manual of Mental Disorders
DZ	Dizygotic (Twins)
EEA	Equal Environment Assumption
HD	Huntington's Disease
HGP	Human Genome Project
IQ	Intelligence Quotient
MC	Monochorionic (Twins)
MBD	Minimal Brain Dysfunction
MDI	Manic-Depressive Insanity
MZ	Monozygotic (Twins)
N	Number of Subjects
NAAR	National Alliance for Autism Research
NAS-NRC	National Academy of Sciences-National Research Council
NIMH	National Institute of Mental Health
n.s.	Statistically Non-Significant
OOA	Old Order Amish
PKU	Phenylketonuria
SES	Socioeconomic Status
SPD	Schizotypal Personality Disorder
SSD	Schizophrenia Spectrum Disorder
UCLA	University of California, Los Angeles
WMD	"Weapons of Mass Destruction"

Chapter 1. Introduction. The Twin Method: Science or Pseudoscience?

In 2003, American psychiatry heralded the beginning of the "genomics era" by proclaiming that gene discoveries "will change the way we treat psychiatric patients,"[1] will lead to the development of "targeted therapies," and will make possible the creation of "pharmacogenomics," which will use genetic information to predict the efficacy of psychiatric drugs. Finally, "one of the most important consequences of genomics will be to individualize treatment by allowing a clinician to tailor therapy on the basis of the unique genotype of each patient rather than the mean responses of groups of unrelated patients."[2]

A problem with these claims is that, despite having searched for over two decades with increasingly sophisticated technology, researchers have found no genes that cause the major psychiatric disorders. This might have compelled psychiatry to rethink the position that its disorders are caused by faulty genes, but this has not occurred. A major reason is that a belief in the genetic basis of mental disorders is a cornerstone of contemporary psychiatry. Yet, the evidence supporting this position is stunningly weak. In this book I show why this is so, and why there must be a massive rethinking of the role of genetics in causing psychiatric disorders. As we will see, it is very possible that genes for diagnoses such as schizophrenia, ADHD, autism, and bipolar disorder are not, as researchers often write, "elusive," but that they are nonexistent. Thus, the "genomics era" in psychiatry may develop into little more than a historical footnote.

In my previous book, *The Gene Illusion: Genetic Research in Psychiatry and Psychology under the Microscope*, I argued that the foundations of genetic theories —

---

1. Insel & Collins, 2003, p. 618.
2. Ibid., p. 618.

9

family, twin, and adoption studies — are based on poor methodology, bias, and a reliance on unsupported theoretical assumptions. My focus was split between research on psychiatric disorders such as schizophrenia and research into "normally distributed" psychological traits such as IQ and personality. In the present book I focus on genetic research in psychiatry, and cover several areas mentioned only briefly or not at all in *The Gene Illusion*. Moreover, some areas, such as molecular genetic research, have seen many changes and are in constant need of update. Two additional chapters focus on how genetic research in psychiatry has been misreported by influential secondary sources. In fact, the misreporting of psychiatric genetic research is a running theme throughout the entire book.

The early part of the 21rst century is characterized by the *geneticization* of diseases and human differences, a term coined by Abby Lippman in 1992.[1] As Lippman subsequently described it, geneticization "capture[s] the ever growing tendency to distinguish people from one another on the basis of genetics; to define most disorders, behaviors, and physiological variations as wholly or in part genetic in origin."[2] This is particularly evident in psychiatry, where psychological, interpersonal, social, and cultural understandings of human suffering have taken a back seat to biological and genetic theories.

The field of *psychiatric genetics* was founded by Ernst Rüdin and his German colleagues in the early part of the 20th century.[3] German psychiatric geneticists used family and twin studies in an attempt to establish the genetic basis of psychiatric disorders. Their primary goal was to promote the eugenic program (called "racial hygiene" in Germany) of curbing the reproduction of people they viewed as carrying the "hereditary taint of mental illness," and in the process its leaders became willing accomplices of Hitler and the Nazis (see Chapter 6).[4] Rüdin and his colleagues played an important role in training people to conduct psychiatric genetic research, and people from other European countries came to Munich to study under them at the Genealogical-Demographic Department of the Kaiser-Wilhelm Institute of Psychiatry.

Contemporary psychiatric geneticists trace their discipline back to Rüdin's "Munich School," often utilizing the same methods and statistical formulations. Today, psychiatric geneticists investigate the causes of mental disorders in order to better treat and prevent them.[5] Unlike researchers in the previous era, they usually avoid discussions of eugenics in relation to their findings. The implications of their theories, however, are obvious, and they often promote the use of "genetic counseling," or recommending to people with certain psychiatric diagnoses not to conceive and bear children. The field of *behavior*

---

1. Lippman, 1992.
2. Lippman, 1998, p. 64.
3. See Proctor, 1988, Weindling, 1989 for a more detailed account of the birth of German psychiatric genetics.
4. Joseph, 2004b, Chapter 2; Müller-Hill, 1998a; Proctor, 1988.
5. Faraone, Tsuang, & Tsuang, 1999.

genetics uses methods similar to psychiatric genetics (e.g., family, twin, and adoption studies), but focuses on assessing genetic influences on "continuously distributed" psychological traits such as personality and IQ, and to a lesser extent on psychiatric disorders.

In this book I discuss behaviors and mental states classified by psychiatry as "mental disorders" or "mental illnesses." These include "schizophrenia," "bipolar disorder" "attention-deficit hyperactivity disorder," and so on. I use these terms only to show that, even if discrete mental disorders actually exist as valid and reliable (biological) entities, as mainstream psychiatry claims, there is little evidence that they have a genetic basis. A far better way of understanding human suffering and abnormal behavior is captured in psychologist Richard Bentall's 2003 "Post Kraepelinian Manifesto" (Emil Kraepelin was the Swiss pioneer of modern psychiatry). In Bentall's view, "There is no clear boundary between mental health and mental illness. Psychological complaints exist on a continuum with normal behaviors and experiences. Where we draw the line between sanity and madness is a matter of opinion." Bentall continued that there "are no discrete mental illnesses. Categorical diagnoses fail to capture adequately the nature of psychological complaints for either research or clinical purposes."[1]

A BRIEF INTRODUCTION TO THE CHAPTERS

Chapter 2 examines evidence supporting the claim that attention-deficit hyperactivity disorder (ADHD) is influenced by genetic factors. Here, I highlight the invalidating flaws of family, twin, and adoption research in this area, and argue that there is little scientifically acceptable evidence supporting a genetic basis for the condition. I discuss ADHD molecular genetic research in Chapter 11.

Chapter 3 takes a close look a crucial aspect of schizophrenia adoption research: the "schizophrenia spectrum" concept. Expanding the definition of schizophrenia was an essential factor in the famous Danish-American schizophrenia adoption studies' conclusions in favor of genetics. However, we will see that the spectrum concept does not hold up to critical examination. My analysis is relevant today for two main reasons: (1) because these adoption studies remain the most frequently cited evidence in support of the genetic theory of schizophrenia, and (2) as an example of how a close examination of one aspect of genetic research can uncover serious and invalidating flaws.

Chapter 4 engages in a bit of historical speculation by predicting the results of twin and adoption studies of pellagra, an early 20th century disease ultimately discovered to be caused by a vitamin deficiency linked to malnutrition. My analysis suggests that the results from twin and adoption studies, often cited in

---

1. Bentall, 2003, p. 143.

support of genetic influences on psychiatric disorders, can be explained on the basis of environmental factors, in addition to methodological flaws and biases.

Chapter 5 looks closely at 43 psychiatry and psychology textbooks' discussions of schizophrenia adoption research. As I demonstrate, this body of research has been systematically misrepresented in these textbooks, and the original investigators' conclusions usually are taken at face value at the expense of a critical examination of their studies' glaring weaknesses. Chapter 5 will be of particular interest to professionals and academics, who often rely on textbooks for information on genetic research in psychiatry.

Chapter 6 takes a close look at perhaps the most relied upon secondary source in the "Genetics of Schizophrenia" area: Irving Gottesman's award-winning 1991 *Schizophrenia Genesis*. This book contains a diagram, reproduced or referred to in many contemporary psychiatry and abnormal psychology textbooks, listing various kinship risk factors for schizophrenia. I show that this book contains many errors and omissions, all in the direction of leading its readers to the mistaken conclusion that schizophrenia is strongly influenced by genetic factors. Although Gottesman is a well-known and respected psychologist, he provides a potentially misleading account of the topic to a new generation of students and professionals.

Chapter 7 examines the evidence supporting autism as a genetically-influenced disorder. Indeed, autism is often regarded as the psychiatric disorder most strongly influenced by genetics. Strikingly, however, we will see that the evidence, consisting mainly of four small twin studies, is plausibly explained on non-genetic grounds. In contrast to other psychiatric disorders, there is evidence suggesting that autism is caused by biological factors. However, "biological" is not the same as "genetic," and the current emphasis on alleged genetic factors detracts and drains away resources from research on potentially relevant biological causes, such as prenatal and postnatal exposure to mercury and other harmful substances. I discuss autism molecular genetic research in Chapter 11.

Chapter 8 focuses on an astonishing 1942 debate in the *American Journal of Psychiatry*. There, neurologist Foster Kennedy argued in favor of "euthanizing" (that is, killing) "defective" and "feebleminded" people. This had already occurred in Nazi Germany, where, even before the Holocaust, tens of thousands of "hereditarily defective" people were exterminated with the active participation of psychiatrists.[1] In response, child psychiatrist Leo Kanner argued against killing because, among other reasons, there would remain fewer people to perform society's dirty work. Kennedy and Kanner were followed by an anonymous editorial leaning towards Kennedy's position in favor of killing, whose authors called upon psychiatrists to focus their attention on the "morbid" attachment of parents opposed to the "disposal by euthanasia of their idiot offspring."

---

1. Lifton, 1986; Müller-Hill, 1998a.

Chapter 9 surveys a large body of literature cited in support of the equal environment assumption (EEA) of the twin method. Although I show later in Chapter 1 that the EEA is untenable regardless of how twin researchers have defined it, twin method results are accepted without question in mainstream contemporary psychiatry. A major reason is that leading twin researchers argue that the EEA is supported by a number of empirical studies. In Chapter 9, I critically review these studies and conclude that they do little to uphold the validity of the EEA. Moreover, I argue that the twin method is invalidated on its face by the fact that, as most twin researchers now recognize, identical twins experience much more similar environments than fraternals.

In *Chapter 10* I examine evidence put forward in support of genetic influences on bipolar disorder (manic depression). As with schizophrenia, ADHD, and autism, I show that the available evidence from kinship research lends little support to genetic theories of causation. I also show that many authoritative secondary sources have made incorrect or misleading claims about the results of bipolar disorder genetic research.

Chapter 11 consists of an up-to-the-moment assessment of psychiatric molecular genetic research. The most remarkable result of this research is that, despite over two decades of sustained work, genes for the major psychiatric disorder have not been discovered. Virtually all previous claims in favor of gene findings have failed replication attempts in subsequent studies. The standard explanation has been that many genes of small effect cause these disorders, and that we are on the threshold of gene discoveries. In this chapter I offer an alternative explanation, which relates to a major theme of this book: *It is unlikely that genes for the major psychiatric disorders exist.* Having shown in previous chapters that studies of families, twins, and adoptees are faulty, I analyze molecular genetic research in schizophrenia, ADHD, autism, and bipolar disorder. I argue that the fruitless search for genes may be the result of psychiatry's misplaced faith in the results of these previous kinship studies. Molecular genetic research in psychiatry is reaching the crisis stage as negative results continue to pile up, and researchers may be more open to questioning whether the genes they are looking for actually exist.

THE CLASSICAL TWIN METHOD

"The knowledge that certain diseases run in families," observed Joseph Alper, "is thousands of years old."[1] Today, it is widely understood that a condition "running in families" can be explained by any number of environmental factors related to the physical and psychological environments shared by family

---

1. Alper, 2002, p. 17.

members. For this reason, most psychiatric geneticists recognize that family studies are unable to disentangle the possible influences of genes and environment. Thus twin studies, which allegedly *are* able to separate these potential influences, are the most frequently cited evidence supporting genetic theories in psychiatry. However, we will see that the main premise of twin studies, upon which genetic theories are based, is faulty.

The main tool of psychiatric genetics is the "classical twin method," more commonly known as "the twin method." As seen in Figure 1.1, the twin method compares the concordance rates or correlations of reared-together identical twins (sharing 100% genetic similarity), versus the same measures of reared-together same-sex fraternal twins (averaging 50% genetic similarity). Based on the assumption that both types of twins experience the same kinds of environments, known as the "equal environment assumption" or "EEA," twin researchers attribute a statistically significant higher concordance rate among identical versus same-sex fraternals to genetic factors. Twins are said to be *concordant* when both members of a pair are diagnosed with the same disorder and *discordant* when only one is diagnosed.

Perhaps at this point I should briefly outline the concept of "statistical significance," which is central to genetic research. By convention, a result is considered statistically significant when its occurrence would be expected by chance less than 5 times in 100 (probability < 5%, or 0.05). If the probability value (often shortened to "p-value") that a finding occurred by chance exceeds 5% in a comparison between groups, the researcher should conclude that there is no difference between the groups. (The logic here is similar to that which applies when we are flipping a coin 10 times. Although we expect heads to come up 5 times with a fair coin, by chance alone it might come up 7 times, or perhaps 3 times. However, if it comes up heads 9-10 times, or 0-1 times, we might suspect that the coin has been altered in some way.)

Using twin research as an example, suppose a research team performs a schizophrenia twin study and finds that 7/17 (41%) identical pairs are concordant for schizophrenia, versus a same-sex fraternal rate of 2/15 (13%). Although the percentage difference appears large, the researchers use a test of statistical significance (in this case, a Fisher's Exact Test) and find a p-value of .086. They would therefore conclude, having failed to find a p-value below the conventional .05 threshold, that their study found no concordance rate difference between identical and fraternal twins. That is, they failed to uncover evidence that would lead them to reject the hypothesis that identical and fraternal twin concordance rates are the same (known as the "null hypothesis" in statistical language). The practice of using small representative samples of a population as a basis for making claims about the population as a whole uses the concept of *inferential statistics*. If results from the sample are unlikely to have been obtained by chance (statistical significance has been reached), researchers make inferences about this finding to the entire population.

Figure 1.1

THE CLASSICAL TWIN METHOD AND ITS ASSUMPTIONS

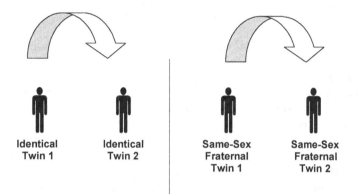

Identical Twin 1	Identical Twin 2	Same-Sex Fraternal Twin 1	Same-Sex Fraternal Twin 2

Identical Twins	*Same-Sex Fraternal Twins*
Share 100% of the same genes.	Share on average 50% of the same genes.
Reared together in the same family.	Reared together in the same family.

The greater resemblance of identical (MZ) versus same-sex fraternal (DZ) twins is attributable to genetic factors and is generalizable to the non-twin population, *assuming that all of the following are true:*

- There are only two types of twins, identical and fraternal.
- Investigators are able to reliably distinguish between these two types of twins.
- The risk of receiving the diagnosis is the same among twins and non-twins (generalizability).
- The risk of receiving the diagnosis is the same among individual identical twins as a population, versus individual fraternal twins as a population.
- Identical twin pairs and same-sex fraternal twin pairs experience the same emotional and psychological bond with each other, as well as experiencing roughly the same social, treatment, and physical environments (known as the "equal environment assumption" or "EEA").

Adapted with revisions from Joseph, 2004b, p. 22.

Returning to our imaginary schizophrenia twin study, although a larger sample showing the same concordance rates would have produced statistically significant results (p < .05), we cannot assume that this would have occurred, and it might be necessary for the researchers to obtain a larger twin sample for the next study. We must also be aware that even if a study finds a statistically significant difference between groups, it has not necessarily shown what *causes* the difference. In the case of twin concordance for psychiatric disorders, a significantly higher identical versus fraternal concordance rate could be caused by identical twins' greater genetic similarity, as twin researchers maintain, or to identicals' greater environmental similarity, as maintained by critics of the twin method. As we will see throughout this book, genetic theories in psychiatry are based in large part on the claim that identical versus same-sex fraternal twin concordance rate differences are caused by genetic factors.

Although the twin method depends on additional assumptions (listed in Figure 1.1), the equal environment assumption has been the main area of contention between twin researchers and their critics. Identical and fraternal twins are usually referred to as monozygotic (MZ) and dizygotic (DZ), respectively, in the scientific literature (as seen in many quotations in this book). I use "identical" and "fraternal," here. Unless otherwise mentioned, all twin studies discussed in this book are of twins reared together in the same family. For an analysis of problems in reared-apart twin research (other than the brief critique below), I refer readers to previous publications.[1]

Twin studies constitute the main evidence cited in support of genetic influences on psychiatric disorders and all types of psychological traits, as counterintuitive as the genetic basis of some traits may seem.[2] In *The Gene Illusion*, I discussed important problems in applying the twin method to psychiatric disorders, which I summarize here.

Methodological problems with studies utilizing the twin method have included (1) the acceptance of unsupported theoretical assumptions; (2) investigator bias in favor of genetic conclusions; (3) the lack of an adequate and consistent definition of the trait or condition under study; (4) the use of nonblinded diagnoses; (5) diagnoses that were made on the basis of sketchy information; (6) inadequate or biased methods of zygosity determination (whether a pair is identical or fraternal); (7) the fact that twins, and identical twins in particular, might have received similar hospital diagnoses because hospital psychiatrists viewed them as sharing a common genetic heritage; (8) the unnecessary use of age-correction formulas; (9) the use of non-representative sample populations; and (10) the lack of adequate descriptions of the methods.

The main problem, however, has been the unwarranted acceptance of the equal environment assumption. From the development of the twin method in the mid-1920s until the early 1960s, twin researchers defined the EEA, without qual-

---

1. See Farber, 1981; Joseph, 2001c, 2004b; Kamin, 1974; Kamin & Goldberger, 2002; Taylor, 1980.
2. In 2004, for instance, a group of researchers claimed that twin studies found important genetic influences on "perfectionism," and another group claimed genetic influences on "breakfast eating patterns" (Keski-Rahkonen et al., 2004; Tozzi et al., 2004).

ification, as the assumption that identical and fraternal twins share the same types of behavior-influencing, physical, and treatment environments. I have called this the "traditional EEA definition."[1] However, most research assessing these environments has shown, as critics have charged since the 1920s, that identical twins spend more time together, more often have the same friends, are treated more similarly by parents and others, and so forth.[2] Moreover, identicals share a closer emotional bond than fraternals, and more often view themselves as being two halves of the same whole (i.e., they experience what some psychologists call identity confusion[3]).

Faced in the early 1960s with mounting evidence against the EEA, twin researchers should have abandoned the twin method as being unable to disentangle possible genetic and environmental influences on psychiatric disorders. Today, most twin researchers concede the fact that identical twins experience more similar environments than fraternal twins. However, they continue to uphold the twin method as a valid instrument for the detection of genetic influences.

Twin researchers accomplished this by *redefining* the EEA. The new definition, which today predominates, has been called the "equal trait-relevant environment assumption" by behavior geneticists.[4] Here, I will shorten this term to read "trait-relevant EEA." An early example of this crucial change in definition was put forward in 1966 by twin researchers Irving Gottesman and James Shields, who wrote that, to invalidate schizophrenia twin studies, the "environments of MZ twins [must be] systematically more alike than those of DZ twins in features which can be *shown* to be of etiological significance in schizophrenia."[5] More recently, psychiatric genetic twin researcher Kenneth Kendler and his colleagues have defined the EEA in the trait relevant sense:

> The traditional twin method, as well as more recent biometrical models for twin analysis, are predicated on the equal-environment assumption (EEA) — that monozygotic (MZ) and dizygotic (DZ) twins are equally correlated for their exposure to environmental influences *that are of etiologic relevance to the trait under study* [emphasis added].[6]

Trait-relevant EEA proponents recognize that identical twins experience more similar environments than fraternals, but argue or imply that the burden of proof for demonstrating that identical and fraternal twins experience dissimilar trait-relevant environments falls not on twin researchers but on twin method *critics*.

---

1. Joseph, 2004b.
2. Ibid.
3. Jackson, 1960.
4. Carey & DiLalla, 1994.
5. Gottesman & Shields, 1966a, pp. 4-5. Emphasis in original.
6. Kendler, Neale, et al., 1993, p. 21.

Although now widely used, the trait-relevant EEA rests on shaky theoretical foundations. Proponents of the twin method are rarely able to pinpoint the environmental factors relevant to the condition they are studying, in spite of their belief that most psychiatric disorders require an environmental "trigger" in combination with a genetic predisposition. As mentioned, twin researchers get around this problem by charging critics with responsibility for showing that identical and fraternal twin environments differ on trait-relevant dimensions (see the discussion below).

What twin researchers fail to recognize, however, is that the trait-relevant EEA has transformed the twin method into little more than a special type of family study. The reason is that the comparison groups in both family studies and the twin method (in family studies, the general population or a control group in the twin method, fraternal twins) are acknowledged to experience environments different to those of the experimental groups (in family studies, the families of people diagnosed with the disorder in question; in the twin method, identical twins). Although contemporary twin researchers retain the trait-relevant requirement for the twin method, but *not* for family studies, virtually every argument they make in defense of drawing genetic inferences from identical-fraternal comparisons they could also make in defense of drawing genetic inferences from family studies. Yet, strangely, these investigators uphold the validity of the EEA and the twin method even as they admit that genetic conclusions based on family studies are confounded by environmental factors. Although most psychiatric geneticists recognize that the results of family studies prove nothing about genetics because "all mechanisms that could lead to a familial clustering of disease should be considered,"[1] they arbitrarily fail to consider all mechanisms that could lead to higher identical versus fraternal twin concordance rates for psychiatric conditions.

The results of "twins reared-apart" (TRA) studies are sometimes put forward in support of the EEA and the twin method, based on the claim that these studies have shown that the status of being reared together in the same home does not lead to much greater identical twin resemblance than if identical twins had been reared apart in different homes. However, these studies are plagued by invalidating problems which include (1) the dubious "separation" of twins, who in many cases grew up together and had quite a bit of contact over much of their lives; (2) the similarity bias of the samples; (3) researchers' failure to publish or share raw data and life history information for the twins under study (Thomas J. Bouchard, Jr.'s Minnesota TRA study, for example[2]), and (4) the impact that the researchers' bias in favor of genetic explanations had on the interpretation of their results. The main problem with TRA studies such as Bouchard's, however, is that the investigators mistakenly compared reared-apart

---

1. Faraone, Tsuang, & Tsuang, 1999, p. 21.
2. Bouchard et al., 1990.

identical twins ("monozygotic twins reared-apart," or "MZAs") to *reared-together* identicals — thereby failing to control for the fact that both sets share several important environmental similarities. These include common age (birth cohort), common sex, similar appearance, and similar political, socioeconomic, and cultural environments. (In addition to MZAs, the Minnesota study used reared-apart fraternals as a comparison group, but they also share several of the environmental influences experienced by MZAs. The investigators also attempted to correct MZA correlations for age and sex effects, but their adjustments were inadequate and unclear.) Thus, behavior geneticists and TRA researchers such as Bouchard and his colleagues used the wrong control group, leading to their erroneous conclusions in favor of genetics. A scientifically acceptable study would compare the resemblance of a group consisting of MZAs reared apart from birth and unknown to each other, versus a control group consisting not of reared-together identical twins, but of *biologically unrelated pairs of strangers* sharing all of the following characteristics: they should be the same age, they should be the same sex, they should be the same ethnicity, the correlation of their rearing environment socioeconomic status should be similar to that of the MZA group, they should be similar in appearance and attractiveness, and the degree of similarity of their cultural backgrounds should be equal to that of the MZA pairs. Moreover, they should have no contact with each other until after they are evaluated and tested. After concluding such a study, we might find that the biologically-unrelated pairs correlate similarly to MZAs, which would suggest that MZA correlations are the result of environmental influences. Because no study of this type has ever been attempted, and because of the major flaws and biases in the studies that have been undertaken, we can draw no valid conclusions in support of genetic influences on psychological trait variation, or in support of the twin method's validity, from reared-apart twin studies published to date.

Twin researchers' recognition that identical twins experience more similar environments and treatments than fraternals invalidates genetic interpretations of identical-fraternal comparisons, for the exact same reason that genetic interpretations of family studies are invalid. Therefore, there is no reason to accept that the twin method measures anything other than the more similar environments of identical versus fraternal twins (plus error), and all conclusions in favor of genetic influences on psychiatric disorders derived from the twin method must be disregarded.

A CLOSER LOOK AT KENDLER'S DEFENSE OF THE EEA

Since the mid-1980s, Kenneth S. Kendler and his colleagues have published a large quantity of psychiatric genetic twin data derived from their work with the Virginia Twin Registry. Kendler has been one of the world's leading twin researchers and psychiatric geneticists for over two decades, and heads the Virginia Institute for Psychiatric and Behavioral Genetics. He is a highly honored psychiatric investigator, and is currently co-Editor of *Psychological Medicine*, as well as a member of the Editorial Boards of *Archives of General Psychiatry*, and *The American Journal of Psychiatry*. Since the 1990s he has also been involved in molecular genetic studies of psychiatric disorders.

In most of the twin studies with which Kendler has been associated, he and his colleagues concluded in favor of important genetic influences on the psychiatric disorders in question. But as Kendler well understands, these conclusions are based on the acceptance of the equal environment assumption. Kendler published a 1983 article in the *American Journal of Psychiatry*, entitled "Overview: A Current Perspective on Twin Studies of Schizophrenia," where he presented a detailed theoretical and empirical argument in defense of the EEA and the twin method. Although he has made some modifications since then, this article remains one of the few detailed defenses of the EEA ever published.[1]

Kendler began by outlining the traditional EEA definition which states that, because identical and same-sex fraternal twins "share environmental factors to approximately the same extent, differences in concordance between the two twin types must be due to the influence of genetic factors."[2] Although the critics' argument that identical twins "share more of their social environment" than do same-sex fraternal twins is true, wrote Kendler, it would be "premature" to invalidate the twin method.[3] The validity of the EEA, he argued, comes down to a pair of competing hypotheses explaining the causal relationship between identical twins' greater behavioral resemblance and their more similar social environments. The *first hypothesis*, put forward by the critics, "is that the similar phenotypes [e.g., behavior, psychiatric disorders] in monozygotic twins are caused by their similar social environment." The *second hypothesis*, put forward by Kendler and others, "is that the similar phenotypes of monozygotic twins are caused by their genetic similarity. The similar phenotypes of these twins are then responsible for creating their similar social environment."[4]

---

1. Kendler, 1983. Recent books promoting psychiatric and behavior genetic research, such as Jang's (2005) *The Behavior Genetics of Psychopathology*, and Carey's (2003) *Human Genetics for the Social Sciences*, provide little in the way of new arguments in support of the EEA.
2. Kendler, 1983, pp. 1413-1414.
3. Ibid., p. 1414.
4. Ibid., p. 1414.

Two "Competing Hypotheses" Discussed by Kendler

Hypothesis #1 (Critics)

The greater behavioral resemblance and greater concordance for psychopathology of identical vs. fraternal twins is caused by identicals' more similar treatment, and by the greater similarity of their social environments.

Critics' Conclusion: The twin method is unable to disentangle possible genetic and environmental influences on behavior and psychopathology.

Hypothesis #2 (Twin Researchers)

The greater behavioral resemblance and greater concordance for psychopathology of identical vs. fraternal twins is caused by identicals' greater genetic similarity. Therefore, identical twins create more similar environments for themselves than do fraternals.

Twin Researchers' Conclusion: The twin method is a valid instrument for the detection of genetic influences on behavior and psychopathology.

Although Kendler recognized that "if the first hypothesis is correct, then critics of twin studies are justified in their criticism," he argued that the second hypothesis is more plausible and that the twin method is, therefore, a valid instrument for the detection of genetic influences on psychiatric disorders. In support of the second hypothesis, Kendler cited a body of empirical "EEA-test" literature. (I review these and subsequent studies in Chapter 9, where I show that they do little to uphold the EEA.) Thus, for Kendler, the validity of the EEA and the twin method comes down to the position that the more similar physical, social, and treatment environments experienced by identical versus fraternal twins are caused by the former's greater genetic similarity.

But, wait! No further analysis is necessary to invalidate the twin method! Apparently, Kendler did not realize that the *reason* identical twins experience more similar environments than fraternal twins — be it environmental or genetic — *is completely irrelevant in assessing the validity of the EEA.* The only relevant question is *whether* — not why — identical twins experience more similar environments.[1]

Kendler attempted to bypass this obvious problem, however, writing that the evidence suggests that:

> the behavioral similarity of monozygotic versus dizygotic twins cannot be ascribed to differences in treatment of the twins by the social environment.... The behavioral similarity of monozygotic twins appears not to result from the similarity in social environment of the twins. Rather, the available evidence suggests that the similarity

---

1. Joseph, 2004b.

of the social environment of monozygotic twins is the result of the behavioral similarity of the twins.[1]

Thus, Kendler acknowledged that identical twins experience more similar environments than fraternals, which he believed is caused by their greater genetic similarity. In other words, for Kendler, identical twins are treated more alike because they behave more alike (second hypothesis), as opposed to the critics' position that they behave more alike because they are treated more alike and experience more similar environments (first hypothesis). The validity of the twin method, for Kendler, rests on the direction of causality: "Although the similarity in environment might make MZ twins more similar, the similarity in behavior of MZ twins might *create* for themselves more similar environments."[2] Kendler approaches these two statements as if they were counterposed, when in fact they are potentially complementary because identical twins could still "create" environmental conditions or exposure leading to greater concordance for psychiatric disorders or physical disease. For example, if identical twins are more genetically predisposed to enjoy sunbathing than fraternal twins, identical twins may well show much higher concordance for skin cancer than fraternals. However, this does not mean that skin cancer is a genetically-based disease.

Another problem with Kendler's argument, which I have called the "twins create their environment theory,"[3] is that it ascribes to parents — but not to children — the ability to alter behavior on the basis of the actions of other family members.[4] Moreover, Kendler's argument is circular because the evidence for twins' behavioral similarity being caused by genetics is implicitly derived from the results of previous twin studies.

Finally, Kendler maintained that "the behavioral similarity of monozygotic versus dizygotic twins cannot be ascribed to differences in treatment of the twins by the social environment," even though his second hypothesis allows that more similar treatment might create more similar twins, but that the twin method remains valid because twins supposedly "are responsible for creating their own environment." In truth, there is little difference between the "first" and "second" hypotheses, because *both* imply that greater identical versus fraternal concordance is related to the more similar environments experienced by identical twins. Moreover, the second hypothesis renders the "trait-relevant" argument irrelevant in the sense that, even if critics were able to demonstrate that identical twins experience more similar trait-relevant environments than fraternals, Kendler and his co-thinkers would still argue for the twin method's validity on the basis of identical twins having "created" their more similar environments.

---

1. Kendler, 1983, p. 1416.
2. Kendler, 1987, p. 706. Emphasis in original.
3. Joseph, 1998.
4. Joseph, 1998, 2004b.

Another example of how genetically oriented researchers use "the twins create their environment" argument in support of the EEA is found in the introductory chapter of the 2002 edition of *The Genetic Basis of Common Diseases*. According to King, Rotter, and Motulsky, the authors of the chapter and editors of the volume, "A higher concordance rate in MZ than in DZ twin pairs (especially like-sexed DZ pairs) indicates that a significant part of the familial aggregation [of a disease] is due to genetic factors...." However,

> A qualification is that it can sometimes be difficult to disentangle heredity from environment, because MZ twins, likely because of their genetic identity, tend to select similar environments. Coronary heart disease, celiac disease, inflammatory bowel disease, diabetes, and schizophrenia are examples of disorders in which the concordance rate is higher in MZ than in DZ twin pairs.[1]

There are at least three problems with this argument:

1. The argument is circular, because the assumption that identical twins select similar environments due to their more similar genetic makeup is based largely, though implicitly, on the results from previous twin studies of behavior and personality.

2. As mentioned above, the idea that identical twins are able to select similar environments ascribes to parents, but not to twins, the ability to change and adjust their behavior to the needs of others.

3. Even if identical twins' more similar environments are due to their greater genetic identity, higher identical versus fraternal concordance for the diseases, syndromes, conditions, and putative diseases listed by King et al. could still be completely caused by environmental factors (see the example of skin cancer I mentioned earlier).

Although King and colleagues devoted only a few sentences to the validity of the equal environment assumption, the next 1,000+ pages of their edited volume cited twin studies as the main evidence supporting a genetic basis for the medical and psychiatric disorders in question.

Let's return to Kendler's position that identical twins' greater behavioral resemblance "cannot be ascribed to differences in treatment of the twins by the social environment." Another problem that Kendler overlooked is that his theory must generalize to mean that no one's behavior — neither twins' nor non-twins' — is influenced by their social environment. The reason, of course, is that each twin is an individual human being receiving treatment by the social environment. But, according to the logic of Kendler's EEA theory, the social environment experienced by each *individual* twin cannot alter his or her behavior.

The position that social and cultural influences *do* influence human behavior is so obvious that I will refrain from burdening myself and readers with a discussion of research supporting it. It's far simpler to cite some concrete examples that don't even address the importance of family (psychodynamic)

---

1. King et al., 2002a, p. 12.

rearing environment, since there are untold behavior-modifying influences in modern society. For example, what is a *law* if not a means of controlling human behavior through a system of rewards and punishment? Whether or not you break a particular law, such as speeding on an open highway or stealing an MP3 player, is at least partly influenced by whether you think it is worth the chance of receiving an expensive traffic ticket or being arrested for shoplifting. And what about income tax? Do people file honest returns solely because of moral virtue or patriotism, or do they also fear an audit and possible punishment if they don't? A small fraction of the behavior-influencing agents in society might include abortion laws, apartment rules, athletic scholarships, being a crime victim, billboards, billy clubs, birth control pills, bubble gum flavored tooth-paste, bullies, cell phones, child labor laws, condoms, corporal punishment, cosmetic surgery, coupons, court marshals, culture, cruise missiles, diplomatic immunity, ethics codes, fashion, fences, freedom of speech, God, guns, "gunboat diplomacy," health insurance, heaven, hunger, income, jaywalking tickets, military dictators, muggers, mutual fund performance, newspaper headlines, oil prices, paychecks, peer pressure, penicillin, police, promotional opportunities, racism, radio, report cards, security guards, sexism, social security, spies, stock market crashes, surveillance cameras, strikes, strikebreakers, teachers, television, the devil, the Internet, the price of gasoline, the size of the lottery jackpot, the Ten Commandments, timecards, "time outs," traffic lights, traffic signs, trends, warnings by the Surgeon General, word of mouth, and zero percent financing.

However, for Kendler's theory to hold, we must deny that these factors influence human behavior. We must believe that humans are hardwired at birth to behave in a predetermined way, uninfluenced by the social environment and by society's system of rewards and punishments.

When Kendler says that critics must demonstrate that twins' environments are "trait-relevant," what he is really saying is that *everyone's* environment is trait-*ir*relevant.

Let's look to the advertising industry to illustrate another example of this untenable position. Advertising agencies and their clients understand that messages created in certain ways for particular audiences can indeed "influence behavior." If Coca-Cola pays $2 million for a thirty-second Super Bowl television commercial, they certainly expect that thousands or millions of people will change their cola buying *behavior*. Furthermore, modern advertising not only targets behavior but seeks to change the way people feel about themselves in society. It's not enough to tout the effectiveness of the latest dandruff shampoo; one also needs to make people feel embarrassed, if not humiliated, that they have dandruff. Self evidently, if the environment does not influence twins' and others' behavior and psychology, as Kendler implies, then corporations would not spend untold billions of dollars each year on advertising. Twin researchers and their EEA test studies (see Chapter 9) resemble the proverbial "blind men" who cannot see the massive "elephant" of environmental influences on the behavior of both twins and non-twins alike.

Kendler went on to assert that:

> Differential treatment of twins by their social environment does not appear to be responsible for the greater similarity of monozygotic versus same-sex dizygotic twins for such characteristics as intelligence, personality, and language and perceptual skills. Therefore, it seems unlikely that such differential treatment could be responsible for the greater concordance for schizophrenia in monozygotic versus same-sex dizygotic twins.[1]

But what if we keep non-twins in mind and rephrase this statement to read: "Differential treatment of people by their social environment does not appear to be responsible for differences in characteristics such as intelligence, personality, and language and perceptual skills. Therefore, it seems unlikely that differences in treatment could be responsible for schizophrenia." The logic in both statements is the same, but the first statement's folly becomes obvious when we strip away the confusion created by the special situation of twins. Moreover, although many genetically-oriented commentators would allow for environmental effects playing some role in schizophrenia when combined with a genetic predisposition, the logic of Kendler's argument implies that the environment has no influence on schizophrenia. However, by 2005 Kendler would recognize that "a large body of descriptive literature shows convincingly that cultural processes affect psychiatric illness,"[2] and that the "impact of genetic risk for psychiatric disorders or drug use can be modified by the rearing environment...by stressful life experiences...and exposure to cultural forces."[3]

According to Kendler's EEA theory, identical twins experience similar environments because they share an identical genetic makeup. However, if identical twins have the same genetic makeup and experience necessarily equal within-pair environments, and psychopathology is the result of genetic plus environmental influences (or of their interaction), identical twin pairs should approach 100% concordance for psychiatric disorders. Moreover, they should be virtual "personality clones" of each other. Yet, as most people are aware, in many cases identical twins differ substantially in values, interests, tastes, psychiatric status, and so on. Indeed, we will see in Chapter 6 that the pooled identical twin schizophrenia concordance rate from the more methodologically sound studies is only about 20-22%.[4]

Kendler missed another essential point in his assessment of the relationship between twins' treatment and schizophrenia. It is not simply a matter of parents and others treating a pair of twins more similarly in a potentially psychologically

---

1. Kendler, 1983, p. 1416.
2. Kendler, 2005c, p. 436.
3. Ibid., p. 437.
4. See also Joseph, 2004b.

harmful manner, but that identical twins are socialized to experience a much stronger emotional bond with each other than are fraternal twins.[1] From the environmental perspective, the twin relationship is a more important aspect of concordance than identical twins merely being treated in a more similar etiologically-relevant manner than fraternals. Thus, twin studies of psychosis may have revealed little more than identical twins' greater propensity to experience folie à deux (shared psychotic disorder) than fraternals (see Chapter 6).[2]

In previous publications I have pointed to a 1967 survey by Norwegian schizophrenia twin researcher Einar Kringlen, who, in addition to computing concordance rates for schizophrenia, assessed his identical and same-sex fraternal twin pairs for emotional closeness and environmental similarity. The results of this survey, which to my knowledge have never been mentioned by anyone other than Kringlen and myself, are presented in Table. 1.1.

Table 1.1

IDENTICAL VS. SAME-SEX FRATERNAL TWIN RELATIONSHIP
AND TREATMENT: KRINGLEN'S 1967 TWIN STUDY

	Identical – 75 Pairs				Same-Sex Fraternal – 42 Pairs				
	F	M	Total	%	F	M	Total	%	p*
Identity confusion in childhood									
Yes	37	31	68	90	3	1	4	10%	<.0000000001
No	0	7	7	9%	17	21	38	90%	
Mistaken for each other by									
Parents and/or sibs	10	6	16	21%	0	0	0	0%	.0004
Teachers	16	10	26	35%	0	0	0	0%	.000001
Strangers	11	15	26	35%	3	1	4	10%	.002
No one	0	7	7	9%	17	21	38	90%	<.000000001
Considered as									
Alike as two drops	31	26	57	76%	0	0	0	0%	<.000000001
More alike than	3	5	8	11%	2	1	3	7%	
Alike as sibs	2	5	7	9%	18	21	39	93%	<.000000001
Uncertain	1	2	3	4%	0	0	0	0%	
Dressed									
Alike	3	28	61	81%	10	18	28	67%	

1. Ainslie, 1985.
2. Jackson, 1960; Joseph, 2004b.

Not alike	2	2	4	5%	4	1	5	12%	
Uncertain	2	8	10	13%	6	3	9	21%	
Brought up "as a unit"									
Similar	2	25	54	72%	3	5	8	19%	.00000003
Dissimilar	2	4	6	8%	11	6	17	40%	
Uncertain	6	9	15	20	6	11	17	40%	
Inseparable as children									
To an extreme	31	24	55	73%	2	6	8	19%	.0000002
Partly inseparable	2	3	5	7%	6	7	13	31%	
Not inseparable	2	7	9	12%	12	8	20	48%	
Uncertain	2	4	6	8%	0	1	1	2%	
Inseparable as adults									
To an extreme	2	12	14	18%	0	0	0	0%	.001
Partly inseparable	2	11	35	47	4	6	10	24%	
Not inseparable	7	12	19	25	14	13	27	64%	
Uncertain	4	3	7	9%	2	2	4	10%	
Mutual friends as children									
Mutual friends	2	23	49	65	7	12	19	45%	.022
Partly	5	3	8	11%	8	5	13	31%	
Different friends	2	5	7	9%	5	4	9	22%	
Uncertain	3	7	10	13%	0	1	1	2%	
GLOBAL EVALUATION OF TWIN CLOSENESS									
Extremely strong	2	21	49	65	2	5	7	17%	.0000003
Moderately strong	5	5	10	13%	8	8	16	38%	
As siblings	1	6	7	9%	7	7	14	33%	.002
Uncertain	3	6	9	12%	3	2	5	12%	

*Source: Adapted from Kringlen, 1967, p. 115. * Probability value determined by Fisher's Exact Test, one-tailed—identical total vs. fraternal total. F = Female, M = Male. For all pairs, one or both were diagnosed with a psychotic disorder.*

As seen in Table 1.1, identical twins were much more likely than fraternals to have been "inseparable" as children and as adults, to have had mutual friends as children, and to have been mistaken for each other. More important are the assessments of twin closeness and "identity confusion." Regarding the latter, 90% of identical twins — but only 10% of fraternal twins — experienced "identity confusion" as children. Moreover, identicals were more often mistaken for each other, considered as "alike as two drops of water," and "brought up as a unit." The final "evaluation of twin closeness" showed that 65% of identical twins had an "extremely strong" level of closeness, which was true for only 19% of fraternals.

Kendler and other defenders of the EEA simply ignore these results. They choose instead to focus on confirmatory data which they cite in support of the EEA and the twin method (see Chapter 9). Still, results such as Kringlen's support the validity of objections made by critics since the twin method's inception in the mid-1920s. Moreover, Kendler's EEA theory stands in direct contrast to the views of most schizophrenia twin researchers, who recognized that at least a portion of identical-fraternal concordance differences can be attributed to environmental factors.[1] Even a psychologist and behavior geneticist as prominent as Michael Rutter has acknowledged that "there will be violations of the Equal Environments Assumption (EEA), which is fundamental to the twin design...The violation arises because part of the difference between MZ and DZ pairs will stem from environmental effects..."[2]

It is widely understood both inside and outside psychiatry that identical twins experience much more similar environments than do fraternals. Moreover, qualifications regarding trait-relevance, or twins "creating their own environments," do nothing to alter the conclusion that the twin method is, like a family study, unable to disentangle possible genetic and environmental influences on psychiatric disorders. Although popularizers of genetic research such as William Wright, in his 1998 book *Born That Way*,[3] claimed that Kendler "refuted" the critics' objections in his 1983 "Overview" article, Kendler's theoretical defense of the EEA is, as we have seen, rather easy to refute.

THE TWIN METHOD AS PSEUDOSCIENCE

Because it rests on at least one clearly false theoretical assumption, one could argue that the twin method can be understood within the framework others have created to separate science from pseudoscience.

In the opening chapter of their 2003 edited volume, *Science and Pseudoscience in Clinical Psychology*, psychologists Scott Lilienfeld, Steven Lynn, and Jeffrey Lohr outlined ten "warning signs" of pseudoscience. "The more such warning signs a discipline exhibits," they argued, "the more it begins to cross the murky dividing line separating science from pseudoscience."[4] The ten warning signs outlined by Lilienfeld and his colleagues (who did not address the twin method, or list it as a

---

1. Joseph, 2004b, Chapter 6.
2. Rutter, 2003, pp. 935-936.
3. Wright, 1998.
4. Lilienfeld et al., 2003, p. 5. All subsequent quotations from these authors are taken from Lilienfeld et al., 2003, pp. 3-10.

pseudoscience) are "an overuse of *ad hoc* hypotheses designed to immunize claims from falsification," "absence of self-correction," "evasion of peer review," "emphasis on confirmation rather than refutation," "reversed burden of proof," "absence of connectivity," "over reliance on testimonial and anecdotal evidence," "use of obscurantist language," "absence of boundary conditions," and "the mantra of holism." Let's take a look at how the twin method and the arguments of its defenders fit into Lilienfeld and colleagues' pseudoscience framework.

An overuse of ad hoc *hypotheses designed to immunize claims from falsification.* An *ad hoc* hypothesis has been defined as "any hypothesis or hypothetical explanation developed to explain a particular set of data that does not fit into an existing theoretical framework. An *ad hoc* hypothesis is one developed after the data have been collected."[1] According to Lilienfeld et al., "The repeated invocation of *ad hoc* hypotheses to explain away negative findings is a common tactic among proponents of pseudoscientific claims. Moreover, in most pseudosciences, *ad hoc* hypotheses are simply 'pasted on' to plug holes in the theory in question."

The twin method is not as vulnerable to negative findings as it is to alternative explanations of its results. As we have seen, in the face of overwhelming evidence that identical twins experience much more similar environments than fraternals, Kendler and others, instead of abandoning the twin method, attempted to "plug holes" created by obvious environmental confounds with ad hoc hypotheses such as the "twins create their environment theory" and the trait-relevant stipulation. Clearly, Kendler and others "pasted on" these internally contradictory theories in order to salvage psychiatric geneticists' main research method.

Absence of self-correction. Pseudosciences tend to avoid eliminating errors from their methods and theories. The only significant changes in psychiatric twin research (apart from more complicated statistical formulas) occurred after psychiatrist Don Jackson, who was not a twin researcher, published a cogent critique of schizophrenia twin research in 1960.[2] Thus, twin researchers made corrections only after environmental confounds in twin research had been exposed by a well-respected critic from outside of their discipline.

Evasion of peer review. There is no reason to believe that twin researchers of the past few decades have evaded the peer review process. This has not been necessary, given the largely uncritical acceptance of twin research in mainstream psychiatry and psychology. Thus, twin researchers submitting their findings to psychiatry and psychology journals do not fear that their work will be rejected because their findings are explainable on environmental grounds. Twin studies are cited frequently in support of the claim that mental disorders are biologically based, a position vitally important to psychiatry and to its partners in the psychopharmaceutical industry. It therefore is unfortunate that so many twin

---

1. Reber, 1985, p. 12.
2. Jackson, 1960. For an analysis of Jackson's argument, see Joseph, 2001b, 2004b.

studies could pass the psychiatric journal peer review process without being rejected on the grounds that their authors' conclusions are based on an unsupported, yet critical, theoretical assumption.

Emphasis on confirmation rather than refutation. This is a hallmark of Kendler's defense of the twin method and the EEA. In Chapter 9, we will see how Kendler has focused mainly on research claiming to uphold the EEA, while ignoring "real world" examples running counter to his argument. "Pseudoscientists," wrote Lilienfeld and colleagues, "tend to seek only confirming evidence for their claims.... a determined advocate can find at least some supportive evidence for virtually any claim..." Moreover, "most pseudosciences manage to reinterpret negative or anomalous findings as corroborations of their claims." An example is found in a 1998 adoption study by prominent behavior geneticists Robert Plomin and colleagues, who, despite having found no personality-test score correlation between birth parents and their 245 adopted-away biological offspring, concluded that a "nonadditive genetic influence, which can be detected by twin studies but not by adoption studies, is a likely culprit."[1]

Reversed burden of proof. We have seen that Kendler's defense of the EEA and the twin method implies that the burden of proof for showing that identical and fraternal environments differ on trait-relevant dimensions falls not on twin researchers themselves but on critics. As Lilienfeld et al. point out, however, "a basic tenet of science is that the burden of proof always falls squarely on the claimant, not the critic...Consequently, it is up to the proponents of these techniques to demonstrate that they work, not up to the critics of these techniques to demonstrate the converse." But since the twin method would indeed be relegated to the museum of pseudosciences if *twin researchers* bore the burden of proof for showing that identical and fraternal twins experience equal (trait-relevant) environments, they place this burden on critics.

In his 1983 "Overview" article, Kendler wrote,

> For a familial-environmental bias in twin studies of schizophrenia to be a tenable hypothesis, nongenetic familial factors must be shown to be of major etiologic importance in the disorder.[2]

When he wrote that familial-environmental factors "must be shown," Kendler clearly meant that these factors must be shown *by critics*. Kendler does this even though (1) shifting the burden of proof to critics runs counter to a "basic tenet of science," (2) family environment is a major factor in environmental explanations of schizophrenia and other psychiatric disorders, and (3) he takes the opposite position when he discusses potential environmental confounds in psychiatric *family* studies. In this case, Kendler does not require critics

---

1. Plomin et al., 1998, p. 211.
2. Kendler, 1983, p. 1416.

to show that "familial factors must be shown to be of major etiologic importance in the disorder" to invalidate conclusions in favor of genetics.

According to twin researcher Michael Lyons and his colleagues, "it would seem that the burden of proof rests with critics of the twin method to demonstrate that 'trait-relevant' environmental factors are more similar for identical than same-sex fraternal twins."[1] Although they wrote, without any justification, that the burden of proof "seems" to fall on critics, a stronger case can be made that twin researchers bear this burden. After all, aren't *they* responsible for demonstrating the validity of their research method?

Similarly, according to twin researcher Thomas J. Bouchard, Jr.,

> It is certainly true that MZ twins experience more similar environments than do DZ twins, but it is also true, if perhaps surprising, that no one has been able to show that such imposed similarities in treatment are trait-relevant.[2]

Elsewhere Bouchard argued, in response to criticisms of the EEA by psychologist Louise Hoffman,

> The equal environment assumption is required only for trait relevant features of the environment; features of the environment that have causal status. Causal status must be demonstrated, not assumed.... It is absolutely mandatory *that Hoffman demonstrate* that the differential treatments she cites have a causal influence on the traits whose similarity she is trying to explain. This is a very difficult task [emphasis added].[3]

And in a response to my criticism of the EEA, Faraone and Biederman wrote in 2000,

> The second claim made by Joseph is that twin studies of ADHD are flawed by the equal environment assumption, which holds that the trait-relevant environments of identical and fraternal twins are the same. He finds this assumption untenable for two reasons. First, several studies have shown that, compared with fraternal twins, identical twins are treated more alike, spend more time together, have more common friends, and experience greater levels of identity confusion. Second, he *infers* from these data that identical twins are more likely to be similarly exposed to "trait-relevant" environmental factors. Notably, Joseph presents no data to support his inference. Thus, readers should view it as a hypothesis to be tested rather than a conclusion to be accepted [emphasis in original].[4]

Thus, in addition to reversing the burden of proof, twin researchers commit the "*ad ignorantium* fallacy," which Lilienfeld et al. described as "the mistake of assuming that a claim is likely to be correct merely because there is no compelling evidence against it."

---

1. Lyons et al., 1991, p. 126.
2. Bouchard, 1997, p. 134.
3. Bouchard, 1993, p. 33.
4. Faraone & Biederman, 2000, p. 570.

Absence of connectivity. By this it is meant that pseudoscience proponents tend to disconnect their field from other disciplines. Kendler's citations in defense of the EEA come mainly from the fields of psychology and psychiatry, and even then only from genetically-oriented researchers in these fields. Findings from neuroscience, criminology, sociology, anthropology, and other areas of psychiatry and psychology demonstrating the importance of the environment on behavioral differences, hardly exist for Kendler (at least until recently) and other leading proponents of the twin method. This is, according to Lilienfeld, consistent with pseudoscientific practices, which "often purport to create entirely new paradigms out of whole cloth rather than to build on extant paradigms. In doing so, they often neglect well-established scientific principles or hard-won scientific knowledge."

Over reliance on testimonial and anecdotal evidence. Although the twin method does not rely on anecdotal evidence, it was used to a great extent in Bouchard and colleagues' reports of allegedly separated twin pairs in their famous Minnesota reared-apart twin studies.[1] In turn, the claimed similarity of these "reared-apart" pairs strengthened the twin method, since twin researchers can and do argue that being reared together does not cause twins to resemble each other much more than if they had been reared apart. In 1987, critic Val Dusek described eight years of published Minnesota anecdotes in the absence of peer reviewed studies as "bewitching science," and compared them "to the sort of evidence often offered as proof for astrology or parapsychology such as extrasensory perception (E.S.P.)."[2]

Use of obscurantist language. This has been a common feature in twin research published since the early 1990s. In many cases, concordance rates or correlations are downplayed in favor of complex path analysis and model fitting diagrams, once described by psychologist Richard Lerner as "dazzling statistical pyrotechnics," accompanied by language difficult to understand by people not directly involved in twin research.[3] For example, in a 1992 twin study of phobias in women, Kendler and his colleagues wrote, in reference to their statistical formulations,

> In univariate analysis, information regarding the causes of variation is obtained by comparing the resemblance of MZ and DZ twin pairs for a single variable. In the multivariate case, the correlation between two or more variables is the primary unit of analysis. By comparing the cross-twin, cross-variable correlation in MZ and DZ twins, and contrasting that to the cross-twin within-variable and within-twin cross-variable correlations, the covariation of two or more variables can be partitioned into its genetic and environmental components.[4]

---

1. See Joseph, 2004b, Chapter 4.
2. Dusek, 1987, p. 21.
3. Lerner, 1995, p. 148.
4. Kendler et al., 1992b, p. 275.

Even if one could decipher this not atypical passage, all statistical formulations, path analyses, model fitting, multiple regression coefficients, and so on depend on the validity of the EEA as a prerequisite for concluding in favor of genetics. The expression "Garbage in, garbage out" could be modified for twin research to read, "False assumptions in, false conclusions out." Twin research publications turn readers' attention away from untenable assumptions and in the direction of impressive looking scientific language and diagrams which, in the words of Lilienfeld and colleagues, "may be convincing to individuals unfamiliar with the scientific underpinnings of the claims in question, and may therefore lend these claims an unwarranted imprimatur of scientific legitimacy."

Absence of boundary conditions. "Most well-supported scientific theories," wrote Lilienfeld et al., "possess boundary conditions, that is, well-articulated limits under which predicted phenomena do and do not apply." This relates to twin research in cases where the investigators insist that their findings can be generalized to the non-twin population. It is even more problematic in adoption research, where little attention is given to whether results can be generalized to the non-adoptee population (see Chapter 6). This is not to say that the results of twin and adoption studies (assuming that their assumptions are valid) are never generalizable, but that researchers, in their haste to conclude in favor of universal genetic principles, pay little attention to this crucial question.

"The mantra of holism."[1] According to Lilienfeld and colleagues, pseudoscience proponents "typically maintain that scientific claims can be evaluated only within the context of broader claims and therefore cannot be judged in isolation." An example is found in psychiatric genetic investigators Faraone, Tsuang, and Tsuang's 1999 *Genetics of Mental Disorders*. Following a discussion of family, twin, and adoption studies, and the limitations of each, they wrote,

> Although some methodological problems limit the effectiveness of twin studies, there is no conclusive evidence that these limitations substantially bias twin study results. Instead, there is a consensus among psychiatric geneticists that the twin method provides an informative source of converging evidence in determining the importance of genetic factors in psychiatric disorders.[2]

Moreover,

> Because each method has its limitations, we cannot rely on either a single study or class of studies to draw conclusions about the effects of genes and environment on mental illness. Instead, from an examination of many studies we seek a pattern of converging evidence that consistently confirms genetic and/or environmental hypotheses about the familial transmission of the disorder.[3]

While Faraone et al. recognized that, due to its "limitations," we "cannot rely" on twin studies to provide evidence in support of genetics, they put these

1. This phrase was coined by John Ruscio (2001).
2. Faraone, Tsuang, & Tsuang, 1999, p. 39.
3. Ibid., p. 45.

studies forward in support of genetics in the context of the supposed "converging evidence" from *other* types of studies. In a similar vein, twin researcher Irving Gottesman (whose work is the topic of Chapter 6) wrote that although schizophrenia family, twin, and adoption studies each "contribute to the genetic argument.... No one method alone yields conclusive proof or disproof."[1] Gottesman's argument is noteworthy in the sense that, if the EEA were truly sound, the twin method would indeed provide "conclusive proof" in favor of genetics.

In 2000, Faraone and Biederman used the "holism" argument in response to my criticism of ADHD twin and adoption studies, claiming that genetic theories of ADHD make better "predictions" and are more "parsimonious."[2] As we have seen, in defense of the EEA they placed the burden of proof on me by invoking the trait-relevant condition for ADHD twin studies.

In their 1998 adoption study of personality, Plomin and colleagues utilized the "holism" argument to explain away results consistent with environmental explanations of personality differences. Rather than highlight their study's failure to find a significant correlation between personality and genetic relationship, the investigators discussed their results in the context of previous twin studies of personality: "The most obvious implication of these results is that other family and adoption studies are needed to triangulate with twin studies on the estimation of genetic influence for personality as assessed by self-report personality questionnaires."[3]

* * *

The framework provided by Lilienfeld and others suggests that the twin method may belong in the pseudoscience category. Indeed, twin researchers' defense of the EEA has done little if anything to dispel the idea that the greater environmental similarity experienced by identical twins explains their greater concordance for psychopathology when compared to same-sex fraternal twins.

THE "HERITABILITY" FALLACY

The *heritability* concept is widely used in reference to genetic influences on psychiatric disorders and psychological trait variation. However, heritability estimates falsely claim to approximate "how much" genetic influence there is. As dissident behavior geneticist Jerry Hirsch has pointed out for many years, a numerical heritability estimate (coefficient) is not a "nature/nurture ratio" of the relative contributions of genes and environment.[4]

---

1. Gottesman, 1991, p. 93.
2. For my original article, Faraone and Biederman's response, and my response to them, see Faraone & Biederman, 2000; Joseph, 2000c, 2000e.
3. Plomin et al., 1998, p. 215.
4. Hirsch, 1997, 2004; McGuire & Hirsch, 1977; Joseph, 2004b.

Contrary to popular belief, whether heritability is 10% or 90% says nothing about the potential efficacy of a particular environmental intervention, nor does a heritability estimate greater than 50% imply that genes are more important than the environment. An example is phenylketonuria (PKU), a genetic disorder of metabolism which, without a specific environmental intervention, causes mental retardation. Although the population variance for PKU susceptibility is completely explained by genetic factors (heritability = 1.0, or 100%), the administration of a low phenylalanine diet to the at-risk infant during a critical period prevents the disorder from appearing. PKU is an excellent example of biologist Richard Lewontin's observation that a "trait can have a heritability of 1.0 in a population at some time, yet could be completely altered in the future by a simple environmental change."[1]

Approaching this question from a different angle, although the human trait of having two arms is inherited, the heritability of humans having two arms is *zero*. The reason is that the heritability statistic describes *variation* in a population attributable to genes. Because virtually everyone is born with two arms, and because people with one arm become that way because of an environmental occurrence, 100% of the "armedness" variation in a population is caused by the environment, and 0% of the variation is caused by genes. At the same time, of course, having two arms is a genetically programmed human trait. Thus, a trait could be 100% *inherited*, yet have a *heritability* of 0%. Hirsch has reminded us that although "heritable" and "inherited" are very different concepts, many people wrongly believe them to be synonymous because they sound alike.[2] Unfortunately, the genetic literature does little to help people avoid such confusion.

A heritability estimate, which is applicable only in a specific population, in a specific environment, and at a specific point in time, was developed in agriculture as a means of predicting the results of a selective breeding program for economically desirable traits.[3] Unfortunately, the invalid extension of the heritability statistic from a breeding predictor to a quantification of the genetic contribution to psychiatric disorders and psychological trait variation has led to a great deal of misunderstanding about the role of genetic influences on these traits and disorders. Moreover, heritability estimates are based on rarely-met assumptions about humans. According to Hirsch,

> Heritability estimation assumes both random mating in an equilibrium population (including therein the equally likely occurrence of every culturally tabooed form of incest) and the absence of either correlation or interaction between heredity and environment. In fact, when one or more of those assumptions are violated, that

---

1. Lewontin, 1974, p. 400.
2. Hirsch, 1997.
3. Lush, 1949. According to Kendler, "contrary to common usage, 'heritability' does not designate a characteristic of a disorder or a trait but only of a disorder or trait in a specific population at a specific time" (Kendler, 2005b, p. 5).

is, random mating in an equilibrium population, correlation or interaction, heritability is undefined.[1]

Heritability estimates are dubious for the additional reason that they are derived from twin and adoptions studies, which are subject to the invalidating environmental confounds and biases I discuss in this book.[2]

Thus, while it is theoretically possible that genetic factors underlie psychiatric disorders, it is inappropriate and misleading to use the heritability statistic to estimate the magnitude of this possible component. Behavior geneticist Richard Rende has written that the heritability statistic serves as "a useful statistical indicator to some, a rather meaningless index to others, and a potentially harmful, biased, and even blatantly incorrect calculation to the harshest critics."[3] Clearly, my views are similar to other "harsh critics" of the heritability concept.

While other terms are preferable, it may be acceptable to use the word "heritable" to indicate that a disorder is influenced by genetics. However, as the example of PKU shows, a heritability percentage or estimate says nothing about the ability to treat or prevent a disorder, or about the magnitude of genetic influences, and its use should therefore be discontinued in psychiatry and psychology.

* * *

Psychiatric geneticists and others view attention-deficit hyperactivity disorder, the most frequently diagnosed psychiatric condition among school-age children, as being strongly influenced by genetic factors. In the following chapter I examine the evidence they put forward in support of this claim.

---

1. Hirsch, 2004, p. 137.
2. Joseph, 2004b.
3. Rende, 2004, p. 112.

Chapter 2. ADHD Genetic Research: Activity Deserving of Attention, or Studies Disordered by Deficits?[1]

Many years ago, Jordan W. Smoller wrote a brilliant satirical psychiatric journal article entitled "The Etiology and Treatment of Childhood."[2] "Childhood," he quipped, "is a syndrome which has only recently begun to receive serious attention from clinicians." Current thinking in child psychiatry, as codified in the APA's Diagnostic and Statistical Manual (4th ed., Text Revision), illustrates that life has an unfortunate tendency to imitate art.[3] The opening diagnostic section of the DSM-IV-TR contains some socially disapproved behaviors transformed into "mental disorders" presumably in need of medical attention, epitomized by diagnoses such as "conduct disorder," "oppositional defiant disorder," and "disruptive behavior disorder." Another grouping of behaviors is called "attention-deficit hyperactivity disorder" (ADHD). The transformation of these behaviors into "ADHD" has been greatly assisted by the psychiatric establishment's acceptance of genetic research. In this chapter, I assess this research from a critical perspective.

According to mainstream psychiatric sources, ADHD has a prevalence rate of 3-7% of American school-age children, although studies find varying results depending on the diagnostic criteria used.[4] I use the term "ADHD" throughout this chapter in place of previous diagnoses in psychiatry such as "hyperactive child syndrome," "minimal brain dysfunction," "hyperactivity," "attention deficit

---

1. This chapter is based on an expanded and updated version of three previous publications (Joseph 2000c, 2000e, 2002a).
2. This essay is widely available on the internet (e.g., http://www.pfc.org.uk/satire/smoller.htm).
3. APA, 2000.
4. APA, 2000; Barkley, 2003.

disorder (ADD)," and so on. I will focus on the two research methods most often cited in support of the genetic basis of ADHD: twin studies and adoption studies. In Chapter 11, I examine ADHD molecular genetic research, which thus far has failed to uncover genes predisposing to ADHD.

In the opinion of Russell Barkley, a leader in the field, ADHD is a "developmental failure in the brain circuitry that underlies inhibition and self-control."[1] Barkley cited studies claiming to show that a portion of ADHD children's brains are smaller than those of "normal" children, and he linked that to genetic factors. Subsequent critical reviews of ADHD neuroimaging studies by Jonathan Leo and David Cohen, however, show that this body of research is greatly flawed.[2] Most mainstream reviews of ADHD research stress the importance of alleged genetic factors.[3] Some, such as *Running on Ritalin* author Lawrence Diller, have taken a middle ground approach, while others, such as critical psychiatrist Peter Breggin, have stressed environmental factors and have questioned the validity of the ADHD diagnosis itself, seeing it as a label justifying the use of drugs to control children's behavior.[4] According to Breggin, the DSM-IV definition of ADHD is "simply a list of behaviors that require extra attention from teachers."[5]

ADHD FAMILY STUDIES

Research suggests that ADHD tends to cluster in families.[6] Although several ADHD family studies contain methodological problems such as non-blinded diagnoses, partisans of both the genetic and environmental positions would be surprised if they did *not* find a familial clustering of ADHD. Despite the formal acknowledgement that family studies are unable to disentangle genetic and environmental influences, however, some ADHD researchers have concluded otherwise. For example, Barkley wrote in 2003 that "ADHD clusters significantly among the biological relatives of children or adults with the disorder, strongly implying a hereditary basis to this condition."[7] And according to ADHD family researchers Nichols and Chen, the "greater risks to relatives of the severely affected children and to relatives of girls, the less frequently affected sex, provided some evidence that the familial association was determined partly by

---

1. Barkley, 1998a, p. 67.
2. Leo & Cohen, 2003, 2004.
3. For example, Cook, 1999; Tannock, 1998; Thapar et al., 1999; Wilens et al., 2002.
4. Diller, 1998. Those questioning the validity of the ADHD concept include Breggin, 1998, 2001a, 2001b; DeGrandpre, 1999; Leo, 2002; Stein, 1998.
5. Breggin, 2001a, p. 203.
6. For example, Biederman et al., 1986; Biederman et al., 1995; Biederman et al., 1990; Cantwell, 1972; Faraone et al., 1991; Morrison & Stewart, 1971; Nichols & Chen, 1981; Welner, et al., 1977.
7. Barkley, 2003, p. 116.

polygenic inheritance."[1] Finally, in their 1995 family study, Biederman and colleagues wrote as follows: "Additional lines of evidence from second-degree relative, twin...adoption, and segregation analysis studies suggest that the familial aggregation of ADHD has a substantial genetic component."[2] However, a method that cannot support the genetic position alone does not constitute evidence when combined with the results of other types of studies. As Diller and his colleagues commented, "Familial clustering, as noted in the [Biederman et al. 1995 family study] article, cannot distinguish between potential genetic and environmental etiologies. While the authors are careful to describe the new data as familial, they nevertheless discuss them only in the context of a genetic etiology."[3]

ADHD TWIN STUDIES

According to Barkley, twin studies furnish "the most conclusive evidence that genetics can contribute to ADHD."[4] ADHD twin studies (all of which used the twin method) have found consistently that identical twins are more concordant for ADHD, or correlate higher for ADHD-related behaviors, than fraternals.[5] The main question, of course, is whether identical twins' greater behavioral resemblance can be attributed to their greater genetic resemblance, as twin method proponents maintain.

The Equal Environment Assumption and ADHD Twin Studies

Because the EEA's validity is crucial to their conclusions, we might expect ADHD twin researchers to have addressed its merits and to have provided evidence or citations in support of the traditional or trait-relevant definitions of the EEA. For the most part, however, this has not been the case. To the extent that they discuss the EEA at all, most ADHD twin researchers adhere to the traditional EEA definition despite the fact that, in the words of twin researchers Scarr and Carter-Saltzman, "the evidence of greater environmental similarity for MZ

---

1. Nichols & Chen, 1981, p. 276.
2. Biederman et al., 1995, p. 432.
3. Diller et al., 1995, p. 451.
4. Barkley, 1998a, p. 68.
5. ADHD twin studies include Cronk et al., 2002; Eaves et al., 1993; Edelbrock et al., 1995; Gilger et al., 1992; Gillis et al., 1992; Gjone et al., 1996; Goodman & Stevenson, 1989a, 1989b; Holmes et al., 2002; Hudziak et al., 1998, 2003; Levy et al., 1997; Lopez, 1965; Nadder et al., 1998; Rietveld et al., 2004; Saudino et al., 2005; Sherman et al., 1997; Silberg et al., 1996; Steffensson et al., 1999; Stevenson, 1992; Thapar et al., 1995; Todd et al., 2001; van den Oord et al., 1996; Willcutt et al., 2000; Willerman, 1973.

than DZ twins is overwhelming."[1] Let's examine how this critical theoretical assumption has been addressed (or ignored) by ADHD twin researchers:

(1) Lopez (1965) — The EEA was not discussed.

(2) Willerman (1973) — The author correctly stated that ADHD family studies are biased because "genetic and environmental influences are not distinguishable," while recognizing that the twin method is only "somewhat less subject to this bias."

(3) Goodman & Stevenson (1989a, 1989b) — The EEA was not discussed.

(4) Gilger et al. (1992) — The EEA was not discussed.

(5) Gillis et al. (1992) — The researchers acknowledged that a violation of the EEA could explain higher identical within-pair correlations, but concluded that the evidence suggested that "MZ and DZ environmental influences are similar."

(6) Stevenson (1992) — The EEA was not discussed.

(7) Eaves et al. (1993) — The EEA was not discussed.

(8) Edelbrock et al. (1995) — The EEA was not discussed.

(9) Thapar et al. (1995) — The authors wrote, "assuming that MZ and DZ twins share environment to the same extent, MZ twins will be more alike than DZ twins for traits that are under genetic influence." Although Thapar and colleagues found that their identical twins scored higher than fraternals on an "index of environmental sharing," which "suggests that MZ twins share environment to a greater extent than DZ twins...," they claimed that environmental similarity "was not associated with MZ twin similarity for hyperactivity scores."

(10) Gjone et al. (1996) — The EEA was not discussed.

(11) Silberg et al. (1996) — The EEA was not discussed.

(12) van den Oord et al. (1996) — The EEA was not discussed.

(13) Levy et al. (1997) — The authors stated that "the assumption of equal environments between MZ and DZ twins [is] often raised as a potential complication in twin studies. If MZ twins have more similar environments than DZ twins, this could be a reason for heritability being overestimated. The consistency here of resemblance to their non-twin siblings to both MZ and DZ twins justifies the conventional equal environment assumption."

(14) Sherman et al. (1997) — The EEA was not discussed.

(15) Nadder et al. (1998) — The EEA was not discussed.

(16) Steffensson et al. (1999) — The EEA was not discussed.

(17) Willcutt et al. (2000) — The EEA was not discussed.

(18) Todd et al. (2001) — The EEA was not discussed.

(19) Cronk et al. (2002) — The purpose of this study was to test the EEA as it related to ADHD and other conditions among a sample of adolescent female twin

---

1. Scarr & Carter-Saltzman, 1979, p. 528.

pairs. The investigators concluded that measures of environmental similarity had little impact on twin correlations for the conditions under study. Therefore, "These results lend support for the validity of the EEA and suggest that estimates of genetic and environmental influences obtained from twin studies of mother-reported child and adolescent emotional and behavioral problems are not unduly biased by the greater environmental similarity of monozygotic than dizygotic twins."[1] (See Chapter 9 for further discussion of this study.)

(20) Holmes et al. (2002) — The EEA was not discussed.

(21) Hudziak et al. (2003) — The EEA was not discussed.

(22) Rietveld et al. (2004) — The EEA was not discussed.

(23) Saudino et al. (2005) — The EEA was not discussed.

As we see, the authors of only three of the last eighteen published ADHD twin studies addressed the EEA, and only Cronk et al. claimed that twins' environments must be equal only as they pertain to the trait-relevant environmental factors of ADHD. Surprisingly, no ADHD twin study other than Cronk et al. cited previous research or publications supporting the validity of the EEA. Thus, implicitly or explicitly, all but one group of ADHD twin researchers based their conclusions on the traditional assumption that the environments of identical and fraternal twins are equal, yet only Gillis and associates claimed that these environments are in fact equal! It appears that for ADHD twin researchers the validity of the EEA is taken for granted...or is indefensible.

According to psychiatric geneticist Stephen Faraone, "The twin study is well known for its ability to separate genetic and environmental sources of etiology.... [The] genetic features of twinning provide a straightforward means of quantifying the impact of environmental and genetic factors on psychopathology." Thus, a statistically significant identical-fraternal correlation difference "must be due to genetic factors." He did add that "this conclusion will be wrong if MZ twins have environments that are more similar than those of DZ twins."[2] It appears that inferring genetic influences from identical-fraternal comparisons is not as "straightforward" as Faraone believes, because most twin researchers (though not, as we have seen, most ADHD twin researchers) now recognize that identical twins experience much more similar environments than fraternal twins.

Echoing Kendler (see Chapter 1), ADHD twin researchers Hay, McStephen, and Levy wrote in 2001 that identical twins "may well be treated more similarly than fraternal twins, but this is far more a consequence of their genetic similarity in behaviour (and of ensuing responses by parents and others) than a cause of such similarity."[3] Like Kendler, Hay and associates failed to

---

1. Cronk et al., 2002, p. 829.
2. Faraone, 1996, p. 596.
3. Hay et al., 2001, p. 12.

understand that the *reason* identical twins experience more similar environments than fraternals is not relevant in determining whether the greater observed resemblance of identical versus fraternal twins for ADHD is caused by genetic factors.

As seen in Chapter 1, the validity of the twin method's most critical theoretical assumption, as Kendler, Faraone, Hay and colleagues and others have defined it, is not supported by the evidence. If identical and fraternal family environments are equal, how do we explain the remarkable finding by Kringlen (Table 1.1) that 91% of identical twins — but only 10% of fraternal twins — experienced identity confusion in childhood, and that identicals were more likely to be "inseparable as children" (73% vs. 19%)? [1] Since the evidence overwhelmingly suggests that identical twins are treated more alike, spend considerably more time together, and experience greater levels of identity confusion and closeness, we would expect identical twins — on purely environmental grounds — to correlate higher than same-sex fraternals on ADHD-related measures.

ADHD twin studies, therefore, are based on an unsupported theoretical assumption and, like family studies, offer only a clue about *possible* genetic influences on ADHD-type behavior. It is quite possible, and even likely, that these studies have recorded nothing more than the greater psychological bond, more similar treatment, and greater environmental similarity experienced by identical versus fraternal twins, plus error.

ADHD ADOPTION RESEARCH

Overview

Another method used to assess genetic influences on ADHD is the study of adopted individuals. In theory, an adoption study is able to disentangle possible genetic and environmental influences on psychiatric disorders because adoptees receive their genes from one family, but are raised in the environment of another family. The Danish-American schizophrenia adoption studies were performed by Seymour Kety, David Rosenthal, Paul Wender, and their Danish associates (see Chapters 3 and 5). These researchers came together on the basis of a common belief that the twin method was unable to satisfactorily separate genetic and environmental influences. Towards the end of his career Rosenthal concluded, in my view correctly, that "in both family and twin studies, the possible genetic and environmental factors are confounded, and one can draw conclusions about them only at considerable risk."[2] According to Wender, who was

---

1. Kringlen, 1967.
2. Rosenthal, 1979, p. 25.

active in the ADHD arena as well, "The roles of 'heredity' (nature) and 'environment' (nurture) in the etiology of ADHD (as with other psychiatric disorders) cannot be determined by adding data from twin studies to the data from family studies." As a champion of psychiatric adoption studies, Wender concluded that the roles of heredity and environment in ADHD "can, however, be more conclusively separated by adoption studies, in which the parents providing the genetic constitution (the biological parents) and those who provide the psychological environment (the adoptive parents) are different people."[1]

While the logic of adoption studies might appear straightforward, the most important psychiatric adoption studies, such as the schizophrenia studies of Kety, Rosenthal, Heston, and Tienari, contained important methodological problems and were subject to selective placement bias.[2] However, despite their numerous flaws, the schizophrenia adoption studies possessed two qualities not found in most ADHD adoption studies: (1) the researchers made diagnoses blindly, and (2) the researchers studied or had psychiatric records for adoptees' biological relatives.

The "Adoptive Parents" Method

As of this writing, ADHD adoption studies have been published by Alberts-Corush et al., Cantwell, Morrison and Stewart, Safer, Sprich et al., and van den Oord et al.[3] Because of the difficulty in obtaining the carefully guarded records of adoptees' biological parents, which the Danish-American researchers were able to obtain through their access to national registers, the authors of four of the six ADHD adoption studies had to rely on the "Adoptive Parents" model, which Wender and colleagues developed in the 1960s.[4] The Adoptive Parents method compares the psychiatric status of three (and sometimes four) types of families as follows:

BH (Biological Hyperactive). This group consists of non-adopted children diagnosed with ADHD who are reared in the homes of their biological parents. (Because these biological parents raised their own ADHD child, this group should not be confused with schizophrenia adoption investigations, which studied the biological families of *adopted-away* diagnosed individuals.)

---

1. Wender, 1995, p. 93.
2. Heston, 1966; Kety et al., 1968; Kety et al., 1975; Kety et al., 1994; Rosenthal et al., 1971; Tienari et al., 1987, 2003, 2004; Wender et al., 1974. For critical reviews of schizophrenia adoption research, see Boyle, 2002b; Jackson, 2003; Joseph, 1999a, 1999b, 2001a, 2004b; Lewontin et al., 1984; Lidz, 1976; Lidz & Blatt, 1983; Lidz et al., 1981; Pam, 1995.
3. Alberts-Corush et al., 1986; Cantwell, 1975; Morrison & Stewart, 1973; Safer, 1973; Sprich et al., 2000; van den Oord et al., 1994.
4. Wender et al., 1968.

AH *(Adoptive Hyperactive)*. This group consists of adopted children diagnosed with ADHD who are raised by adoptive parents with whom they share no genetic relationship.

BN *(Biological Normal)*. This group typically consists of non-adopted normal (non-ADHD) children who are raised by their biological parents, and is designated as a control group.

AN *(Adoptive Normal)*. The AN control group consists of adoptees having no record of ADHD or related diagnoses, who are reared by their adoptive parents. Only Alberts-Corush and colleagues utilized this group.

In Figure 2.1, the Adoptive Parents design is seen alongside the two most popular designs in schizophrenia adoption research.

Figure 2.1

COMPARISON OF THE ADHD ADOPTIVE PARENTS METHOD WITH TWO METHODS USED IN SCHIZOPHRENIA ADOPTION RESEARCH

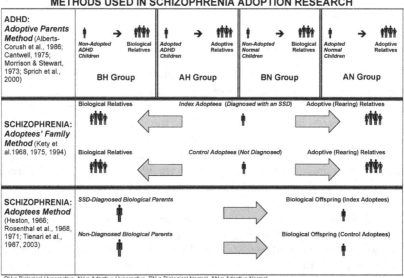

BH = Biological Hyperactive, AH = Adoptive Hyperactive, BN = Biological Normal, AN = Adoptive Normal.
SSD = Schizophrenia Spectrum Disorders as defined by the original investigators. Relatives in *italics* indicate the study's first identified relatives (known as a "probands" in the psychiatric genetic literature). Arrows point towards relatives identified and diagnosed after the identification of the probands.

The authors of the four Adoptive Parents studies (Alberts-Corush et al.; Cantwell; Morrison & Stewart; Sprich et al.) compared the ADHD rate among the relatives of 3 (or 4) different types of families, but had no information on their ADHD adoptees' biological relatives. In fact, *no ADHD adoption study has investigated the biological relatives of adopted-away children.* Therefore, their authors made no direct comparisons between the biological and adoptive families of the same child in any of these studies. In striking contrast, Kety and colleagues' schizophrenia adoption studies diagnosed adoptees' biological and adoptive relatives. (I have omitted siblings from Figure 2.1 in order to simplify differences in the study designs.)

Unfortunately, ADHD genetic researchers usually fail to discuss the severe limitations of the Adoptive Parents design unless compelled to do so by critics.[1] Too often, they obscure the fact that adoptees' biological relatives were not studied. For example, Faraone and Biederman wrote that a "testable psychosocial theory" must be able to explain "the elevated rates of ADHD and associated traits among the biological relatives of adopted away ADHD children," falsely implying that data on these biological relatives were gathered.[2] And in a subsequent review article in which he covered ADHD adoption research, Faraone wrote, "By examining both the adoptive and biological relatives of ill probands, one can disentangle genetic and environmental sources of familial transmission.[3]" This was the logic of Kety's schizophrenia adoption studies. However, no ADHD adoption study examined the "adoptive and biological relatives" of the same "ill" adoptees. Faraone continued that the

> Early [ADHD adoption] studies showed that the adoptive relatives of hyperactive children are less likely to have hyperactivity or associated disorders than are the biological relatives of hyperactive children..."[4]

These lines are technically correct, but Faraone's readers might incorrectly assume that ADHD adoption studies investigated the biological and adoptive relatives of the same group of adoptees.

Other potentially misleading accounts include the following:

Thapar et al., 1999

The "fundamental assumption" of adoption studies is that a disorder "will show greater prevalence among biological relatives of affected children, than among adoptive relatives.... Early [ADHD] adoption studies...found significantly higher rates of hyperactivity among biological parents of children with hyperactivity (7.5%) compared with adoptive parents (2.1%)."[5] [Again, Thapar's description is correct, but she does little to help her readers understand the crucial difference between the Adoptive Parents method and other adoption study models.]

Sprich et al., 2000

The title of Sprich and colleagues' 2000 study, "Adoptive and Biological Families of Children and Adolescents with ADHD," could be misinterpreted as

---

1. For example, see Faraone & Biederman, 2000.
2. Ibid., p. 570.
3. Faraone, 2004, pp. 305-306.
4. Ibid., p. 306.
5. Thapar et al., 1999, p. 108.

meaning that these investigators studied the biological and adoptive relatives of the same adoptees.

Schachar & Tannock, 2002

In adoption studies, the "biological parents of AD-HKD [sharing features of ADHD and hyperactivity disorder] individuals are more likely to exhibit AD-HKD or related disorders than are adoptive parents."[1]

Ogdie et al., 2003

"Adoption studies demonstrate an increased frequency of ADHD diagnoses in biological relatives of probands... [which is] consistent with genetic underpinnings."[2]

Waslick & Greenhill, 2004

"Adoption studies have confirmed that ADHD tends to occur more frequently in first-degree biological relatives of adopted probands than in families in which adopted probands cohabitate."[3]

Biederman, 2005

"Early studies showed that the adoptive relatives of hyperactive children are less likely to have hyperactivity or associated disorders than are the biologic relatives of hyperactive children."[4]

Faraone et al., 2005

"Two studies found that biological relatives of hyperactive children were more likely to have hyperactivity than adoptive relatives."[5]

And we will see in upcoming chapters that the problem of secondary sources' potentially misleading accounts of psychiatric genetic research is by no means limited to ADHD.

As seen in Figure 2.1, the Adoptive Parents method compares diagnoses in a group consisting of *adopted-away ADHD children* and their adoptive families (AH), versus a group consisting of the families of *other ADHD children* living with their biological parents (BH).

For proponents of the genetic position, a greater rate of disturbance in BH families compared with AH families suggests the operation of genetic factors.

---

1. Schachar & Tannock, 2002, p. 405.
2. Ogdie et al., 2003, p. 1268.
3. Waslick & Greenhill, 2004, p. 485.
4. Biederman, 2005.
5. Faraone et al., 2005.

According to Morrison and Stewart, whereas family studies cannot establish whether familial clustering is due to genetic or environmental factors,

> Examining the legal parents of adopted hyperactive children could help decide the issue, for if a similar excess of "personality disorder" were found in the adopting parents, an environmental hypothesis for the transmission of behavior disorder could be sustained. However, if it were found that parents (and their extended families) who have adopted hyperactive children showed no such high prevalence of psychiatric illness, the argument for the genetic transmission of hyperactivity would be strengthened.[1]

Morrison and Stewart, however, overlooked another important limitation of the Adoptive Parents model. Adoptive parents (or at least those who have gone through legal adoption procedures) constitute a population screened for mental health as part of the adoption process. They are — by definition — a group in which we would expect to find fewer psychiatric disorders than in the general population. Thus, as behavior geneticist Michael Rutter and his colleagues pointed out, low rates of psychological disturbance among adoptive parents in ADHD adoption studies "could be no more than an artifactual consequence of the tendency to select mentally healthy individuals as suitable adopting parents."[2] Even Wender recognized, if only in passing, that "one problem with the adoptive parents method is that the prospective adoptive parents have usually been screened by adoption agencies and excluded if they had significant psychopathology."[3] And according to schizophrenia adoption researcher David Rosenthal, "The screening with respect to adopting parents is well known, since adoption agencies have long taken the view that mentally ill people do not make the kinds of parents that serve the best interests of the child."[4] Therefore, the Adoptive Parents method's comparison of diagnoses among two unrelated groups of families — one in which parents are screened for psychopathology (AH), and the other in which parents are not screened for psychopathology (BH) — provides no important information about genetic factors in ADHD.

Keeping this in mind, let's examine the individual studies. (I have marked the four Adoptive Parents studies with an asterisk*.)

Safer, 1973

Daniel Safer studied the siblings and half-siblings of a group of 17 children living in foster care who were diagnosed with ADHD. These foster children had been born into abusive and neglectful families: "In nearly every case, through neglect or cruelty, the natural parents mismanaged the care of these children and

---

1. Morrison & Stewart, 1973, p. 888.
2. Rutter et al., 1990.
3. Wender, 1995, p. 95.
4. Rosenthal, 1971a, p. 194.

subsequently lost custody of them."[1] In addition, 5 of the 14 mothers for whom Safer had information had been in jail, and three others were alcoholics. Half of the fathers had been in prison, and three were alcoholics.

Safer found that the children's full-siblings were diagnosed with ADHD at a significantly higher rate than their half-siblings (based on retrospective, non-blinded "trained professional" diagnoses of 9/19 full-sibs, vs. 2/22 half-sibs). However, he did not state how much time the 17 children lived together with their siblings (although he provided median ages at placement). According to Safer, it is likely that "a genetic proclivity in association with a high rate factor increases the likelihood of MBD [minimal brain dysfunction, now known as ADHD]."[2] However, his study was far too limited and problematic to reach any such conclusion.

*Morrison & Stewart, 1973**

Here, the researchers studied the families of three groups of children: a BH group (N = 59), an AH group (N = 35), and a BN control group (N = 41), basing their diagnoses on non-blinded interviews with relatives. According to Morrison and Stewart, their results supported the genetic position on the basis of two findings. The first was that BH relatives had a significantly greater rate of alcoholism, sociopathy, and hysteria, conditions "for which there appears to be a genetic basis."[3] Because their study and previous family studies found an association between ADHD and these three diagnoses, Morrison and Stewart argued that their results supported the genetic hypothesis. However, evidence for the genetic basis of alcoholism, sociopathy, and hysteria was (and still is) weak, and even if it were strong, one cannot conclude that ADHD is genetically related to them on the basis of association, which can be explained by any number of possible factors (see Chapter 3).

Morrison and Stewart's second line of evidence related to their finding significantly more cases of ADHD among BH versus AH relatives. They based their diagnoses on parents' recollections of whether the relative had been "hyperactive, aggressive, or reckless as a young child; had been involved in antisocial behavior such as lying, cheating, fighting or truancy at home or at the school; had suffered from distractibility, poor concentration, or had specific learning problems or failure in school."[4] However, several researchers have concluded that parental recall is unreliable for research purposes.[5] Moreover, Morrison and Stewart did not adequately describe how they made their (non-blinded) diag-

---

1. Safer, 1973, p. 179.
2. Ibid., p. 184.
3. Morrison & Stewart, 1973, p. 891.
4. Ibid., p. 889.
5. Bradburn et al., 1987; Halverson, 1988; Holmberg & Holmes, 1994; Reuband, 1994; Robbins, 1963; Yarrow et. al., 1970.

noses, which were based on second-hand information. As a critic noted, the non-blinded interviewers could have "unwittingly" encouraged BH relatives to provide information leading to an ADHD diagnosis.[1]

Morrison and Stewart's claim that BH relatives had a significantly higher rate of childhood ADHD is based on data they provided in a table.[2] There was no indication that BH parents, who were the first-degree biological relatives of the hyperactive child, had significantly more childhood ADHD diagnoses than AH parents. The investigators found statistical significance only by combining parents with aunts and uncles. Conspicuously missing from their table were the rates among the grandparents and siblings of the hyperactive children, even though Morrison and Stewart counted grandparents in another table, and BH children had an average "sibship size" of 3.6, while AH children averaged 2.2. Thus, Morrison and Stewart failed to report the diagnostic status of over 190 biological and adoptive siblings. Clearly, their conclusion that the results "clearly favor a genetic hypothesis for transmission" cannot be sustained.[3]

Cantwell, 1975*

Psychiatric researcher Dennis Cantwell investigated the families of 139 boys (50 BH, 39 AH, 50 BN) matched on the basis of age, sex, race, and social class. Like Morrison and Stewart, Cantwell examined the relationship between ADHD and alcoholism, hysteria, and sociopathy in addition to comparing ADHD diagnoses between groups. Cantwell, who made diagnoses non-blinded, found significantly more cases of ADHD among BH relatives compared to AH and BN relatives, and also found significantly more psychiatrically diagnosed people in the BH relative group. He concluded that these results were "strongly suggestive of genetic factors operating" in ADHD, and that the data supported a genetic relationship between ADHD and alcoholism, sociopathy, and hysteria.[4]

Like Morrison and Stewart, Cantwell did not diagnose blindly, and relied on retrospective accounts to make diagnoses. Critic Robert McMahon questioned the method of relying on interviews to make diagnoses:

> It would have been important to attempt to validate these diagnostic procedures using independently rated behavioral observations; medical records; reports of friends, relatives and coworkers; and, perhaps, psychological test data. The need for independent, concurrent validation of diagnoses is especially critical when dealing with demonstrably unreliable parental attempts to assess retrospectively patterns of hyperactive behavior in themselves and in their relatives.[5]

---

1. McMahon, 1980, p. 148.
2. Morrison & Stewart, 1973, p. 890.
3. Ibid., p. 891.
4. Cantwell, 1975, p. 278.
5. McMahon, 1980, p. 148.

As McMahon suggested, there were problems with the reliability and validity of the children's diagnoses.

In a later essay, Cantwell wrote,

> Almost 90% of adopted children are illegitimate. They are at greater potential risk, then, for poor prenatal care and certain types of birth hazard, such as low birth weight. Single mothers, particularly pregnant teenagers, may be exposed to greater social stress during the pregnancy and at the time of the decision to give up the child, so there are both biological and social factors related to the pregnancy that may make the adopted child at greater risk.[1]

If, as Cantwell suggested, social and prenatal factors might place adoptees at higher risk for ADHD than non-adoptees, it is possible that Cantwell's (and Morrison & Stewart's) AH group recorded little more than the damage done to a child before placement, which is consistent with the low rate of psychiatric diagnoses among the AH adoptive parents. And research suggests that, in the general population, adoptees are indeed more likely than non-adoptees to receive an ADHD diagnosis.[2] This finding casts further doubt on the already shaky conclusions of the ADHD adoption studies.

Alberts-Corush et al., 1986*

Utilizing psychological tests, Alberts-Corush and associates assessed attention deficits and impulsivity among the parents in groups BH, AH, BN, and AN. They found significantly more attention deficits among the BH parents, but no differences in impulsivity. While concluding that their results "provide support for an association between childhood hyperactivity and attentional deficits in the biological parents of hyperactives," they drew no conclusions in favor of genetics.[3] The investigators also recognized the limitations of the Adoptive Parents method: "studies involving the biological and adoptive parents of the same hyperactive child would assuredly provide a more definitive analysis of the gene-environment interaction."[4]

van den Oord et al., 1994

Dutch behavior geneticist Edwin van den Oord and his colleagues compared the correlations of two groups of adopted sibling pairs. The participants were international adoptees (mean age = 12.4 years) who had been placed into Dutch adoptive homes. About two-thirds were born in Korea and other Asian countries, and another 18% were born in Colombia. The investigators studied three groups of adoptees: a *biological sib group* consisting of 111 pairs of adopted biological sibling pairs raised in the same adoptive home, a *non-biological sib group*

---

1. Cantwell, 1989, p. 82.
2. Deutsch, 1989; Deutsch et al., 1982.
3. Alberts-Corush et al., 1986, p. 423.
4. Ibid., p. 422.

consisting of 221 pairs of biologically *unrelated* adoptees raised in the same adoptive home, and a group of 94 "only child" adoptees. Adoptees were scored according to their adoptive parents' responses on the Child Behavior Checklist (CBCL, an assessment tool that surveys various aspects of children's behavior).

The investigators reasoned that a greater correlation of ADHD-type behavior among the biological sibs compared to the non-biological sibs would suggest the operation of genetic factors. They found that the biological sib group correlated significantly higher on the CBCL Attention Problems and Externalizing scales, and concluded in favor of important genetic effects on these behaviors.

Although some reviewers have cited this study in support of genetics, it contains several important problems.[1] To begin, the investigators found no statistically significant differences between the biological and nonbiological pairs on the CBCL Total score, or on the Internalizing behavior category. In the Attention Problems category for pairs of boys (who are diagnosed with ADHD more often than girls), the biological siblings correlated at a modest.169, while the nonbiological siblings correlated at 0.089.[2] The authors did not indicate that this difference was statistically significant.

Also problematic was the investigators' assumption that "the common environments were similar for the two groups of siblings."[3] Although all biological pairs had the same country of origin, this was true for only 75% of the non-biological sibling pairs.[4] While most were raised in The Netherlands from an early age, the ethnic composition of the sibling pairs would likely affect their level of mutual interaction. We could reasonably expect greater emotional closeness between a pair of ethnically Korean siblings than a pair consisting of a Korean and an Austrian, yet a pair of the latter type could only have been found in the nonbiological group. It is also likely that non-white "foreigners" living in The Netherlands would have experienced greater levels of discrimination and mistreatment than the European adoptees, who constituted 14.2% of the non-biological group, but only 2.7% of the biological sibs.[5] It therefore is likely that siblings with the same biological heritage (and more similar appearance) would be treated more similarly by parents and the social environment. Other behavior genetic researchers have found evidence that parents' ratings of twins are influenced by their expectations of how similarly identical and fraternal twins should behave.[6] Because parents might expect biologically related siblings to behave more similarly than non-biologically related pairs, it is possible that just such an

---

1. e.g., Barkley, 2003; Tannock, 1998.
2. van den Oord et al., 1994, p. 200, Table 4.
3. Ibid., p. 203.
4. Ibid., p. 195.
5. Ibid., p. 195.
6. Goodman & Stevenson, 1989b.

"expectancy effect" was operating in the parents' CBCL rating of their adopted children.

Another problem was that, although the researchers provided adoptees' mean age, age difference, and age at placement, they did not indicate how much time the pairs lived together, or whether there was a correlation between time living in the same home and behavioral similarity. It seems more likely that the biological sibs would be placed into their adoptive home at the same time. More importantly, the biological siblings were raised in the same pre-placement environments in their native countries, meaning that they were more similarly exposed to potentially behavior-influencing environmental factors than the non-biological group, who lived in more dissimilar pre-placement environments. The researchers assessed adoptees' pre-placement environments by attempting to quantify factors such as "abuse," "caretaking," "health," and "neglect" from mean scores based only on the adoptive parents' knowledge. These figures, however, cannot provide an adequate picture of the adoptees' pre-placement rearing environments.

The investigators also were unable to control for age of placement, which represents another potentially confounding factor in this study. The biological sibs averaged a much later age of placement (43.5 months) than the non-biological sibs (20.7 months). An earlier Dutch study by Verhulst and associates, based on a larger sample of international adoptees, found that for 10- to 15-year-old adoptees, "the older the child at placement the greater the probability that the child will develop behavioral/emotional problems and/or will perform less well in school."[1] Because van den Oord and colleagues' biological sibs were, on average, two years older than the non-biological sibs at the time of placement, an important difference in potentially disturbance-creating environments existed between the two groups.

Another study by Verhulst and associates found that among 12- to 15-year-old boys, adoptees were three times more likely than non-adoptees to score in the deviant range on the CBCL Hyperactivity scale, and twice as likely to score in the Externalizing scale deviant range.[2] Regardless of the cause, it is clear that the international adoptees were at greater risk for hyperactive behavior than non-adoptees. It is therefore unlikely that the results of an adoption study of this type can be generalized to the non-adoptee population, although van den Oord and colleagues might argue that they controlled for the effects of the adoption process by their use of a non-biological adoptee group. However, we have seen that they were unable to control for the differing environments of the two groups of adoptee pairs.

---

1. Verhulst et al., 1990b, p. 104.
2. Verhulst et al., 1990a.

Sprich et al., 2000*

Susan Sprich and colleagues published their ADHD Adoptive Parents study in 2000.[1] They identified an AH group of 25 ADHD adopted children and their 62 first-degree adoptive relatives, a BH group of 101 ADHD children and their 310 first-degree biological relatives, and a BN group of 50 non-ADHD children and their 153 first-degree biological relatives. The results showed that the BH relative group had a significantly higher ADHD diagnosis rate versus the AH and BN relatives. Sprich and colleagues concluded that their results "add to mounting evidence from multiple lines of research strongly supporting the genetic hypothesis for ADHD."[2]

This study suffers from most of the invalidating flaws of the earlier Adoptive Parents investigations, and the authors, who discussed several limitations of their study, failed to point out that AH parents typically are screened for mental health as part of the adoption process. In addition, their diagnoses were based on retrospective questionnaires. On the positive side, this was the first-ever ADHD adoption study to make diagnoses blindly.

Another limitation discussed by Sprich and colleagues involved their failure to establish an AN group of adopted non-ADHD children: "It would have been interesting to include a group of adopted non-ADHD [AN] probands [i.e., children] for comparison."[3] Although the investigators viewed this omission as only a minor flaw, a 1999 psychiatric genetics textbook by Faraone, Tsuang, and Tsuang *required* an AN group for psychiatric adoption studies: "The increased risk for psychiatric disorders among adoptees limits generalizabilty and *demands* that any psychiatric study of adoptees use an adoptee control group [emphasis added]."[4] Thus, by this one standard established by leading psychiatric geneticists (one of whom, Faraone, was Sprich's collaborator), the study is methodologically unsound. Yet another problem relates to the fact that adoptees were placed into their adoptive homes as late as one year after birth, and that "only a minority of available subjects had been adopted at birth or shortly thereafter as we required for inclusion in this study."[5] If only a minority of adoptees met the investigators' stated requirements, it would seem that they were unable to control for the confounding features introduced by late placement. Again, this change of criteria limits genetic inferences on the basis of Faraone and colleagues' description: "If a child has lived with a parent for even a short period of time prior to adoption, the biological relationship will have been 'contaminated' by the environment created by the child's biologic parents."[6]

---

1. Sprich et al., 2000.
2. Ibid., p. 1436.
3. Ibid., p. 1436.
4. Faraone, Tsuang, & Tsuang, 1999, p. 42.
5. Sprich et al., 2000, p. 1436.
6. Faraone, Tsuang, & Tsuang, 1999, p. 43.

To summarize, apart from having made blind diagnoses, this study is sus-ceptible to all of the methodological problems of the earlier Adoptive Parents studies, and carries the added dimension of violating standards established by leading psychiatric geneticists (one of whom co-authored the study) near the time the study was published.

Summary and Discussion of ADHD Adoption Research

The Adoptive Parents method, used in four of the six ADHD adoption studies, cannot provide evidence in favor of genetics because, among other reasons, it does not assess the status of adoptees' biological relatives. Even in otherwise methodologically sound studies (which the ADHD studies certainly were not), the Adoptive Parents method offers, at best, only a clue that genetic factors might be operating. As behavior geneticists Plomin and colleagues wrote in 1997, ADHD "adoption studies to date have been quite limited methodologi-cally."[1] And Faraone and Biederman recognized that ADHD adoption studies' failure to study an adoptee's biological relatives, in addition to "relatively minor methodological problems...limit the strength of any inferences we can draw from these studies."[2] The methodological problems they dismissed as "minor," however, are actually *massive*. To review, these problems include (1) the failure to study adoptees' biological relatives; (2) the researchers' use of non-blinded diag-noses, which they sometimes made on the basis of relatives' recollections; (3) inadequate definitions of ADHD; (4) the inability to control for environmental confounds; (5) the inability to control for the status of adoptive parents as a pop-ulation screened for psychiatric disorders; (6) potential researcher bias; and (7) the use of late-separated adoptees. While Faraone and Biederman recognized that these studies contain flaws which limit their usefulness in assessing for pos-sible genetic factors, their claim that genetic theory correctly predicts that "ADHD should be transmitted through biological, not adoptive family relation-ships" rests on these flawed adoption studies.[3]

The bias introduced by genetically-oriented ADHD adoption researchers failing (other than Sprich et al.) to perform blind diagnoses is reason enough to invalidate their conclusions in favor of genetics. Faraone and colleagues have written that "a control group will not serve its purpose if the diagnosticians know which study participants are related to cases and which to controls."[4] And David Rosenthal wrote, "With respect to all such research, in which the dependent variable is the diagnosis of relatives, it is essential that the diagnos-tician not know whether the individual examined is related to an index or control proband...because it is easy to be swayed by knowledge regarding index

1. Plomin et al., 1997, p. 189.
2. Faraone & Biederman, 2000, p. 570.
3. Ibid., p. 569.
4. Faraone, Tsuang, & Tsuang, 1999, p. 19.

or control status."[1] Thus, for Rosenthal, who had intimate knowledge of how these studies were performed, blind diagnoses are "essential" because it is "easy" for researchers to be influenced by their knowledge of a relative's group status.

Summarizing the evidence in 1995, Wender wrote, "What have these adoption studies added to the data on ADHD from the family and twin studies? First, they have provided more solid data showing that 'hyperactivity' (broadly defined) has genetic contributions."[2] Because Wender believed that both family and twin studies are contaminated by environmental factors, one might ask what "solid data" he was referring to. Another important finding from ADHD adoption research, according to Wender, was that it has "shown that some psychiatric disorders associated with conduct disorder — 'alcoholism,' Antisocial Personality Disorder ('psychopathy,' 'sociopathy'), somatization disorder ('Briquet's Syndrome,' 'hysteria') — are associated with hyperactivity and are also genetically transmitted."[3] Like the authors of the original ADHD adoption studies, such as Cantwell, and Morrison and Stewart, Wender saw a genetic link between ADHD and alcoholism, sociopathy, and hysteria on the basis of what can only be characterized as very weak evidence. His position was based on two mistaken viewpoints: (1) that the evidence in favor of the genetic basis of alcoholism, sociopathy, and hysteria is solid; and (2) that the *associative* relationship of psychiatric disorders is evidence of their *genetic* relationship. The most outstanding example of Wender's embrace of viewpoint #2 was his support of the controversial Danish-American "schizophrenia spectrum" concept (see Chapter 3).

A GENETIC PREDISPOSITION FOR ADHD?

Psychiatric geneticists see children as inheriting a predisposition which will develop into ADHD in the presence of (possibly unknown) environmental triggers, which might include psychological factors, viruses, toxins, etc. Unfortunately, the public as well as some professionals mistakenly equate "genetic predisposition" with "it's genetic."

Theoretically, the knowledge that a child carries a genetic predisposition (or actual genes) is helpful to the extent that genetically vulnerable children can be helped to avoid environmental factors associated with ADHD. Thus, behavior geneticists Hay and Levy argued that if "early behaviour genetic markers" or "molecular markers" are discovered, "they will only be of real use if acceptable interventions are available,"[4] while genetic researcher Ed Cook wrote that "as the genetic risks are determined, it may become more feasible to determine spe-

---

1. Rosenthal, 1975, p. 20.
2. Wender, 1995, p. 99.
3. Ibid., p. 99.
4. Hay & Levy, 2001, p. 221.

cific environmental risk factors in the context of identified genetic risk."[1] However, as with all psychiatric disorders, "early intervention" strategies are complicated by the potential impact of knowing that a child carries genes for ADHD. This knowledge could, in itself, be a life altering event, affecting the way parents, classmates, teachers, and others treat a child.[2] Thus, even in the unlikely event that presumed ADHD genes are found in the future (see Chapter 11), society might still decide to concentrate on eliminating environmental factors contributing to ADHD-type behavior. In doing so, society's goal will be to aid *all* children in the same way that anti-smoking campaigns help reduce tobacco use. Is it necessary to identify people at greater genetic risk for nicotine addiction before initiating an anti-smoking campaign? Clearly, it is not.

HERITABILITY

Twin researchers have put the heritability of ADHD at between 0.39 and 0.91, while more recent estimates put the figure at between 0.88 and 1.0.[3] They arrive at these figures by doubling the identical-fraternal twin correlation or concordance rate difference. For example, if identicals correlate at 0.90, and fraternals correlate at 0.50, twin researchers would estimate heritability at 0.80 (80%). Aside from the fact that these estimates are derived from the flawed studies and methods discussed earlier, in the previous chapter we saw that heritability estimates for psychiatric conditions and psychological traits are inappropriate. Thus, we cannot determine "how much" of the "ADHD phenotype" variation is attributable to genes because, like PKU, a timely (and possibly simple) environmental intervention could prevent a condition with a stated heritability as high as 1.0.

CONCLUSIONS

The presumed genetic basis of ADHD rests on the results of family, twin, and adoption studies. Although research seems to indicate that ADHD is familial, the fact that families share a common environment as well as common genes permits no valid conclusions in support of genetics.

The twin method is no less confounded by environmental factors than family studies because, as most people clearly understand, identical twins share more similar environments than fraternals. Therefore, the greater resemblance of

---

1. Cook, 1999, p. 196.
2. See Joseph, 2004b, Chapter 10.
3. Wilens et al., 2002.

identical versus same-sex fraternal twins for ADHD, or on ADHD-related tests, can be completely explained on environmental grounds, plus error. Typically, ADHD twin researchers discuss the EEA briefly or not at all, and only one group used the trait-relevant EEA definition. Most ADHD genetic researchers' conclusions, therefore, are based on the simple assumption that identical and fraternal environments are equal, when clearly they are not.

ADHD adoption studies are greatly inferior to the flawed schizophrenia adoption studies which preceded them, and offer no scientifically acceptable evidence in favor of the genetic position. The fact that most made non-blind diagnoses, and that none assessed adoptees' biological relatives, invalidates any inferences in support of genetic factors. Finally, despite concerted worldwide efforts, no genes for ADHD have been found (see Chapter 11). An examination of the total weight of evidence in favor of a genetic basis or predisposition for ADHD leads to the conclusion that a role for genetic factors is not supported, and that future research should be directed towards psychosocial causes.

Nevertheless, Faraone and Biederman argued that the genetic theory of ADHD "has consistently made predictions which turn out to be correct."[1] However, there are non-genetic explanations for most of these "correct predictions," and their claim is based on the predictions they decide to make, often with the data in hand. An alternative analysis would review facts about ADHD not typically discussed in the context of genetics, but which make genetic explanations unlikely: (1) ADHD-type behavior is often exhibited by an individual in some situations but not in others.[2] According to Barkley, "all the primary symptoms of ADHD show significant fluctuations across various settings and caregivers."[3] In other words, children with alleged genetic defects and shrunken brain areas[4] are often fine when playing baseball and video games, but display "symptoms" in boring and unstimulating environments. (2) ADHD symptoms typically do not persist into adulthood, or in the words of the DSM-IV-TR, "In most individuals, symptoms (particularly motor hyperactivity) attenuate during late adolescence and adulthood..."[5] (3) ADHD is diagnosed from three to ten times more often in boys than in girls.[6] (4) ADHD has been widely recognized as a problem only since the early 1970s.[7] (5) Aspects of ADHD-like behavior are found in a large percentage of "normal" children.[8] (6) Over four million children in the United States consume stimulants, whereas in a country like France (pop-

---

1. Faraone & Biederman, 2000, p. 570.
2. APA, 1994.
3. Barkley, 1998b, p. 73.
4. Barkley, 1998a.
5. APA, 2000, p. 90.
6. Barkley, 1998b.
7. Arnold & Jensen, 1995.
8. Barkley, 1998b.

ulation 60 million), about 7,000 children receive these drugs.[1] While individually none of these points rules out genetic factors, together they argue against the idea that genes are involved. This position is strengthened by the evidence in this chapter showing that ADHD family, twin, and adoption studies do not establish the genetic position.

To the extent that children do exhibit ADHD-type behaviors, we should look more closely at environmental factors as the cause. Psychiatrists L. E. Arnold and Peter Jensen, who hold many mainstream views on ADHD, noted the "probable interaction between the complexity of environmental demands and manifestation of the symptoms of ADHD." They pointed out that "it is...possible that today's complex environments are overstimulating" and that,

> Children who assimilate a steady diet of video games, television, multiple after school activities, harried parents, and interchangeable caretakers may have their attentional systems down-regulated as a means of reducing the noise. They may become used to many novel, complex stimuli, and their attentional systems may not respond to the lower-level stimuli involved in academic work.[2]

Even as an oversimplified thesis, I find this a plausible explanation for the apparent increase of ADHD-type behavior in late 20th and early 21st century North America.

---

1. David Cohen, personal communication, 6/10/2005.
2. Arnold & Jensen, 1995, p. 2300.

CHAPTER 3. A CRITIQUE OF THE SPECTRUM CONCEPT AS USED IN THE DANISH-AMERICAN SCHIZOPHRENIA ADOPTION STUDIES[1]

Studies validating the dominant positions in psychiatry usually are not subjected to in-depth critical examination by those who defend them. This is particularly true about research cited in support of genetic and biological influences on the major psychiatric disorders. Yet, conclusions in favor of biology and genetics frequently depend on the investigators' decision-making process during the course of their studies. Whom should they count as cases? How should they define the disorder in question? What statistical procedures should they use? Which comparisons should they emphasize? Too often, studies have been published in which the methods, results, and conclusions appear together for the first time, allowing researchers to present the study as a neat package.[2]

This can occur because there is no procedure in psychiatry requiring researchers to submit and/or publish their methods *before* they collect data. Thus, even highly ethical investigators might be tempted to pick and choose results enabling them to find statistically significant results, which are often a prerequisite for having their study published.[3] These problems could be reduced if a system were established requiring researchers to submit a description of their methods prior to the collection and analysis of data. Although it is understood that there "is a cardinal rule in experimental design that any decision regarding the treatment of data must be made *prior* to an inspection of the data,"[4] accountability in psychiatric research, as well as research in other fields, is inadequate.

---

1. This chapter is based on a revised version of a previous publication (Joseph, 2000a).
2. Joseph & Baldwin, 2000.
3. Hubbard & Armstrong, 1997.
4. Walster & Cleary, 1970, p. 18, emphasis in original.

59

The Danish-American adoption studies, the subject of this chapter, helped establish schizophrenia as psychiatry's paradigmatic genetic disorder. Yet, although the results depended on greatly expanding the definition of schizophrenia, I am aware of no evidence that Seymour Kety, David Rosenthal, Paul Wender, and their Danish associates agreed on this definition before they collected and analyzed their data. Moreover, we will see that these investigators made faulty calculations of their *published* data, which led them, mistakenly, to conclude in favor of genetic influences on schizophrenia.

In this chapter I will show that a detailed analysis of a key premise of one body of research — in this case some of the most famous studies in the history of psychiatry — can change its results. I know that looking in-depth at one issue, particularly where numbers are involved, can test readers' patience. However, it is necessary in this case because the Danish-American results have had a tremendous impact on how the public and professionals view schizophrenia and other psychiatric disorders.

As I and others have argued elsewhere, the Danish-American schizophrenia adoption studies contained numerous flaws.[1] These include, but are not limited to, (1) inconsistent and biased methods of counting relative diagnoses; (2) the frequent failure to adequately describe the basis upon which a schizophrenia diagnosis was arrived at; (3) counting first- and second-degree relatives with the same weighting; (4) the lack of case history information, which would allow reviewers to assess the environmental conditions experienced by adoptees and relatives; (5) the bias introduced by counting relatives individually in the Adoptees' Family studies, which violates an assumption of the statistical measures used; (6) the use of late-separated or late-placed adoptees, with the accompanying rupture in parent-child attachment; (7) evidence that the investigators' bias in favor of genetic explanations had an important influence on their methods and conclusions; and (8) selective placement bias.[2]

Moreover, the statistically significant results Kety and colleagues reported depended on their decision to greatly expand the definition of schizophrenia to include what they called "schizophrenia spectrum disorders" (referred to here as "SSDs"). Kety and associates' schizophrenia spectrum included chronic schizophrenia (designated "B1") and several other diagnoses supposedly related to B1. These consisted of "acute schizophrenia" (B2), "borderline schizophrenia" (B3), "uncertain chronic schizophrenia" (D1), "uncertain acute schizophrenia" (D2), "uncertain borderline schizophrenia" (D3), and "schizoid or inadequate personality" (C).[3] Kety and colleagues' 1968 description of these diagnoses is seen in Figure 3.1.

---

1. Boyle, 2002b; Breggin, 1991; Cassou et al., 1980; Joseph, 1999b, 2001a, 2004b, 2004c; Lidz, 1976; Lidz & Blatt, 1981; Lidz at al., 1983; Pam, 1995.
2. Joseph, 2004b, Chapter 7.
3. Kety et al., 1968, p. 352.

Figure 3.1
THE 1968 DANISH-AMERICAN SCHIZOPHRENIA SPECTRUM AS DEFINED BY KETY, ROSENTHAL, WENDER AND COLLEAGUES
B1. "Chronic Schizophrenia ('chronic undifferentiated schizophrenia,' 'true schizophrenia,' 'process schizophrenia'). *Characteristics:* (1) Poor pre-psychotic adjustment; introverted; schizoid; shut-in; few peer contacts; few heterosexual contacts; usually unmarried; poor occupational adjustment. (2) Onset—gradual and without clear-cut psychological precipitant. (3) Presenting picture: presence of primary Bleulerian characteristics; presence of clear rather than confused sensorium. (4) Post-hospital course—failure to reach previous level of adjustment. (5) Tendency to chronicity."
B2. "Acute schizophrenic reaction (acute undifferentiated schizophrenic reaction, schizo-affective psychosis, possible schizophreniform psychosis, [acute] paranoid reaction, homosexual panic). *Characteristics:* (1) Relatively good premorbid adjustment. (2) Relatively rapid onset of illness with clear-cut psychological precipitant. (3) Presenting picture: presence of secondary symptoms and comparatively lesser evidence of primary ones; presence of affect (manic-depressive symptoms, feelings of guilt); cloudy rather than clear sensorium. (4) Post-hospital course good. (5) Tendency to relatively brief episode(s) responding to drugs, EST, etc."
B3. "Border-line state (pseudoneurotic schizophrenia, border-line, ambulatory schizophrenia, questionable simple schizophrenia, 'psychotic character,' severe schizoid individual). *Characteristics:* (1) Thinking: strange or atypical mentation; thought shows tendency to ignore reality, logic and experience (to an excessive degree) resulting in poor adaptation to life experience (despite the presence of normal IQ); fuzzy, murky, vague speech. (2) Experience: brief episodes of cognitive distortion (the patient can, and does, snap back but during the episode the idea has more the character of a delusion than an ego-alien obsessive thought); feelings of depersonalization, of strangeness or unfamiliarity with or toward the familiar; micropsychosis, (3) Affective: anhedonia—never experiences intense pleasure—never happy; no deep or intense involvement with anyone or anybody. (4) Interpersonal behavior: may appear poised, but lacking depth ('as if' personality); sexual adjustment: chaotic fluctuation, mixture of hetero- and homosexuality. (5) Psychopathology: multiple neurotic manifestations which shift frequently (obsessive concerns, phobias, conversion, psychosomatic symptoms, etc.); severe widespread anxiety."
C. "Inadequate personality. *Characteristics:* A somewhat heterogeneous group consisting of individuals who would be classified as either inadequate or schizoid by the APA *Diagnostic Manual* [II]. Persons so classified often had many of the characteristics of the B3 category, but to a considerably milder degree."
D1, 2, or 3. "Uncertain B1, 2 or 3 either because information is lacking or because even if enough information is available, the case does not fit clearly into an appropriate B category."
Source: Kety et al., 1968, p. 352.

These were the days before the publication of the 1980 DSM-III and its "operationalized" diagnostic criteria. Although the investigators made "global diagnoses" using the descriptions presented in Figure 3.1, it remains unclear how many of these characteristics were necessary in order to diagnose someone with a spectrum disorder, or at what point in the investigative process Kety and colleagues agreed upon these descriptions.

Three methods of studying adoptees were used in the Danish-American series: (1) Kety and colleagues' "Adoptees' Family method,"[1] which studied the biological and adoptive relatives of Danish adoptees diagnosed with a schizophrenia spectrum disorder, versus the relatives of controls (see Figure 2.1); (2) Rosenthal and colleagues' "Adoptees method,"[2] which studied the adopted-

---

1. Kety et al., 1968; Kety et al.,1975; Kety, Rosenthal, Wender, Schulsinger, & Jacobsen, 1978; Kety et al., 1994.
2. Rosenthal et al., 1968, 1971.

away biological offspring of schizophrenia spectrum and manic-depressive parents versus controls (see Figure 2.1); and (3) Wender and colleagues' "Cross-fostering method," which investigated the adopted-away biological offspring of non schizophrenia-diagnosed parents. (These crossfostered adoptees were raised by an adoptive parent diagnosed with an SSD, and were compared with controls.[1]) Although the investigators counted all SSDs as "schizophrenia," the validity of their schizophrenia spectrum, even on the basis of their stated criteria for how a diagnosis qualified as a spectrum disorder, is questionable.

Throughout this chapter I refer to the most important and influential Danish-American investigations. The first, published by Kety et al. in 1968, made diagnoses on the basis of information found in institutional and governmental records. The 1975 Kety et al. study used the same group of adoptees and relatives, many of whom the investigators personally interviewed.[2] They made many diagnoses on the basis of blind analyses of these interviews. In this chapter I make frequent references to the Kety et al. 1968 study (diagnoses based on records), and the Kety et al. 1975 study (diagnoses based on interviews). Rosenthal and colleagues' first publication appeared in 1968, and they published their final results in 1971.[3] Wender and colleagues' Crossfostering paper appeared in 1974.[4] Unless otherwise noted, "1968," "the 1968 study," "1975," or "the 1975 study" refer to the Kety et al. Adoptees' Family investigations.

ORIGINS OF THE SPECTRUM

The Danish-American researchers counted all SSDs equally as "schizophrenia" in comparisons between various groups of relatives, regardless of whether they were the adoptee's first- or second-degree relatives. Although the spectrum concept predates the Danish-American studies, Kety and colleagues made it a central aspect of their work. According to Kendler, they are "widely and justly credited with having stimulated modern interest in the schizophrenia spectrum...."[5] A stated goal of the investigators was to test the validity of the spectrum. As Kety et al. wrote in 1976,

> American psychiatrists...have broadened the concept [of schizophrenia] to include two additional syndromes — "latent" or "borderline" schizophrenia and "acute schizophrenic reaction" — on the assumption that these are variants of the original concept. In the hope that ultimately we might be able to examine the genetic relatedness of these three syndromes, we decided to use them (grouped under "definite schizophrenia") in the selection of index cases and the diagnosis of

1. Wender et al.,1974.
2. Kety et al., 1968; Kety et al., 1975.
3. Rosenthal et al., 1968; Rosenthal et al., 1971.
4. Wender et al., 1974.
5. Kendler, 2003a, p. 1550.

relatives.... We decided to test the hypothesis that there was an even wider "schizo-phrenia spectrum" that would include "uncertain schizophrenia" and "schizoid or inadequate personality."[1]

The schizophrenia spectrum, whose existence was first made public in 1967, originally was described as follows:

> We had recognized certain qualitative similarities in the features that charac-terized the diagnoses of schizophrenia, uncertain schizophrenia, and inadequate personality, which suggested that these syndromes formed a continuum; this we called the schizophrenia spectrum of disorders.[2]

In previous publications I have touched on the most important aspects of the spectrum's origins, which I summarize here.[3]

The American investigators Kety, Rosenthal, and Wender began a collabo-ration with Danish psychiatric investigators in 1963. Together, they obtained the records of a large number of adoptees through the extensive population and psy-chiatric registers then existing in Denmark.

The investigators obtained information on 5,483 adoptees from the greater Copenhagen area. Based on known population rates, they apparently expected to identify enough cases of chronic (B1) schizophrenia to be able to conduct their study. Because the lifetime expectancy rate for schizophrenia is usually stated as 0.8-1.0%,[4] the investigators probably expected to identify 50-55 B1 adoptees from this population (5,483 x .01= 54.8). Based on lower rates reported for Denmark (0.69%), they would have expected about 38 B1 cases (5,483 x 0.0069 = 37.8).[5] However, they found only 16 cases of chronic B1 schizophrenia among these 5,483 adoptees, which Rosenthal acknowledged was a "lower than expected yield."[6] This rate of about 3/1000 is one-half to one-third of the expected general population rate.

This unexpectedly low adoptee schizophrenia rate suggests that rearing by parents screened by adoption agencies for psychopathology had reduced the schizophrenia rate by over 60%. Thus, at the outset of their work the investi-gators uncovered, yet overlooked, important evidence supporting a link between family rearing environment and schizophrenia.

One of the investigators, David Rosenthal, stated clearly that he and his colleagues would not have found statistically significant results had they decided to define schizophrenia only in its chronic form:

---

1. Kety et al., 1976, p. 414.
2. Kety et al., 1968, p. 353.
3. Joseph, 2004b.
4. Rosenthal, 1970; Slater & Cowie, 1971.
5. Slater & Cowie, 1971, p. 13.
6. Rosenthal, 1972, p. 65.

If we had relied only on hard-core, process [B1] cases, we would have found no significant difference between our index and control subjects.[1]

And elsewhere he wrote,

The second [important] feature [of the research] has to do with the fact that we have included a broad spectrum of disorders in the ones I am calling schizophrenic. These include not only the classical chronic, process types of cases, but patients called doubtful schizophrenic, reactive, schizo-affective, borderline or pseudo-neurotic schizophrenic, or schizoid or paranoid. If we dealt only with hardcore [B1] schizophrenia, our ns [number of subjects] would be too small to make any of these studies meaningful.[2]

Had Kety and colleagues decided to count only chronic B1 cases, as schizophrenia was defined in Denmark,[3] they would have had no possibility of achieving statistically significant results in the genetic direction.

The evidence suggests that the Danish-American investigators created the spectrum in order to have enough cases to conduct their study and not, as they maintained in their major publications, in order to test the hypothesis that the SSDs were related to chronic schizophrenia.[4] Even Rosenthal understood that "It seems somewhat ironic that...Paul Wender and I...in concert with Seymour Kety [were] in effect broadening the concept of schizophrenic disorder as widely as it may have ever been reasonably conceived before." Rosenthal further explained that the investigators "strained to encompass all disorders that shared salient clinical and behavioral manifestations with process schizophrenia and to group these as a spectrum of schizophrenic disorder."[5] Although Kety, Rosenthal and colleagues described the spectrum in their original 1968 publication (see Figure 3.1), I am aware of no evidence that they created this expanded definition of schizophrenia before they searched the Danish adoptee and relative records.

THE INCLUSION OF "BORDERLINE SCHIZOPHRENIA" (B3) IN THE SCHIZOPHRENIA SPECTRUM

The B3 "borderline schizophrenia" diagnosis (which Kety et al. sometimes called "latent schizophrenia") was a crucial component supporting the investigators' claims of statistically significant findings in the Danish-American

---

1. Ibid., pp. 73-74.
2. Rosenthal, 1971b, p. 194.
3. For example, Kety, Rosenthal, & Wender (1978, p. 214) wrote that chronic B1 schizophrenia "is the only syndrome which merits the designation of schizophrenia in Denmark."
4. This was suggested by Lewontin and colleagues in 1984, and Pam in 1995.
5. Rosenthal, 1975, p. 19.

studies. As seen in Figure 3.1, Kety and colleagues used symptoms such as "strange or atypical" thinking, "feelings of depersonalization," "anhedonia — never experiences intense pleasure," "lacking in depth," "mixture hetero- and homosexuality," and "multiple neurotic manifestations...severe widespread anxiety" to make a B3 diagnosis. They believed that a condition was genetically related to chronic schizophrenia if it affected "a significant concentration in the biologic index relatives compared with their controls," whereas they believed that conditions were unrelated to chronic schizophrenia if cases clustered in statistically non-significant numbers among index versus control biological relatives.[1] According to the investigators, B3 met the criteria for inclusion in the spectrum.

Contrary to Kety and colleagues' claims, however, we will see that B3 diagnoses *did not* cluster in statistically significant numbers among index versus control biological relatives, casting doubt on their conclusion that this diagnosis is associated with, or genetically related to, B1. In the 1980s, several commentators noted the lack of support for Kety and colleagues' position, which I elaborate upon here.[2]

To test Kety and colleagues' claim, I will count the B3 biological relatives of B1 adoptees, plus the B1 biological relatives of B3 adoptees. I shall call this the *B1/B3 relationship* as it relates to individual relatives. In the 1970s, critic Lorna Benjamin pointed out that in order to assure that the assumption of independent observations is not violated, only differences between affected biological *families* should be considered.[3] When making this comparison, I define the B1/B3 relationship as the number of B3 index adoptees' biological families with at least one B1 member, plus the number of B1 index adoptees' biological families with at least one B3 member.

Diagnostic comparisons between index biological relatives or families versus controls, which are displayed in Tables 3.1 through 3.4, do not support Kety and associates' contention that B1 and B3 are related. In Table 3.1, I calculate the 1968 B1/B3 relationship as it pertains to individual relatives. If we look at the B1/B3 relationship from the standpoint of affected 1968 *families*, the results found in Table 3.2 are obtained.

---

1. Kety, Rosenthal, Wender, Schulsinger, & Jacobsen, 1978, pp. 29-30.
2. E.g., Lidz & Blatt, 1983; Lewontin et al., 1984.
3. Benjamin, 1976. See also Joseph, 2004b.

Table 3.1

THE B1/B3 RELATIONSHIP—INDIVIDUALS: 1968

Number of B3 Diagnoses Among the Biological Relatives of B1 Index Adoptees, Plus B1 Diagnoses Among the Biological Relatives of B3 Index Adoptees—vs. Controls: Diagnoses Based on Records

	Total Adoptees	B1 Index Adoptees	B3 Index Adoptees	B1 Biological Relatives of B3 Index Adoptees	B3 Biological Relatives of B1 Index Adoptees	Index B1/B3 Relationship/ Control B3 Biological Relatives
Index	33	16	10	0/38 (0%)	3/82 [a] (3.6%)	3/120 (2.5%)
vs. Matched Controls	26					1/121 (0.8%)
Probability*						.31 (ns)
vs. All Controls	33					3/156 (0.6%)
Probability*						.22 (ns)

Based on figures from Kety et al. (1968, pp. 354-355). B1/B3 Relationship defined as the number of B3 biological relatives of B1 index adoptees, plus the number of B1 biological relatives of B3 index adoptees.
ns = statistically non-significant at the 0.5 level.
*Fisher's Exact Test, one-tailed.
[a] All three B3 relatives were paternal half-siblings from the same family.

Table 3.2

THE B1/B3 RELATIONSHIP — FAMILIES: 1968

Number of B1 Index Adoptees' Biological Families with at least One B3 Diagnosis, Plus Number of B3 Index Adoptees' Biological Families with at least One B1 Diagnosis — vs. Controls: Diagnoses Based on Records

	Total Adoptees	B1 Index Adoptees	B3 Index Adoptees	B3 Adoptee Biological Families with at least One B1 Member	B1 Adoptee Biological Families with at least One B3 Member	Index B1/B3 Relationship/ Control Families with at least One B3 Member
Index	33	16	10	0/10 (0%)	1/16 (6.3)	1/26 (3.8%)
vs. Matched Controls	26					1/26 (3.8%)
Probability*						(ns)
vs. All Controls	33					1/33 (3.0%)
Probability*						(ns)

Based on figures from Kety et al. (1968, pp. 354-355). The B1/B3 relationship is defined as the number of B3 index adoptees' biological families with at least one B1 member, plus the number of B1 index adoptees' biological families with at least one B3 member.
ns = statistically non-significant at the .05 level.
*Fisher's Exact Test, one-tailed.

If we look at Kety and associates' 1975 study, we find the same results. In Table 3.3, I calculate the B1/B3 relationship in terms of individual biological relatives.

Table 3.3

THE B1/B3 RELATIONSHIP—INDIVIDUALS: 1975

Number of B3 Diagnoses Among the Biological Relatives of B1 Index Adoptees, Plus B1 Diagnoses Among the Biological Relatives of B3 Index Adoptees—vs. Controls: Diagnoses Based on Interviews

	Total Adoptees	B1 Index Adoptees	B3 Index Adoptees	B1 Biological Relatives of B3 Index Adoptees	B3 Biological Relatives of B1 Index Adoptees	*Index* B1/B3 Relationship/ *Control* B3 Biological Relatives
Index	33	17	9	0/38 (0%)	5/102 [a] (4.9%)	5/140 (3.6%)
vs. Matched Controls [b]	26					3/121 (2.5%)
Probability*						.44 (ns)
vs. Screened Controls	23					1/113 (0.9%)
Probability*						.16 (ns)
vs. All Controls	34					3/174 (1.7%)
Probability*						.25 (ns)

All biological relative diagnoses based on figures from Kety et al. (1975, pp. 158-161).
B1/B3 Relationship defined as the number of B3 biological relatives of B1 index adoptees, plus the number of B1 biological relatives of B3 index adoptees. ns = statistically non-significant at the .05 level.
*Fisher's Exact Test, one-tailed.
[a] Three potential fathers were named in the adoption report of index adoptee S-22, counted here as one father.
[b] Number of matched control B1 + B3 index and relative cases based on Kety et al., (1968, pp. 354-355). Matching status was not provided in the 1975 study.

If we examine the B3 figures pertaining to affected families in the 1975 study, the results are as follows.

Table 3.4

THE B1/B3 RELATIONSHIP—FAMILIES: 1975

Number of B1 Index Adoptees' Biological Families with at least One B3 Diagnosis, Plus Number of B3 Index
Adoptees' Biological Families with at least One B1 Diagnosis — vs. Controls: Diagnoses Based on Interviews

	Total Adoptees	B1 Index Adoptees	B3 Index Adoptees	B3 Adoptee Biological Families with at least One B1 Member	B1 Adoptee Biological Families with at least One B3 Member	*Index* B1/B3 *Relationship/ Control* Families with at least One B3 Member
Index	33	17	9	0/9 (0%)	5/17 (29%)	5/26 (19.2%)
vs. Matched Controls [a]	26					2/26 (7.6%)
Probability*						.21 (ns)
vs. Screened Controls	23					1/23 (4.3%)
Probability*						.13 (ns)
vs. All Controls	34					3/34 (8.8%)
Probability*						.21 (ns)

All biological relative diagnoses based on figures from Kety et al. (1975, pp. 158-161). The B1/B3 relationship is defined as the number of B3 index adoptees' biological families with at least one B1 member, plus the number of B1 index adoptees' biological families with at least one B3 member. ns = statistically non-significant at the .05 level.
*Fisher's Exact Test, one-tailed.
[a] Number of matched control index and relative cases based on Kety et al. (1968, pp. 354-355). Matching status not provided in the1975 study.

Thus, Kety and colleagues should have concluded that B3 was unrelated to
B1. As they stated, "the schizophrenia spectrum was and still is a hypothesis or
group of hypotheses on which we hoped our continuing studies might cast some
light."[1] However, the 1968 and 1975 results I have outlined in Tables 3.1 to 3.4
clearly show that the sum total of the B3 biological relatives of B1 adoptees, plus
the B1 biological relatives of B3 adoptees, was statistically non-significant when
compared to the B3 rate among the control group biological relatives.

Readers may wonder why, if according to Tables 3.1 to 3.4, B3 was unrelated to B1, Kety and associates did not also recognize this finding. The answer
relates to how they decided to count biological relatives when assessing the relationship between B1 and B3. Looking specifically at B3 in Tables 3.1 to 3.4, I
counted only the B3 biological relatives of the B1 index adoptees, plus the B1 biological relatives of the B3 index adoptees. The reason I counted only these diagnoses is that we learn nothing about the relationship between B1 and B3 by
counting the B3 biological relatives of the B2 and B3 adoptees, the B1 biological
relatives of the B1 and B2 adoptees, or the B2 biological relatives of B1, B2, or B3
adoptees. Kety and associates, on the other hand, counted all these (real or

---

1. Kety et al., 1976, p. 417.

potential) diagnoses together, along with D relatives, and claimed that all were genetically related to B1 on the basis of a greater *combined* clustering of diagnoses among index versus control biological relatives. Table 3.5 outlines (1) the B1/B3 relationship, (2) Kety and colleagues' method of counting and combining spectrum diagnoses, and (3) the counterposed conclusions that flow from each.

Table 3.5

THE B1/B3 RELATIONSHIP VS.
KETY AND COLLEAGUES' CRITERIA

B1/B3 RELATIONSHIP	KETY ET AL. CRITERIA
TOTAL OF:	TOTAL OF:
B3 Biological Relatives of B1 Adoptees ＋	**B3 Biological Relatives of B1 Adoptees** ＋
B1 Biological Relatives of B3 Adoptees	B1 Biological Relatives of B3 Adoptees ＋
	B1 Biological Relatives of B1 Adoptees ＋
	B1 Biological Relatives of B2 Adoptees ＋
	B3 Biological Relatives of B2 Adoptees ＋
	B3 Biological Relatives of B3 Adoptees ＋
	B2 Biological Relatives of B1 Adoptees ＋
	B2 Biological Relatives of B2 Adoptees ＋
	B2 Biological Relatives of B3 Adoptees ＋
	D1 Biological Relatives of B1 Adoptees ＋
	D1 Biological Relatives of B2 Adoptees ＋
	D1 Biological Relatives of B3 Adoptees ＋
	D2 Biological Relatives of B1 Adoptees ＋
	D2 Biological Relatives of B2 Adoptees ＋
	D2 Biological Relatives of B3 Adoptees ＋
	D3 Biological Relatives of B1 Adoptees ＋
	D3 Biological Relatives of B2 Adoptees ＋
	D3 Biological Relatives of B3 Adoptees
VERSUS B3 DIAGNOSES AMONG CONTROL BIOLOGICAL RELATIVES	VERSUS SPECTRUM (SSD) DIAGNOSES AMONG CONTROL BIOLOGICAL RELATIVES

NOTE: *Biological relative diagnoses in* **bold type** *indicate those Kety et al. diagnosed and counted. Those not in bold type would have been counted by Kety et al. on the basis of their criteria, but they made no such diagnoses.*

My Conclusions Regarding B3	Kety and Colleagues' 1975 Conclusions Regarding B3
B3 does not cluster significantly among index vs. control biological relatives. For this and other reasons, it cannot be counted as schizophrenia, nor does it belong in a "schizophrenia spectrum of disorders."	B3 and other spectrum diagnoses, when combined, cluster significantly among index vs. control biological relatives. Therefore, these diagnoses are genetically related to B1 and should be counted as schizophrenia when calculating schizophrenia rates among adoptees and relatives.

For diagnostic codes, see Figure 3.1. Based on data and conclusions in Kety et al. (1975). Although Kety et al. (1975, p. 154) listed C diagnoses (schizoid/inadequate personality) as "schizophrenia spectrum" disorders, they did not count C in some statistical comparisons, nor did they show which adoptees had C-diagnosed relatives.

According to Kety and colleagues' own 1975 figures, there was no signif-icant B3 clustering among their index versus control biological relatives.[1] They found statistically significant results only by combining B1/B3 relationship diag-noses with other spectrum diagnoses, such as the ones I have outlined in Table 3.5, and concluding that *all* were genetically related to chronic B1 schizophrenia. In doing so, they skipped a crucial step in determining the relationship between B1 and B3. That is, they decided that individual diagnoses, standing alone, did not have to cluster significantly among their index versus control biological rela-tives. In addition, they counted all spectrum diagnoses among the biological rel-atives of B2 and B3 adoptees, even though the hypothesis they sought to test (that each SSD is related to B1) required them to count only diagnoses exem-plified by my B1/B3 relationship formula. Yet, had they decided to assess each diagnosis individually, they would have been compelled to remove B3 from the schizophrenia spectrum.

In a discussion of the results of his 1968 study, Kety noted that, of his B3 index adoptees' 38 biological relatives, 3 were diagnosed as B3, 1 as D1, 1 as D3, and 1 as C. He concluded that "this finding supports the notion that borderline schizophrenia is a form of schizophrenia and is related to chronic schizo-phrenia."[2] These results, however, permit no such conclusion. Because Kety found no significant B1 clustering among these relatives, he cannot conclude in favor of a relationship between B1 and B3. In fact, as he described it, Kety found *zero* B1 diagnoses among the 38 biological relatives of his B3 adoptees. This should have led him to draw the opposite conclusion from the one he actually drew. Here and elsewhere, Kety committed the logical fallacy of *assuming* that B1 and B3 were genetically related in order to *conclude* that they were genetically related.[3]

---

1. Kety et al., 1975, p. 154, Table 3. Index 6/173 = 3.5%, vs. control 3/174 = 1.7%, p = .25.
2. Kety, 1970, p. 242.
3. The Kety et al. 1994 final report on the Danish-American Provincial study (from "the rest of Denmark") reported a significantly higher rate of "latent schizophrenia" (comparable to B3) among index versus control biological relatives (which was *not* statistically significant if the comparison is limited to first-degree relatives). However, Kety et al. removed 18 of the 42 control adoptees (and their relatives) from the study — over a decade after it began — for insupportable reasons (Joseph, 2004b). Of these 18 excluded control adoptees, 13 were diagnosed with a "serious or confounding mental illness" (typically a major affective disorder), and five could not be interviewed (Kety & Ingraham, 1992). According to Kendler and Diehl, who viewed this study's results while in preparation, there was no significant difference in latent schizophrenia diagnoses *before* the reduction of the control group: "Latent and uncertain schizophrenia was not found to be significantly more common in the biologic relatives of the schizophrenia adoptees than in those of the control adoptees (6.5% vs. 5.5%, respectively)" (Kendler & Diehl, 1993, p. 265). Thus, the Provincial study results do not support a significant "latent schizophrenia" clustering among the biological relatives of adoptees diagnosed B1, vs. control biological relatives.

According to Rosenthal, "in adoption studies, biological relatives separated from schizophrenic family members also developed borderline schizophrenia or schizoid personality."[1] However, in addition to the evidence I have presented thus far, this did not occur in Rosenthal's own Adoptees study. Table 3.6 shows the results of that investigation.

Table 3.6

PREVALENCE OF "BORDERLINE SCHIZOPHRENIA" AND SCHIZOID DIAGNOSES
AMONG THE ADOPTED-AWAY OFFSPRING OF PARENTS DIAGNOSED WITH A SCHIZOPHRENIA
SPECTRUM DISORDER VS. CONTROLS:
ROSENTHAL AND COLLEAGUES' 1971 DANISH-AMERICAN ADOPTEES STUDY

	Total Adoptees	Adoptees Diagnosed Schizoid	Adoptees Diagnosed Borderline Schizophrenia	TOTAL Adoptees Diagnosed Schizoid or Borderline Schizophrenia
Index	52 [a]	2 [b] (4%)	3 [c] (6%)	5/52 (10%)
vs. Control	67	2 (3%)	3 (4.5%)	5/67 (7.5%)
Probability*		⬇ .59 (ns)	⬇ .53 (ns)	⬇ .46 (ns)

All diagnoses above, and diagnostic descriptions below, from Rosenthal et al. (1971, pp. 309-310). ns = statistically non-significant at the .05 level.
*Fisher's Exact Test, one-tailed.
[a] Excludes adoptees whose biological parents were diagnosed with "manic depressive psychosis."
[b] Index diagnoses: "Schizoid: schizophreniform borderline?" and "Probably borderline paranoid; schizoid." Control diagnoses: "Moderately schizoid" and "Schizoid; beginning schizophrenia?"
[c] Index diagnoses: "Borderline schizophrenia" (3). Control Diagnoses: "Borderline schizophrenia" (2), and "Schizophrenic-like border case" (1).

Looking back on the spectrum's formation, Kety wrote,

> The investigators did not necessarily believe that all of these disorders would be found to be related to schizophrenia, but it would have been inappropriate to exclude any prematurely. Furthermore, if the different components were kept separate, it might eventually be possible to evaluate the relationship of each to paradigmatic [B1] schizophrenia.[2]

However, Kety decided *not* to evaluate B3 separately, arguing that "the number of these illnesses which we found in the relatives was too small to permit a further breakdown of the schizophrenia spectrum."[3] Elsewhere, he wrote that "there were insufficient cases to permit testing individual components of the spectrum with any reliability."[4] On the other hand, Kety and col-

---

1. Rosenthal, 1971a, p. 96.
2. Kety, 1988, p. 218.
3. Kety, 1975, p. 21.
4. Kety, 1987, p. 424.

leagues recognized that B2 acute schizophrenia did not cluster significantly among their index versus control biological relatives, concluding in 1978 that "most of the acute psychoses which in America have been labeled 'acute schizophrenia reaction' are not subtypes of schizophrenia." In striking contrast to B3, they chose to remove B2 from their schizophrenia spectrum.[1] It seems that the investigators' decisions about whether a diagnosis should be excluded, and when it would be "inappropriate" to do so, had a major impact on their results as well as on the conclusions that flowed from them. And these results were sometimes known to them when they made these crucial decisions.

How can such arbitrary, after-the-fact reasoning be accepted in scientific research, where two statistically non-significant comparisons (B3 index vs. control biological relatives; B2 index vs. control biological relatives) are treated differently even though, statistically speaking, *both results are the same*? And we will see in Chapter 5 that controversial research methods and biased decision making are rarely questioned in mainstream psychiatry and psychology textbooks.

The Danish-American investigators' procedures illustrate the need for the establishment of research registers to help ensure against possible post-data collection manipulations. Elsewhere, I described a research register as follows: "Researchers would be expected to submit a written description of how they will obtain participants, how they will define and measure the variables of interest, how they will perform group comparisons, and what conclusions they will draw from the possible results they obtain.... A register would create a permanent public record of the intentions and methods of researchers before data collection, analysis, and publication."[2] In this case, Kety and colleagues would have been required to submit this information in the initial stages of their investigation, thereby removing any doubt that they decided to change definitions and comparisons in order to find the desired results.[3] And there is no question that they desired and expected to discover important genetic influences on schizophrenia.

The most explicit rationalization for the investigators' decision to combine diagnoses which, by themselves, showed no statistically significant index/control biological relative difference, is found in a 1988 article by Ingraham and Kety. In reference to the 1975 study, they wrote,

> Latent or borderline schizophrenia was found at a 4-5% prevalence in the biological index relatives and 1% to 1.5% in the biological relatives of controls. This is also true where the symptoms are less distinct and the diagnosis is designated uncertain.[4]

---

1. Kety, Rosenthal, & Wender, 1978, p. 217.
2. Joseph, 2004b. p. 339.
3. And as I documented in Chapter 7 of *The Gene Illusion*, there is evidence suggesting that the investigators changed group comparisons in the 1968 study when the original comparison did not yield evidence in support of genetics.
4. Ingraham & Kety, 1988, pp. 121, 123.

Ingraham and Kety went on to concede that *none* of these diagnoses was found in statistically significant numbers versus controls: "Since neither in chronic nor in latent schizophrenia the results for the definite or uncertain diagnoses are statistically different," they wrote, "It appears justified to combine them."[1] However, four diagnoses failing tests of genetic relatedness *separately* do not — when combined — constitute a genetically related spectrum of disease. Confronted with the finding that each SSD category fell short of statistical significance when compared with controls, Kety could just as easily have concluded that his findings failed to support the spectrum concept or the genetic basis of schizophrenia. Thus, genetically-oriented researchers' post data-collection *decisions*, and secondary sources' largely uncritical endorsement of these decisions, are the foundation of contemporary theories in psychiatry supporting a genetic basis for schizophrenia and other psychiatric disorders.

Moreover, from the standpoint of their study's viability, Kety et al. had a vital interest in retaining B3 in the spectrum. Had they decided to remove this diagnosis, their index group would have been reduced to the original 23-24 B1 and B2 adoptees, or possibly only the 16-17 B1 adoptees. Thus, despite their claim that they created the spectrum in order to test hypotheses, a decision to remove B3 (and therefore D3) would have either meant the unsuccessful conclusion of their work, or their need to ascertain thousands more Danish adoptees in the hope of finding enough B1 cases to be able to continue the study. In the 1975 investigation, 24 of the 56 biological relative SSDs were either B3 or D3 (24 of 30, excluding category C). Clearly, there was a relationship between Kety and associates' willingness to remove a diagnosis from the spectrum, and that diagnosis's usefulness to them in achieving statistically significant results in the genetic direction.

The Siever and Gunderson Reanalysis

Psychiatrists Larry Siever and John Gunderson's 1979 reanalysis of Kety and colleagues' 1975 study assessed the relationship between B1 and the other B categories.[2] They began by acknowledging the lack of a statistically significant index versus control B3 clustering ($p = 0.25$), while claiming that the B1 index clustering was significant ($p = 0.03$; although we will see in Chapter 5 that this comparison was statistically significant only because Kety et al. failed to count a B1 control biological relative). "This suggests," wrote Siever and Gunderson, "that genetically transmitted factors specifically related to schizophrenia may play a more important role in the etiology of chronic [B1] schizophrenia than in the etiology of the borderlines [B3], although the small numbers preclude any definite conclusions."[3] Like Kety, Siever and Gunderson concluded that B2 did

1. Ibid., p. 123.
2. Siever & Gunderson, 1979.
3. Ibid., p. 63.

not qualify as an SSD but, also like Kety, they decided to include B3 in the spectrum even though — like B2 — it showed no statistically significant clustering among the biological relatives of index adoptees versus controls.

Siever and Gunderson counted B3 diagnoses among the biological relatives of B1 and B3 index adoptees, and compared this total to the control group: "In this analysis, the prevalence rate of borderlines in relatives of the B1 and B3 index cases (8/142 or 5.6 percent) significantly exceeded the rate in relatives of screened controls (1/113 or 0.9 percent) (p = .039)."[1] However, there at least five reasons why this conclusion is unconvincing. *First*, 5 of the 8 index B3 diagnoses were given to half-siblings, which runs counter to genetic predictions. *Second*, the B3 rate among the biological relatives of B3 adoptees tells us nothing about the relationship between B1 and B3, yet Siever and Gunderson figured three such diagnoses into their comparison (see Table 3.5). *Third*, Siever and Gunderson removed the biological relatives of the B2 adoptees from their comparison. However, there was equal justification for removing both B2 *and* B3. *Fourth*, 2 of 8 index B3 diagnoses were record-based only, and were not counted by Kety et al. in 1975. One relative had emigrated, and the other received no 1975 diagnosis. Had Siever and Gunderson decided to apply the same criteria to B1 as they had to B3, their index/control B1 comparison, like the B3 comparison, would have been statistically nonsignificant. And *fifth*, they arbitrarily limited their comparison to the 1975 "screened" controls, which reduced the control B3 rate from 1.7% to 0.9%.[2]

Although Siever and Gunderson concluded that "the results all converge in suggesting a genetic relationship between borderlines and chronic schizophrenics," they reached this conclusion on the basis of the five errors I have just described.

Eugen Bleuler and "Latent Schizophrenia"

Kety and associates justified the inclusion of latent schizophrenia (B3) in their spectrum on the grounds that it was identified and described by the creator of the schizophrenia concept, the Swiss-German psychiatrist Eugen Bleuler. (In the Danish-American series, "latent schizophrenia" and "borderline schizophrenia" were used interchangeably.) For example, Kety wrote that in addition to the DSM-II description of latent schizophrenia,

> We also took into account Bleuler's description of the symptoms of latent schizophrenia as he observed them in the relatives of overt schizophrenia patients.

1. Ibid., p. 63.
2. See Joseph, 2004b, pp. 239-241. The 1975 "screened" controls had been interviewed by the researchers and judged free from schizophrenia or related diagnoses. The *entire* Kety et al. 1968 control group was "unscreened," which did not prevent mainstream psychiatry from pointing to the allegedly significant results it produced. By Siever and Gunderson's logic, we should completely disregard the 1968 study's results because the control group was not screened.

Bleuler's description of latent schizophrenia actually was the most useful guide since only those observations, like ours, had been made on individuals not hospitalized or seeking treatment.[1]

Bleuler did indeed identify the condition in his 1911 book on schizophrenia: "There is also a latent schizophrenia, and I am convinced that this is the most frequent form, although admittedly these people hardly ever come for treatment."[2] However, Kety, Wender, and Rosenthal overlooked a critical element in the way Bleuler proposed that this condition be diagnosed. Had they remained true to Bleuler's teachings, they would not have included latent schizophrenia in their spectrum.

Bleuler believed that latent schizophrenia could be diagnosed only *after* a person had manifested more serious symptoms. In the following passage, he described the difference between chronic and "milder cases of schizophrenia":

> As in every other disease, the symptoms must have reached a certain degree of intensity if they are to be of any diagnostic value. Yet in milder cases of schizophrenia we find a number of prominent manifestations, which strongly fluctuate within the limits of what is regarded, if not as healthy, at least as "not mentally ill." Character anomalies, indifference, lack of energy, unsociability, stubbornness, moodiness, the characteristic for which Goethe could only find the English word, "whimsical," hypochondriacal complaints, etc., *are not necessarily symptoms of an actual mental disease;* they are, however, often the only perceptible signs of schizophrenia. *It is for this reason that the diagnostic threshold of schizophrenia is higher than that of any other disease...* [emphasis added].[3]

Thus, according to Bleuler, one cannot reliably distinguish between "milder cases of schizophrenia" and merely "whimsical" people. (It is beyond the scope of this chapter to challenge his concept of schizophrenia as a "disease."[4]) He therefore demanded a high diagnostic threshold for schizophrenia to eliminate the possibility that people not suffering from an "actual mental disease" would be diagnosed with schizophrenia. "Only a few isolated psychotic symptoms can be utilized in recognizing the disease," wrote Bleuler, "and these too, have a very high diagnostic threshold value."[5]

For Bleuler, mild or "simple" schizophrenia (a B3 equivalent; see Figure 3.1) was a retrospective diagnosis to be made on the basis of a patient's *later* difficulties:

> Such mild cases are often considered to be "nervous" or "degenerated" individuals, etc. But if we follow the anamnesis of those who are admitted to the hospital in later years because of an exacerbation of their difficulties, a criminal charge, a pathological drinking bout or some such episode, we can usually find throughout

---

1. Kety, 1985b, p. 592.
2. Bleuler, 1950, p. 239.
3. Ibid., p. 294.
4. See Szasz (1976, Chapters 1 & 3) for an excellent critical discussion of this topic.
5. Bleuler, 1950, p. 294.

the entire past history of the individual mildly pathological symptoms which *in the light of their recent illness* unquestionably have to be considered as schizophrenic [emphasis added].[1]

Moreover, according to Kety, Bleuler saw "latent schizophrenia" as a "broader and milder" form of "simple schizophrenia"[2]

Basing his disease model on the then recently discovered etiology of neurosyphilis,[3] Bleuler viewed schizophrenia as "remain[ing] latent until an acute pathological thrust produces prominent symptoms, or until a psychic shock intensifies the secondary symptoms."[4] Kety, who claimed allegiance to Bleuler's theories, ignored his warning that mild cases should be diagnosed as schizophrenia only "in the light of [a patient's] recent illness." On the other hand, Kety cited Bleuler in support of removing B2 from the spectrum, arguing that the DSM-II had "deviated" from Bleuler's teachings by including "'acute schizophrenic reaction'...despite Bleuler's admonition that these are 'partial phenomena of the most varied diseases [whose] presence is often helpful in making the diagnosis of a psychosis, but not in diagnosing the presence of schizophrenia.'"[5] In this case Kety cited Bleuler's high diagnostic threshold in support of his decision to remove B2 from the spectrum, while ignoring a comparable threshold Bleuler established for B3, a diagnosis Kety decided to retain. Thus, Kety's differing standards on the statistical plane were paralleled by his selective citation of Bleuler on the historical plane.

Bleuler insisted on a "very high diagnostic threshold" for diagnosing schizophrenia, in direct contrast to Kety, Rosenthal, and associates, who may have applied the schizophrenia label to dozens of "whimsical" relatives in their studies. Two contrasting quotations show how far Kety and colleagues strayed from Bleuler's teachings:

Bleuler

> Only a few isolated psychotic symptoms can be utilized in recognizing [schizophrenia], and these too, have a very high diagnostic threshold value.[6]

Rosenthal

---

1. Ibid., p. 239.
2. Kety, 1980, p. 423.
3. Szasz, 1976.
4. Bleuler, 1950, p. 463.
5. Kety, 1985a, p. 6. Although Kety gave the reference for Bleuler's quotation on acute schizophrenia as "Bleuler, 1911/1950, p. 204," Bleuler did not mention acute schizophrenia on page 204, and I have not been able to find this quotation elsewhere in Bleuler's 1911 monograph.
6. Bleuler, 1950, p. 294.

Paul Wender and I were working downstairs, in concert with Seymour Kety, in effect broadening the concept of schizophrenic disorder as widely as it may have ever been reasonably conceived before...[1]

Thus, B3 not only failed to cluster significantly among the biological relatives of index adoptees, it also lacked the historical and theoretical legitimacy claimed by Kety and associates.

OTHER SCHIZOPHRENIA SPECTRUM DISORDERS (SSDS)

Kety et al. made 16 biological relative "uncertain" D diagnoses in the 1975 study (13 index and 3 control), which represented 28% of all SSD diagnoses. However, they described D as a "highly subjective and as yet nonexplicit category," and "vague and subjective...which hardly qualifies as schizophrenia according to our own or other criteria."[2] Clearly, "uncertain" diagnoses should not be counted in schizophrenia epidemiological research. American psychologist Nicholas Pastore came to a similar conclusion in his 1949 analysis of psychiatric geneticist Franz J. Kallmann's celebrated 1938 German schizophrenia family study.[3] In that study, Kallmann counted "doubtful" cases of schizophrenia with the same weight as "definite" schizophrenia cases, and some of his major conclusions depended on counting these doubtful relatives. In Pastore's view, "there is no justification" for counting doubtful cases as schizophrenia, "unless it is assumed that the 'doubtfuls' will eventually become definite schizophrenics, in which case the 'doubtfuls' should have been so classified at the outset of the investigation."[4]

I quoted Kety earlier to the effect that he wanted to "test the hypothesis" that "uncertain schizophrenia" was related to chronic schizophrenia. His fundamental error, however, was his belief that a *lack of information constitutes a diagnosis.* To illustrate this point, when oncologists lack sufficient evidence, they do not diagnose their patients with "uncertain cancer." Instead, they investigate further and either diagnose cancer or rule it out.

Kety wrote on several occasions to the effect that, "in the case of the relatives, questionable or uncertain schizophrenia had to be added if relatives with less certain diagnoses were not to be lost."[5] Keeping track of relatives is one thing. It is quite another to count "questionable or uncertain" relatives as "schizophrenia" in statistical comparisons. Elsewhere, Kety wrote that "a category of uncertain schizophrenia was necessary because a relative could not be

---

1. Rosenthal, 1975, p. 19.
2. Kety et al., 1976, p. 420.
3. Pastore, 1949.
4. Ibid., p. 290.
5. Kety, 1987, p. 424.

rejected from the study as a candidate proband [adoptee] could be if the diagnosis were less than definite."[1] Although Kety believed that these relatives "could not be rejected," he provided no explanation for *why* they could not be rejected, other than to prevent them from becoming "lost." In the Danish-American series, "uncertain borderline schizophrenia" (D3) second-degree relatives carried the same statistical weight as chronic schizophrenia (B1) first-degree relatives.

In 1983, Kety recognized that Category C (schizoid and inadequate personality) did not belong in the schizophrenia spectrum:

> Our diagnoses of schizoid and inadequate personality in this study [Kety et al., 1975] did not discriminate between the genetic relatives of schizophrenic adoptees and their controls; in fact, the prevalence was the same in both. There was thus no justification for believing that schizoid and inadequate personality, as we had diagnosed them in the interview study, were related to schizophrenia, and were therefore excluded from the subsequent analyses.[2]

Noteworthy is Kety's decision to remove C on the grounds that its prevalence "was the same in both" groups. From the standpoint of inferential statistics, which (as we saw in Chapter 1) usually requires a p-value below .05 to claim statistical significance, he could have made the exact same statement about B3, which he decided to *retain* in the spectrum. Arbitrarily, Kety did not apply the same criteria to B3 as he did to C (or B2).

There is, in fact, a substantial overlap between the C, B3, and D3 diagnoses.[3] This is seen in the following comparison between (SSD) B3 and (non-SSD) C. In a 1985 article, Kety acknowledged that his 1975 B3 interview diagnoses were based on milder symptoms than the hospitalized 1968 B3 record-based diagnoses.[4] Table 3.7 presents a side-by-side comparison of Kety's 1975 B3 cases, and the DSM-II definition of schizoid personality. As it turns out, the 1975 B3 relatives were barely distinguishable from those diagnosed as C.

The similarity between Kety's 1975 latent (borderline) schizophrenia and DSM-II schizoid personality is apparent in Table 3.7. Both diagnoses were made on the basis of (1) flat affect/inability to express feelings, (2) poor contact/seclusiveness, (3) poor interpersonal relationships/avoidance of close relationships, and, (4) bizarre thinking/eccentricity.

---

1. Ingraham & Kety, 1988, p. 121.
2. Kety, 1983a, p. 723.
3. Lidz & Blatt, 1983.
4. Kety, 1985b. Only 1 of the 7 record-based B3 diagnoses was sustained by interview, meaning that 8 of 9 B3 diagnoses in the 1975 study were new.

Table 3.7

DIAGNOSTIC DESCRIPTIONS OF SCHIZOPHRENIA SPECTRUM BORDERLINE (LATENT) SCHIZO-
PHRENIA, AND NON-SCHIZOPHRENIA SPECTRUM SCHIZOID PERSONALITY

Borderline (Latent) Schizophrenia Relative Diagnoses in the Kety et al. 1975 Adoptees' Family Study	Complete DSM-II Definition of Schizoid Personality
"For our diagnoses made from interviews in the 1975 study....We used the DSM-II description of latent schizophrenia, schizoid and inadequate personality.... Our diagnoses of latent and uncertain schizophrenia in the relatives, therefore, included a majority with flat affect, bizarre thinking, poor contact, and poor interpersonal relationships rather than the positive symptoms which appeared to characterize the [1968] hospitalized group" (Kety, 1985b, p. 592).	"This behavior pattern manifests shyness, over-sensitivity, seclusiveness, avoidance of close or competitive relationships, and often eccentricity. Autistic thinking without loss of capacity to recognize reality is common, as is daydreaming and the inability to express hostility and ordinary aggressive feelings. These patients react to disturbing experiences and conflicts with apparent detachment" (APA, 1968, p. 42).
Summary of Kety's 1975 B3 symptoms (schizophrenia spectrum):	Summary of DSM-II schizoid personality symptoms (non-spectrum):
(1) Flat affect. (2) Poor contact. (3) Poor interpersonal relationships. (4) Bizarre thinking.	(1) Inability to express feelings. (2) Seclusiveness. (3) Avoidance of close relationships. (4) Eccentricity, autistic thinking.

In 1979, psychiatric investigators Robert Spitzer and Jean Endicott[1] examined the Danish-American records in an effort to define diagnostic categories for DSM-III.[2] They created a new diagnosis, schizotypal personality disorder (SPD), from eight symptoms distinguishing Kety and colleagues' SSD and non-SSD relatives. (Kety subsequently described schizotypal personality disorder as "comparable to our diagnosis of latent schizophrenia."[3]) Consistent with the evidence I have presented in this chapter, Spitzer and Endicott could not adequately separate schizoid and schizotypal personalities into two discrete diagnostic categories, concluding that SPD "was merely a subdivision of what has for years been referred to as Schizoid Personality Disorder."[4] The similarity between these two diagnoses was codified in the DSM-III, which differentiated schizoid personality disorder and schizotypal personality disorder on the sole basis of the latter's "eccentricities of communication or behavior."[5] It is therefore understandable that Kety, Rosenthal, and Wender could write, in reference to B3, D3, and C, that "it is doubtful that we could demonstrate a significant differentiation between these categories."[6] That is, they recognized their inability to differentiate between an SSD and other diagnoses or behavioral constellations.

---

1. Spitzer & Endicott, 1979.
2. APA, 1980.
3. Kety & Ingraham, 1992, p. 250.
4. Spitzer & Endicott, 1979, p. 98.
5. APA, 1980, p. 310.
6. Kety, Rosenthal, & Wender, 1978, p. 220.

Yet, the validity of their study depended on an ability to reliably make such differentiations.

Although schizoid personality is widely known in psychiatry, "inadequate personality" is little known and was not included in the DSM-III and subsequent editions. The complete DSM-II description of inadequate personality reads as follows:

> This behavior pattern is characterized by ineffectual responses to emotional, social, intellectual and physical demands. While the patient seems neither physically nor mentally deficient, he does manifest inadaptability, ineptness, poor judgment, social instability, and lack of physical and emotional stamina.[1]

While few psychiatrists would diagnose such as person with schizophrenia, Kety wrote that the investigators "had great difficulty making a clear and satisfying distinction between 'inadequate personality' and mild 'borderline schizophrenia.'"[2] Years later, Kety wrote that the 1968 B3 relatives had more "negative symptoms" than the 1975 B3 group.[3] If Kety and associates had "great difficulty" distinguishing "mild" 1968 B3 cases from DSM-II inadequate personality, one can safely assume that they had extremely great difficulty distinguishing inadequate personality from the 1975 B3 cases.

Does Association Imply Genetic Relatedness?

Thus far, we have found little evidence in the Danish-American studies linking chronic B1 schizophrenia to the other SSDs, even on the basis of the investigators' stated criteria (although not their method) for determining genetic relatedness. However, even if B3 or any other diagnosis had been significantly concentrated among the biological relatives of B1 index cases, it would not necessarily indicate that the diagnosis is related, genetically or otherwise, to B1 chronic schizophrenia.

According to Kety and colleagues, a mere correlation between diagnoses suggests genetic relatedness. It is well known, however, that correlation is not the same as causation. In the following passage Rosenthal concluded, on the basis of a correlation, that diagnoses are genetically related to each other. Referring to the non-B1 SSDs, he wrote,

> These combined diagnoses occurred about six times more frequently among the biological relatives of our index cases than among the biological relatives of the controls. For this reason we feel justified in having broadened the range of schizophrenic disorders studied to include those that we thought might be genetically related to process [B1] schizophrenia.... [the B2 diagnosis] may have to be elimi-

---

1. APA, 1968, p. 44.
2. Kety, 1970, p. 238.
3. Kety, 1985b.

nated from the spectrum. Among the 30 biological relatives of 7 index cases who had this diagnosis, we did not find a single instance of schizophrenic spectrum disorder.[1]

Thus, Rosenthal believed that a diagnosis is genetically related to schizophrenia because, when combined with other diagnoses, it is found more frequently among index versus control biological relatives.[2] Discussing the investigators' decision to broaden the definition of schizophrenia, psychologist Alvin Pam observed that "with a bigger net, they caught more fish."[3] Expanding on this analogy, one might note that, according to Kety and colleagues' logic, seagulls and crabs caught up in the net would be classified as fish, if the total catch were significantly greater than that of a control group of fisherman.

In his 1938 family study, Kallmann found an "enormous increase" in mortality from tuberculosis among the children of schizophrenia patients.[4] In addition, he found that almost 22% of these patients had themselves died of tuberculosis. This led him to conclude that schizophrenia and tuberculosis were genetically related to each other, and that "a very particular significance must be assigned to tuberculosis in the entire heredity-circle of schizophrenia."[5] Kallmann's correlation, however, was spurious because he failed to recognize that the high rate of tuberculosis among schizophrenia patients and their relatives was the result of environmental conditions common to both schizophrenia and tuberculosis patients. Had Kallmann created a "schizophrenia spectrum" in 1938, tuberculosis would have likely been a part of it — and with greater justification than any of the non-B1 Danish-American SSDs. Tuberculosis is a physical disease found in significantly greater numbers among the families of Kallmann's hospitalized schizophrenia patients versus the general population rate. Conversely B3, which Rosenthal described as a diagnosis "so difficult to make with good reliability in many cases,"[6] was found in *statistically non-significant* numbers among the biological relatives of SSD index adoptees versus controls. It is possible that the greater level of psychological distress among these relatives was the result of differences in rearing environments due to, among other possibilities, the selective placement of adoptees (see Chapter 4).[7] Thus, it is likely that Kety and colleagues' conclusion that the SSDs are genetically related to schizophrenia (and to each other) was as spurious as Kallmann's.

---

1. Rosenthal, 1972, pp. 68-69.
2. Rosenthal's claim that non-B1 SSDs were found "six times more frequently" among index biological relatives requires clarification. In the 1968 Kety et al. study, 9 of 13 non-B1 index biological relative SSDs were given to half-siblings. Among first-degree relatives versus controls, the rate was a non-significant 4 index versus 2 control.
3. Pam, 1995, p. 29.
4. Kallmann, 1938a, p. 82.
5. Ibid., p. 86.
6. Rosenthal, 1971a, p. 95.
7. See Joseph, 2004b.

The 1968 schizophrenia spectrum included conditions ranging from possibly "psychotic" individuals (B1), to non-psychotic people who shun social contact or whose personalities are deemed "inadequate" (C). The B3 diagnosis, which Kety and colleagues admitted they made "in the absence of frank delusions or other psychotic symptoms," does not describe people commonly regarded as psychotic.[1] They diagnosed individuals B3 because they saw them as strange or reclusive people who don't conform to societal norms. Furthermore, B3 individuals don't think the same way that "we" do; they don't understand "our" sense of reality or logic; they can't adapt to "our" society; and they may have sexual relations with people not of the opposite sex, which "we" find distasteful. In fact, Kety et al. included homosexuality in all SSD diagnostic formulations (other than C; see Figure 3.1). One could imagine them diagnosing someone with an SSD on the basis of abnormal behavior plus homosexuality. And there is, in fact, at least one documented example of Kety, Rosenthal, and Wender diagnosing a person with an SSD on the basis of suspected homosexuality. The complete "diagnostic statement" of this individual read, "Schizophrenia borderline or perverse (homosexual, transvestite). Could break down with schizophrenia episode."[2]

Although the American Psychiatric Association removed homosexuality (or at least "ego-syntonic homosexuality") from its list of mental disorders in 1974,[3] the Danish-American investigators made no mention of this in their publications. Moreover, homosexuality was not mentioned in the three DSM-II diagnostic descriptions from which Kety et al. derived B1, B2, and B3, meaning that they — but not the DSM-II — used homosexuality as a diagnostic criterion for schizophrenia. (Homosexuality was not mentioned in the DSM-I schizophrenia descriptions, either.) Rosenthal cited studies claiming that 75% of people diagnosed with schizophrenia are "overt homosexuals" or have had homosexual experiences.[4] Because Rosenthal, Kety and colleagues associated homosexuality with their SSDs, one might ask why, in light of Rosenthal's belief in a large "clinical overlap between schizophrenia and homosexuality," they did not count this "disorder" as an SSD as well.[5]

Even if he did not explicitly include homosexuality, Rosenthal defined "schizophrenia-related illness" more broadly than anyone else:

> For every hospitalizable schizophrenic, there are many more people in the community who have a schizophrenic-like type of disorder which is not severe enough to require hospitalization. These individuals are called borderline or pseudoneurotic schizophrenic, schizoid, paranoid, or simply cold, distant and inadequate, or odd and eccentric.[6]

1. Kety et al., 1994, p. 445.
2. Wender et al., 1974, p. 124.
3. Kutchins & Kirk, 1997.
4. Rosenthal, 1979, p. 28.
5. Ibid., p. 28.
6. Ibid., p. 23.

Disregarding Bleuler's admonitions, for Rosenthal it apparently was enough to be "simply cold, distant and inadequate," or simply "odd and eccentric," to deserve the schizophrenia label. Thus the spectrum, which Rosenthal and his colleagues created to enable them to continue their investigations, became transformed in Rosenthal's mind into evidence supporting a very broad definition of schizophrenia. In essence, Rosenthal widened the definition of schizophrenia in order to conduct his studies, and then reasoned backwards that a major finding of his studies had been that the SSDs are genetically related to chronic schizophrenia. This is not a scientific finding; it is a self-fulfilling prophecy.

Had Rosenthal decided against expanding the schizophrenia definition he would have been compelled, by his own admission, to conclude that he and his colleagues had found no evidence in support of genetic influences on schizophrenia:

> It should be apparent now that if we had included in our comparisons of index and control relatives only those who clearly had process [chronic B1] schizophrenia, we would have found no difference between the two groups of relatives.[1]

It is difficult to discard deeply held beliefs — entailing the negation of one's life work and endangering personal and professional ties related to this work — in the face of evidence that these beliefs are unfounded. However, a belief system can be maintained, in the face of evidence against it, by redefining the theory and premises upon which it is built by inventing ad hoc hypotheses (see Chapter 1). On a personal level this is understandable. What is more difficult to accept is how clearly erroneous conclusions flowing from investigators' personal beliefs, professional affiliations, and possibly economic interests are transformed into a discipline's core knowledge.

Wender's conception of schizophrenia was similar to Rosenthal's:

> The heterogeneous group of individuals who are believed by some to be biological — as well as familial — relatives of schizophrenics merges with the group described as having schizoid personality disorder. Called schizotypal, these people vary from the shy, timid and unsociable to the callous, cold, harsh, and distant, from the quiet, empty, and intelligent to the sensitive and poetic or to the militant, rigid, and fanatic (political or religious).[2]

One might ask how many "shy, timid and unsociable," "odd and eccentric," "sensitive and poetic," or "rigid" people Rosenthal and Wender diagnosed as "schizophrenic" (B3, D3, or C) in the Danish-American series.[3]

---

1. Rosenthal, 1972, p. 68.
2. Wender & Klein, 1981, p. 127.
3. In their publications, the investigators referred to B1, B2, and B3 as "definite schizophrenia." This is puzzling because they intended to *test the hypothesis* that B2 and B3 were related to B1. Because the investigators eventually removed B2 from their spectrum, this diagnosis was not so "definite" after all. Regarding B3, Lidz and Blatt (1983, p. 430) commented wryly that "many of the subjects placed in this category were not definitely schizophrenic but, rather, were definitely not schizophrenic."

Psychologist Mary Boyle's comments on the broad definition of schizo-phrenia utilized by some twin researchers are relevant to Kety and colleagues' schizophrenia spectrum as well. Using a medical example, Boyle observed that diabetes twin studies "looked at concordance for signs and symptoms and for *signs* only."[1] She argued that if a subject was found to have had an abnormal response to a glucose-tolerance test, severe thirst, sugar in the urine, weight loss, tiredness, etc., it would be wrong to consider a co-twin concordant who had complained only of weight-loss and tiredness in the absence of signs. Fur-thermore,

> The behavior of twin researchers who use the spectrum concept is, in fact, not dissimilar to that of medieval medicine men who tried to group physical complaints in ignorance of their antecedents and who created classificatory chaos. It must be added that in "schizophrenia" research the error is compounded because concor-dance is not judged by degree of similarity between twins but by the extent to which each twin independently satisfies the researchers' criteria for "schizophre-nia" or the "schizophrenia spectrum." Because these criteria are so broad and sub-jective it is quite possible to call concordant a twin pair who have none or very few of the behaviors of interest in common...[2]

In the Danish-American schizophrenia spectrum, for example, a person diagnosed B2 would share few if any diagnostic criteria with a schizoid indi-vidual. Yet, Kety and colleagues counted B2 and C cases as "schizophrenia" in statistical formulations.

CONCLUSION

The evidence suggests that Kety, Rosenthal, and Wender created their schizophrenia spectrum more by necessity than on the basis of theoretical or empirical soundness. We have seen that in the Danish-American studies, the B3, B2, and C categories are not related to chronic B1 schizophrenia by statistical, theoretical, or empirical evidence. Moreover, it was improper for the investi-gators to count "uncertain" cases in general, and "uncertain borderline schizo-phrenia" in particular, as schizophrenia. Because the Danish-American investigators subsequently removed several SSDs, there remain only two dis-puted diagnoses: B3 and D3.

Looking specifically at B3, we saw that Kety and associates made a crucial error in the way they counted relatives with this diagnosis. Contrary to their methods of calculation, the "B1/B3 relationship" formulation is the only valid method to determine if B3 clustered in statistically significant numbers among index versus control biological relatives. According to my calculations, there was

---

1. Boyle, 2002b, p. 165.
2. Ibid., p. 166.

no significant B3 clustering in either the Kety et al. 1968 or 1975 studies (or in Rosenthal's 1968 and 1971 studies). We also saw that, even if such clustering had occurred, one still could not conclude that B1 and B3 are genetically related. The reason is that mere association does not necessarily indicate that diagnoses share a common genetic etiology. We also saw that Eugen Bleuler, inventor of the schizophrenia concept and an important source of Kety's justification for the spectrum, most likely would not have diagnosed Kety's B3 adoptees and relatives with a milder form of schizophrenia. Bleuler diagnosed these people with schizophrenia retrospectively, on the basis of their subsequently more serious difficulties.

In this chapter I have argued that the schizophrenia spectrum, as Kety, Rosenthal, and Wender defined it, is invalid on several grounds. Thus, B1 chronic schizophrenia is the only diagnosis they should have used in their studies, implying that the Danish-American schizophrenia adoption studies, and their authors' conclusions in favor of genetics, should be reassessed on this basis.

CHAPTER 4. PELLAGRA AND GENETIC RESEARCH

Pellagra is a disease that ravaged poor people in the southern United States during the first half of the 20th century.[1] Before then, it had been known in southern Europe for almost 200 years. The often fatal disease, still found among the world's poor, is characterized by severe skin rash, gastrointestinal problems, and mental disturbance. Between 1730 and 1930, pellagra claimed over half a million lives, including tens of thousands of poor black and white Southerners.[2] Due to the pioneering work of Joseph Goldberger and others, it was firmly established that pellagra has a single cause: a vitamin deficiency (mainly niacin) linked to malnutrition. In other words, pellagra was (and is) a disease of hunger and poverty. Despite the publication of Goldberger's findings in the late 1910s, it took 25 more years to finally wipe out pellagra in the United States.[3] The tragedy of the unnecessary period between discovery and effective action has been detailed elsewhere.[4]

In this chapter I discuss pellagra as the subject of genetic research, both real and hypothetical, since it illustrates how psychiatric genetic research methods are potentially confounded by environmental factors, and how the "genetic predisposition" concept is often irrelevant and potentially harmful. The discussion is limited to a published family pedigree study and to speculation about what pellagra twin and adoption studies might have found. Although there is a current widespread consensus in psychiatry that individuals who suffer from mental disorders carry a genetic predisposition, many of the environmental factors thought to trigger these disorders are either controversial or are unknown. Fortunately, the cause of pellagra *is* known and, as we will see, psy-

---

1. This chapter is based on a revised and expanded version of a previous publication (Joseph, 2000d).
2. 10,000 deaths in 1929 alone; Carpenter, 1981.
3. Goldberger & Wheeler, 1915; Goldberger et al., 1918.
4. Chase, 1980, Chapter 9; Etheridge, 1972; Roe, 1973.

chiatric genetic methods would be expected to point to the operation of genetic factors in a condition known to be caused not by genes, but by a dietary deficiency

A PUBLISHED FAMILY STUDY OF PELLAGRA

We have seen that although the general public often equates "running in the family" with "genetic," most epidemiologists and psychiatric geneticists now recognize that, because families share both common genes and a common environment, genetic conclusions from the results of family studies are confounded by environmental factors.

Before pellagra's true cause was discovered, leading early 20th-century American eugenicists such as Charles Davenport promoted the idea that pellagra had an important genetic component. Although "pellagra is not an inheritable disease in the sense in which brown eye color is inheritable," he wrote, "the course of the disease does depend... on certain constitutional, inheritable traits of the affected individual."[1] In support of a hereditary basis for pellagra, Davenport provided 15 pages of affected family pedigrees from his own "eugenical family study."[2] His pedigree diagrams showed that pellagra clustered in families more often than the general population expectation. According to Davenport, "the constitution of the organism must be held to be the principal cause of the diversity which persons show in their reaction to the same disease-inciting factors. This constitution of the organism is a racial, that is, hereditary factor."[3] Unfortunately, Davenport failed to mention that most pellagrins lived in poverty, or that reports by several epidemiological investigators pointed to nutritional factors in pellagra.

Thus, Davenport's pellagra family pedigrees, and the conclusions he and others drew from them, are a classic example of the fallacy of inferring genetic causation from the results of family studies, but only because we now know that pellagra is caused by a niacin deficiency. This is why we can look back today and recognize the fallacious thinking of Davenport and his associates. Although most psychiatric conditions also run in families, they differ from pellagra in that (1) their causes are either unknown, unproven, or controversial, and (2) they are not physical diseases of the body.

Although pellagra was known to run in families, to my knowledge no pellagra twin or adoption study was ever published. In the rest of this chapter I speculate about the results pellagra twin and adoption researchers would have

---

1. Davenport, 1916, p. 13.
2. Ibid.
3. Ibid., p. 4; Davenport mistakenly believed that African-Americans were constitutionally less susceptible to pellagra.

obtained. I then examine the implications this might have for contemporary psychiatric genetic research.

TWIN STUDIES

Suppose that the twin method had been utilized in early 20th century Spartanburg, South Carolina, a center of pellagra investigation, in order to assess for possible genetic influences on the disease. The investigators would have likely identified subjects either by obtaining the names of hospitalized pellagrins who had a twin (commonly referred to as the "resident hospital method"), or by obtaining the names of all pellagrins admitted to a hospital during a particular time period, and then identifying the twins from this group (commonly referred to as the "consecutive admissions method"). Suppose that researchers had discovered 50 identical and 50 same-sex fraternal twin pairs in this manner. Given that we now know that pellagra is caused by a vitamin deficiency, what results would we expect to find among these twins?

At first it might appear that we would not find a significant identical-fraternal concordance difference on environmental grounds because, as Don Jackson argued was the case with schizophrenia, we would not expect higher identical twin pellagra concordance on the basis of the greater psychological association of identical twins.[1] Association and "ego fusion" might compel a twin to behave like his or her co-twin, as recognized by a schizophrenia genetic researcher as prominent as David Rosenthal, but would not produce more physical disease among identical versus fraternal co-twins.[2] However, we saw in Chapter 1 that identical twins are much more likely to spend time together and experience similar environments than fraternals.

Looking specifically at food choices, one study found that identical twins "show the greatest similarity in food tastes" when compared to same-sex and opposite-sex fraternal twins, and another that identicals had more similar preferences for fruits and vegetables (but not for staple items).[3] More recently, the authors of a large study of 455 pairs (232 identical, 223 fraternal) looked at differences in twin dietary intake and found that "intraclass correlations are consistently and significantly higher for identical twins for every nutrient."[4] Confirming these findings, another group found "a greater similarity in food preference between members of monozygotic twin pairs than between dizygotic twin pairs...."[5]

---

1. Jackson, 1960.
2. Rosenthal, 1960.
3. Wilson, 1934, p. 338; Smith, 1965, p. 56.
4. Fabsitz et al., 1978, p. 19.
5. Falciglia & Norton, 1994, p. 154.

Of course, unlike the largely middle-class twins in these dietary studies, impoverished Southern identical and fraternal twins had few dietary choices and often went hungry. Thus, we cannot be sure that the greater identical twin correlation for food preference and nutrient intake could have been generalized to this population, but to the extent that identical twins spend more time together, they will eat more similar foods than fraternal twins and will therefore be more concordant for pellagra. There is also reason to believe that identical twins would have eaten more similarly within the home. The 2-3 times higher pellagra rate among poor Southern women than men has been explained on the basis that men, as the family breadwinners, were given the best food available to the family, such as meat and eggs. To the extent that identical twins are treated more alike than fraternal twins, it is likely that they would have eaten more similar foods within the family than fraternals.

It is therefore probable that a pellagra twin study (using the classical twin method) would have found a significantly higher identical versus fraternal twin concordance rate on environmental grounds alone, but the difference, though statistically significant, would not have been as great as most psychiatric conditions, where large identical-fraternal differences are the result of common environment *plus* association, ego-fusion, and more the more similar treatment of identicals.

Reared-Apart Twins

The concordance rates of *reared-apart* identical twins have been cited as evidence supporting a genetic basis for psychiatric disorders such as schizophrenia. However, it is important to note that, because samples are too small to yield enough cases, there has never been a systematic study of psychiatric disorders using reared-apart twins. Some researchers have reported on *individual pairs* of twins judged concordant or discordant for various psychiatric disorders. Regarding schizophrenia, reviewers sometimes claim that the concordance rate of a handful of individual reared-apart identical pairs in the literature is around 65%.[1] Rather than dispute this figure, or discuss why these cases do not constitute scientifically acceptable evidence in favor of genetic factors, I will concentrate on the fact that reared-apart identical twins (also known as monozygotic twins reared apart, or MZAs) are usually placed into highly correlated socioeconomic environments, often into different branches of the same family, and many are poor. As Susan Farber noted in her exhaustive 1981 review of the world MZA literature:

> Twins are usually separated because of poverty or death in the family. Though the different cultures and time spans make exact rating of socioeconomic status impossible, it is clear that most of the cases were born to poor families and reared in

---

1. Gottesman & Shields, 1982.

conditions not greatly different. Many were reared by relatives. The sample...is highly biased in the direction of lower and lower-middle-class rearing conditions.[1]

Thus, had investigators searched the southern United States for pellagrins having a reared-apart identical twin, it would not have been surprising if they had found a certain percentage concordant for pellagra for socioeconomic reasons alone.

Although reared-apart twin researchers usually argue, albeit incorrectly, that correlated environments explain only a tiny fraction of MZA personality and IQ correlations,[2] we certainly would expect MZAs' correlated socioeconomic environments to increase their concordance for a disease like pellagra, where malnutrition is the known cause. MZAs placed into separate impoverished families eating a pellagragenic diet would have been at risk for concordance; MZAs living in separate middle- or upper-class homes would have been at very low risk.

In general, because studies of both twins reared together and twins reared apart are contaminated by environmental factors, their use in the study of pellagra probably would have led to conclusions as erroneous as those reached by Davenport and his co-thinkers on the basis of family pedigree studies.

ADOPTION STUDIES

There is little doubt that the results of a pellagra adoption study conducted in the southern United States in the first part of the 20th century would have led researchers to conclude that important genetic factors were implicated. This is due to the likelihood that adoptees' family background would have played a role in determining the types of homes into which they were placed, as well as playing a role in the types of families willing to adopt them.

The Adoptees method (Figure 2.1) is the most frequently used adoption design in psychiatric genetic research. As we recall, this method compares the prevalence of a particular diagnosis (which I'll call "Disorder X" here) among the adopted-away biological offspring of parents (usually mothers) diagnosed with Disorder X, versus a control group consisting of the adopted-away biological offspring of parents never diagnosed with Disorder X. According to contemporary psychiatric geneticists, a statistically significant excess of Disorder X among index (experimental) adoptees versus control adoptees suggests the operation of genetic factors. Conversely, they argue that similar rates in both groups suggests that Disorder X is caused by (possibly unspecified) environmental factors. Unfortunately, most adoption researchers downplay the possibility that

---

1. Farber, 1981, p. 18.
2. e.g., Bouchard, 1993.

adoption agency policies could contribute to a significantly higher index adoptee rate of Disorder X, even if the disorder is caused solely by environmental factors. Therein lays the problem of selective placement, which all adoption studies must address. According to genetic researchers Clerget-Darpoux, Goldin, and Gershon:

> The critical underlying assumption in adoption studies is that the genetic and environmental variables are separated by the adoption process. This is not true if the trait in question is correlated with an environmental variable for which there is some matching between adoptive and biological parents.[1]

Pellagra, which Davenport's contemporaries viewed as being influenced or caused by bad heredity, provides an example of how such a correlation could occur. They key points to remember are, (1) that adoption agencies frequently used socioeconomic status (SES) as a matching criterion, influencing into which families they decided to place children, and (2) that adoption agencies and prospective adoptive families believed that pellagra was caused by hereditary factors.

Let's begin with the first point. Suppose a team of adoption researchers had traveled to South Carolina in 1915 to conduct a pellagra adoption study, and had obtained the records of 50 women who had been diagnosed with pellagra and who also had given up a child for adoption. Because most pellagrins were poor, the adoption agencies' policy of matching adoptees and prospective adoptive families for SES meant that most of these adoptees would have been placed into the homes of poor families, thus becoming more susceptible to consuming a pellagragenic diet than if they had been randomly placed into the full range of available adoptive homes. If the investigators had then located 50 undiagnosed mothers for a control group, without carefully matching both groups of mothers for SES, then clearly we would have expected the index adoptees, who had grown up in lower SES environments, to have more cases of pellagra than the control adoptees. Moreover, matching might have occurred on the basis of other trait-relevant variables. As Clerget-Darpoux and colleagues pointed out, "for most of the disease[s] and traits studied, we do not know the etiology and thus do not know all of the relevant variables to examine for correlation between adoptive and biological parents."[2]

Thus the knowledge of a disorder's environmental trigger can greatly aid, or might even be indispensable to, the design of a well-controlled adoption study. In the case of pellagra, whose cause was not known in the early 20th century, researchers would have needed to control for untold environmental factors in order to ensure that one particular factor did not confound the results of their study. Paradoxically, the adoption study itself may have been unnecessary had the investigators known that pellagra was triggered (if not caused) by

---

1. Clerget-Darpoux et al., 1986, p. 305.
2. Ibid., pp. 307-308.

a niacin deficiency. The reason is that niacin-enriched foods could then have been provided to the entire population (as they eventually were), thereby protecting everyone from pellagra, and the question of whether genes were also involved would have held little social or medical importance.

The second point relates to adoption agencies' and prospective adoptive families' *belief* that pellagra was caused by hereditary factors, which would have had a major impact on the types of families into which the biological offspring of pellagrins would have been placed. Clearly, few early 20th century southern United States adoption agencies would have placed the biological offspring of mothers diagnosed with pellagra into the homes of middle-class or well-to-do families. Moreover, prospective adoptive parents would not have selected these children as long as other "untainted" children were available. It is critical to understand that this was an era in which beliefs in the power of heredity were very strong, and eugenic programs aimed at preventing the reproduction of "hereditarily tainted" people were in force, often the result of state law. And although downplayed or ignored by schizophrenia adoption researchers, adoptees in their studies were placed in an era and location in which the view of schizophrenia (or "insanity") as a hereditary disease was axiomatic.[1]

Thus, it is likely that the biological offspring of mothers diagnosed with pellagra would have been placed into lower SES families, or might have ended up in an orphanage, due to their undesirability as adoptees. However, most psychiatric adoption researchers studying disorders such as schizophrenia understood that knowledge of the mother's condition may have influenced adoption agencies' placement policies. For this reason, most adoptees in their studies were placed before the mother was diagnosed with schizophrenia, theoretically eliminating this potential bias from their studies.

What is never mentioned, however, is that — on either environmental or genetic grounds — the biological relatives of a person diagnosed with a psychiatric disorder would be expected to have an excess of biological relatives diagnosed with the same disorder. Thus we would expect a family history of the disorder even in cases where a birthmother gives up her child for adoption *before* she herself is diagnosed with the disorder. Moreover, it is reasonable to expect that the adoption agencies and prospective adoptive parents (if they were informed by the agencies) would have been aware of this family history, especially in countries maintaining psychiatric registers. Therefore, the fact that adoption agencies and prospective parents did not factor the birthmother's diagnostic status into placement decisions does not exclude the possibility that her *family history* was an important factor influencing placements.

This leads to a crucial aspect of adoption studies of psychiatric disorders believed to be the result of bad heredity. Although researchers using the Adoptees method usually attempt to control for environmental confounds by

---

1. Joseph, 2004b.

matching index and control biological (birth) mothers on variables such as socioeconomic status, they invariably fail to match on the most important environmental variable: the status of these mothers' biological relatives for the disorder under study. This presents a problem in psychiatric genetic adoption research, since the independent variable (the diagnostic status of mothers giving up a child for adoption) is necessarily linked to the most important variable that researchers must control for.

It therefore becomes clear that an early 20th century pellagra adoption study carried out in the southern US would have been plagued by potential environmental confounds relating to an adoptee's biological background. And although we now know that pellagra is caused by a vitamin deficiency, we could have seen a significantly higher pellagra rate among index adoptees for the sole reason that they were more likely to be placed into impoverished homes lacking pellagra preventative diets. Thus, genetic researchers' belief that "environmental factors cannot contribute to the similarity of the adopted children and their biological parents because members of an unrelated family raised the children" is false in many cases.[1]

CONCLUSIONS

We have seen that pellagra genetic partisans erroneously interpreted pellagra family pedigrees as showing that the condition had a genetic basis, and that twin and adoption studies would likely have confirmed this conclusion. But suppose it had been shown that some people *are* genetically vulnerable to pellagra. That is, despite the clear cause of pellagra being a vitamin deficiency linked to malnutrition, suppose it was shown that some individuals, because of their genetic predisposition, were more likely to develop pellagra following malnutrition. Still, the discovery of a genetic predisposition for pellagra would have had little importance. Pellagra was wiped out in the United States by the relief programs of the 1930s, and, more importantly, by a federally-mandated World War II-era program requiring the enrichment of flour and corn meal with the vitamins needed to prevent pellagra. In other words, once the environmental factor was identified and eliminated, any possible genetic predisposition had been rendered unimportant. Perhaps this explains why, to the best of my knowledge, no one ever bothered to perform a pellagra twin or adoption study.

But let us go further. Imagine that a genetic factor for pellagra had been proven, but that the disease remained widespread today because the environmental factor (niacin deficiency) remained unknown. If this were the case, pellagra would still be seen as having an important genetic component. This

---

1. Jang, 2005, p. 21.

"diathesis-stress" model is precisely how disorders such as schizophrenia, bipolar disorder, and ADHD are currently viewed both in an out of psychiatry. This leads to the following point: *For psychiatric conditions believed to carry a genetic predisposition requiring an unknown environmental trigger, the importance researchers give to the genetic predisposition represents little more than a recognition that they have not identified the environmental cause or trigger.* There is an inverse relationship between researchers' belief in the importance of genetic factors for a given disorder and their knowledge of the disorder's environmental component.

Now suppose that, like pellagra, specific environmental causes for psychiatric disorders such as schizophrenia and ADHD are discovered. We could then remove these factors from people's environments, and schizophrenia and ADHD would recede or vanish *along with the idea that they are genetic disorders.* X-linked juvenile retinoschisis is an inherited disorder of the retina, leading to visual impairment. But other genetically-influenced visual disorders, such as astigmatism and farsightedness, are not commonly thought of as genetic disorders. The reason is that astigmatism and farsightedness can be corrected with glasses, whereas retinoschisis can not.

As historian Allan Chase observed, "Pellagra has pretty much disappeared, without any apparent change in anyone's genes."[1] Indeed, pellagra would have disappeared even if some people were genetically more susceptible than others. Chase also showed that the eugenicists' view of pellagra as a hereditary-infectious condition meant that it would take an additional generation to conquer the disease in the United States. Of course, it is far easier to enrich bread with the proper nutrients than it is to "enrich" the frequently neglectful, abusive, or destructive childhood and adult environments experienced by most people diagnosed with psychiatric disorders.

The example of pellagra shows that the genetic predisposition concept can delay discovery of the true causes of a condition at the cost of unnecessary suffering, and can promote the unwarranted stigmatization of diagnosed individuals. For psychiatric disorders, the claim of a hereditary component — even if true — can have similar consequences.

---

1. Chase, 1975, p. 86.

CHAPTER 5. A GENERATION MISINFORMED: PSYCHIATRY AND PSYCHOLOGY TEXTBOOKS' INACCURATE ACCOUNTS OF SCHIZOPHRENIA ADOPTION RESEARCH

Textbooks are valuable tools for transmitting the knowledge and history of various academic fields to students and professionals.[1] Unfortunately, they can also help perpetuate myths. Modern psychiatry is dominated by the biological/psychopharmacology paradigm, which must show that its diagnoses are biologically/genetically based. Furthermore, the prevailing views in psychiatry influence psychology and related fields.

This chapter examines textbooks' reporting of a specific area of psychiatric research: the study of adoptees as a means of testing the hypothesis that genetic factors influence schizophrenia. Unfortunately, we will see that textbook descriptions and interpretations of this body of greatly flawed research more closely match the dominant views in psychiatry than the facts as reported in the actual studies, and critical analysis is largely absent.

Historian of science Diane Paul discussed the findings of her 1985 investigation into how the authors of genetics textbooks presented the "Genetics of IQ" topic. She found these textbooks' discussions of the heritability concept to be "confused in the extreme," and that "the Cyril Burt scandal of the mid 1970s had only trivially affected the content of the texts. While Burt's name had disappeared — except as an example of fraud in science — his data were still reported."[2] Paul noted that "the most astonishing feature of the textbooks was their similarity. The same data were cited in support of the same conclusions, often in practically the same words, in text after text."[3]

---

1. This chapter is based on a revised and updated version of a previous publication (Joseph, 2000b).
2. Paul, 1998, p. 37.
3. Ibid., p. 37.

Paul went on to discuss the "dramatic changes that have taken place over the past quarter-century in the way textbooks are published."[1] Among these changes she mentioned a greater emphasis on packaging as opposed to content, the tendency for textbooks to be similar (or "virtual clones, both stylistic and substantive, of a market leader" as she put it), and the "widespread" practice of borrowing or "cribbing" from other textbooks.[2] Paul concluded that textbooks' similarities can be explained by "publishers who want to maximize profits," and authors who "want to minimize time and effort."[3]

I found the publications I review in this chapter through library searches of textbooks discussing the findings and importance of schizophrenia adoption research. When I found such a book, I included it in the sample. No book was rejected because it presented an accurate description of these studies. I include a brief critique of the studies themselves mainly to inform readers that there are doubts about particular issues which the authors of these textbooks should have discussed. I focus on the Oregon study and the early publications coming out of the Danish-American work because these investigations played a crucial role in turning public and professional opinion in the direction of accepting genetic theories of mental disorders as valid.

Three subsequent adoption studies were the Danish-American Crossfostering study,[4] the Danish Provincial Adoptees' Family study,[5] and the Finnish Adoptive Family Study of Schizophrenia.[6] Due to space limitations, I will not analyze the surveyed textbooks' treatment of these studies here, other than to point out that they were discussed in much the same way as the earlier studies.[7] These investigations, which contain flaws similar to the older studies (see Chapter 3), are generally seen as confirming what is believed to have been established by the early studies discussed here.

* * *

As we have seen, adoption studies are theoretically able to disentangle possible genetic and environmental influences on schizophrenia. However, many academics and clinicians have not read these studies' original publications, and therefore tend to rely on textbooks and secondary sources for accounts of the original investigators' results. Here, I analyze 43 accounts of these studies (listed

---

1. Ibid., p. 39.
2. Ibid., p. 39.
3. Ibid., p. 50.
4. Wender et al., 1974.
5. Kety et al., 1994.
6. Tienari et al., 1987, 2003.
7. For a critique of the Finnish study, see Jackson, 2003; Joseph 1999a, 2004b.

in Table 5.1), which include 15 psychiatry textbooks,[1] 11 abnormal psychology textbooks,[2] 6 books devoted entirely to schizophrenia,[3] 6 books whose authors argue that genes play an important role in determining human behavioral differences,[4] 2 chapters from annual psychiatry reviews,[5] 2 neuroscience textbooks,[6] and the DSM-IV-TR.[7] Simply put, these sources constitute the authoritative texts of psychiatry and abnormal psychology.

Table 5.1

43 TEXTBOOKS SURVEYED IN THIS CHAPTER

Authors/Editors	Title	Year
	Psychiatry Textbooks	
Andreasen	Brave New Brain	2001
D'haenen et al. (Eds.)	Biological Psychiatry (Vol. 1)	2002
Gelder et al.	Oxford Textbook of Psychiatry (3rd ed.)	1996
Gelder et al. (Eds.)	New Oxford Textbook of Psychiatry	2000
Hales et al. (Eds.)	Textbook of Psychiatry (3rd ed.)	1999
Hill et al. (Eds.)	Essentials of Postgraduate Psychiatry	1986
Judd & Groves (Eds.)	Psychiatry: Psychobiological Foundations of Clinical Psychiatry (Vol. 4)	1986
Kaplan & Sadock (Eds.)	Comprehensive Textbook of Psychiatry (6th ed., Vol. 1)	1995
Kolb & Brodie	Modern Clinical Psychiatry (10th ed.)	1982
Maxmen & Ward	Essential Psychopathology and its Treatment (2nd ed., rev. for DSM-IV)	1995
Nicholi (Ed.)	The New Harvard Guide to Psychiatry	1988
Sutker & Adams (Eds.)	Comprehensive Textbook of Psychopathology (2nd ed.)	1993
Tasman et al. (Eds.)	Psychiatry (Vol. 2)	1997
Trimble	Biological Psychiatry	1988
von Praag et al. (Eds.)	Handbook of Biological Psychiatry (Part III)	1980
	Abnormal Psychology Textbooks	
Coleman et al.	Abnormal Psychology and Modern Life	1984
Comer	Abnormal Psychology	1998
Davidson & Neale	Abnormal Psychology	1990

1. Andreasen, 2001; D'haenen et al., 2002; Gelder et al., 1996; Gelder et al., 2000; Hales et al., 1999; Hill et al., 1986; Judd & Groves, 1986; Kaplan & Sadock, 1995; Kolb & Brodie, 1982; Maxmen & Ward, 1995; Nicholi, 1988; Sutker & Adams, 1993; Tasman et al., 1997; Trimble, 1988; van Praag et al., 1980.
2. Coleman et al., 1984; Comer, 1998; Davidson & Neale, 1990; Gottesfeld, 1979; Kazdin et al., 1980; Martin, 1981; Nolen-Hoeksema, 1998; Oltmanns & Emery, 1998; Price & Lynn, 1986; Sarason & Sarason, 1984; van Hasselt & Hersen, 1994.
3. Bellack, 1979; Gottesman & Shields, 1982; Hirsch & Weinberger, 2003; Keefe & Harvey, 1994; Murray et al., 2003; Neale & Oltmanns, 1980.
4. Faraone, Tsuang, & Tsuang, 1999; King et al., 2002b; Pfaff et al., 2000; Plomin et al., 1990; Plomin et al., 1997; Rosenthal, 1970.
5. Byerley & Coon, 1995; Schultz, 1991.
6. Adelman, 1987; Kandel et al., 2000.
7. APA, 2000.

Gottesfeld	Abnormal Psychology: A Community Mental Health Perspective	1979
Kazdin et al. (Eds.)	New Perspectives in Abnormal Psychology	1980
Martin	Abnormal Psychology: Clinical and Scientific Perspectives (2nd ed.)	1981
Nolen-Hoeksema	Abnormal Psychology	1998
Oltmanns & Emery	Abnormal Psychology (2nd ed.)	1998
Price & Lynn	Abnormal Psychology (2nd ed.)	1986
Sarason & Sarason	Abnormal Psychology: The Problem of Maladaptive Behavior (4th ed.)	1984
Van Hasselt & Hersen	Advanced Abnormal Psychology	1994

Schizophrenia

Bellack (Ed.)	Disorders of the Schizophrenic Syndrome	1979
Gottesman & Shields	Schizophrenia: The Epigenetic Puzzle	1982
Hirsch & Weinberger	Schizophrenia (2nd ed.)	2003
Keefe & Harvey	Understanding Schizophrenia	1994
Murray et al. (Eds.)	The Epidemiology of Schizophrenia	2003
Neale & Oltmanns	Schizophrenia	1980

Genetics

Faraone et al.	Genetics of Mental Disorders	1999
King et al. (Eds.)	The Genetic Basis of Common Diseases (2nd ed.)	2002
Pfaff et al. (Eds.)	Genetic Influences on Neural and Behavioral Functions	2000
Plomin et al.	Behavioral Genetics: A Primer (2nd ed.)	1990
Plomin et al.	Behavioral Genetics (3rd ed.)	1997
Rosenthal	Genetic Theory and Abnormal Behavior	1970

APA Annual Reviews

Oldham & Riba (Eds.)	Annual Review of Psychiatry (Vol. 14)	1995
Tasman & Goldfinger	Annual Review of Psychiatry (Vol. 10)	1991

Neuroscience Textbooks

Adelman (Ed.)	Encyclopedia of Neuroscience (Vol. 2)	1987
Kandel et al. (Eds.)	Principles of Neural Science (4th ed.)	2000

DSM-IV-TR

APA	Diagnostic and Statistical Manual of Mental Disorders (4th ed., text rev.)	2000

Undoubtedly, these books have played an important role in shaping students' and professionals' views on the "genetics of schizophrenia" topic. However, their authors' accounts of schizophrenia adoption research are flawed, for reasons which include:

- They emphasize the original researchers' conclusions at the expense of independent critical analysis.
- They often rely on secondary sources.
- They typically do not discuss, or mention only briefly, the views and publications of critics.

• They often misreport studies' methods and results.

• While some surveyed textbooks discuss possible environmental confounds in schizophrenia twin research, few discuss the likelihood that genetic inferences from adoption studies are confounded by the selective placement of adoptees with a family history of mental disorders.[1]

• Few discuss problems with the reliability and validity of a schizophrenia diagnosis in the context of genetic research.

• Few discuss adoption study problems such as late placement and attachment disturbance.

• They sometimes cite studies failing to find statistically significant results in the genetic direction as evidence in favor of genetic factors.

• They usually accept the original researchers' definition of schizophrenia (or "schizophrenia related disorders") without question. In particular, they typically present the Danish-American investigators' results as supporting a "schizophrenia spectrum of disorders," but rarely question the validity of this concept, which was usually necessary in order to find statistically significant results (see Chapter 3).

A detailed description of the adoption studies' methods and results can be found in the original publications and in critiques (in Chapter 3, I briefly outlined problems in this area).[2] My purpose here is to review the way the surveyed textbooks discussed, or failed to discuss, important issues in these studies.

HESTON'S OREGON ADOPTION STUDY

The majority of surveyed textbooks credited Leonard Heston's 1966 schizophrenia adoption study with providing important and groundbreaking evidence in favor of genetic factors. Heston's 47 experimental (index) adoptees were born to mothers residing in Oregon state mental hospitals, all of whom were diagnosed with schizophrenia. No surveyed textbook, however, mentioned that during most of the period in which these adoptees were placed (1915-1945), Oregon state law permitted the forcible sterilization of "the insane" for eugenic purposes, and had even established a State Board of Eugenics to oversee the process.[3] It is therefore likely that the children of these patients, who were the presumed carriers of the "hereditary taint" of schizophrenia, were placed into environments inferior to those experienced by the "untainted" control adoptees. Thus, Heston's conclusions in favor of genetics were confounded by the selective placement of adoptees on the basis of the psychiatric status of their biological

---

1. See Joseph, 2004b.
2. Ibid.
3. Ibid.

mothers. Since the authors of the surveyed textbooks did not provide this information (nor was it mentioned by Heston), their readers were probably led to believe that index and control adoptees were randomly placed into available adoptive homes.

Most textbooks discussing this study also failed to point out that Heston did not diagnose his adoptees blindly. Heston personally collected information on all 97 of them (47 experimental, 50 control), and had conducted interviews with 72. After dossiers were compiled for each adoptee, diagnoses were made, according to Heston, "blindly and independently by two psychiatrists. A third evaluation was made by the author [Heston]."[1] When differences arose between these three raters, "a fourth psychiatrist was asked for an opinion and differences were discussed in conference."[2] Therefore Heston, who compiled the dossiers and was *not* diagnosing blindly, had an important input into the diagnostic process and was able to influence the "blind" raters in conference. Heston already knew from the records that five experimental and zero control adoptees had received a hospital diagnosis of schizophrenia, and that this difference was statistically significant. However, if one fewer experimental or one more control adoptee had received a schizophrenia diagnosis, the experimental-control difference would have been statistically non-significant. If the two "blinded" psychiatrists had diagnosed a control adoptee with schizophrenia, Heston, who was aware of the adoptee's status and who was a partisan of the genetic position, could have influenced them into changing their diagnosis. Thus, it is misleading to state that diagnoses were made by blinded raters.

Although Gottesman and Shields wrote in their textbook, "It is important to note that diagnoses were made blindly by two psychiatrists in addition to the author," they failed to emphasize that the diagnostic process was contaminated because "the author" was *not* blinded.[3] According to Martin, "Five of the 47 children whose biological mothers were schizophrenic were diagnosed, without knowledge of the group to which they belonged, as schizophrenic; none of the children in the control group was so diagnosed."[4] According to R. J. Rose, "The dossier compiled on each subject...was evaluated blindly and independently by two psychiatrists."[5] Rose took this last phrase directly from Heston's 1966 paper, but he left off Heston's next sentence, which explained that a "third evaluation was made by the author." And in their 2003 schizophrenia textbook chapter, Riley and colleagues wrote that "Heston, along with two other psychiatrists, made diagnoses blind to the parental diagnoses."[6]

---

1. Heston, 1966, p. 821.
2. Heston & Denney, 1968, p. 368.
3. Gottesman & Shields, 1982, p. 131.
4. Martin, 1981, p. 298.
5. Rose, 1980, p. 103
6. Riley et al., 2003, p. 255.

The surveyed texts also failed to mention Heston's inadequate criteria for making a diagnosis of schizophrenia. Heston stated only that he and his colleagues made diagnoses using "generally accepted standards,"[1] and that they diagnosed schizophrenia "conservatively."[2] Neale and Oltmanns claimed that diagnoses were made "by standard American criteria based on DSM-II," even though Heston did not state this in his publications.[3] In fact, the DSM-II was not published until 1968.[4]

Other problems in Heston's study which the textbook authors failed to discuss include (1) that Heston had very little information about his index adoptees' biological fathers, or about his control adoptees' biological mothers and fathers; (2) Heston's questionable inclusion of 25 experimental adoptees who were not interviewed; (3) Heston's failure to publish case history information for those under study; and (4) the fact that almost half of the adoptees had spent months or years in an orphanage, and that, as Heston conceded, "none of the subjects were reared in typical or 'normal' circumstances."[5] This did not prevent the editor of the *American Journal of Psychiatry*, Nancy Andreasen, from writing in 2001 that in Heston's study, "adoptees were reared in families that were considered to be 'normal' or 'healthy.'"[6]

Finally, Heston found a greater rate of "psychosocial disability" (e.g., criminality, alcoholism) among his experimental adoptees. Although some textbooks reported this, most failed to consider this finding as evidence that the experimental adoptees experienced more psychologically damaging rearing environments than controls. Instead, some authors followed Heston in speculating about a possible genetic link between schizophrenia and these psychosocial disabilities, and textbook authors such as Neale and Oltmanns, Coleman et al., and others speculated that adoptees might have inherited tendencies toward these behaviors from their biological fathers.

On a final note, Nobel Laureate psychiatrist and neurophysiology researcher Eric Kandel, in his 2000 chapter in *Principles of Neural Science*, provided a mistaken description of Heston's research design. According to Kandel, in Heston's study "the rate of schizophrenia was higher among the biological relatives of schizophrenic adoptees than among relatives of normal adoptees."[7] As we have seen, Heston began with *mothers* diagnosed with schizophrenia and studied the rate of schizophrenia among their adopted-away biological offspring. Kandel apparently believed that Heston used Kety's Adoptees' Family

---

1. Heston, 1966, p. 822.
2. Heston & Denney, 1968, p. 369.
3. Neale & Oltmanns, 1980, p. 194.
4. APA, 1968.
5. Heston & Denney, 1968, p. 374.
6. Andreasen, 2001, p. 199.
7. Kandel, 2000, p. 1194.

method, when Heston actually used the Adoptees method, later used by Rosenthal (see Figure 2.1).[1]

Most surveyed textbooks cited David Rosenthal and associates' 1968 and 1971 Danish Adoptees studies as providing important evidence in favor of the genetic position.[2] Unfortunately, none discussed the likelihood that the adopted-away offspring of index group parents were placed into inferior homes due to the prevalence of psychiatric disorders in their biological families, which as suggested elsewhere, played an important role in the Danish adoption process.[3] Like Oregon, Denmark had compulsory eugenic sterilization laws for most of the period in which Danish adoptees were placed.[4] Moreover, Rosenthal's study contained several glaring methodological problems such as late placement, questionable diagnostic methods, the failure to provide life history information, and generalizabilty issues.[5] In addition, he found no statistically significant schizophrenia spectrum disorder (SSD) difference between the adopted-away biological offspring of SSD parents versus the adopted-away offspring of control parents.[6] In fact, Rosenthal found only one index adoptee with a hospital record of chronic schizophrenia, and only one additional case was diagnosed by interview.

To reach statistical significance, Rosenthal had to include several offspring whose parents were diagnosed by one or more rater with "manic depressive disorder" — a condition Rosenthal believed was "genetically distinct and different" from schizophrenia.[7] In a later textbook chapter, however, Rosenthal claimed that he studied "individuals who had a history of schizophrenic disorder and who were the biological parents of children who had been given away at an early age for adoption by nonrelatives."[8]

In general, the surveyed textbooks either claimed or implied that Rosenthal's study confirmed Heston's (allegedly) significant results, which clearly was not the case. Their authors cited higher — though statistically nonsignificant — index versus control adoptee rates as evidence in favor of genetics. However, a lack of statistical significance means that we cannot reject the null

---

1. For a critique of Kandel's chapter, see Leo & Joseph, 2002.
2. Rosenthal et al., 1968; Rosenthal et al., 1971. For critical reviews of Rosenthal's studies, see Boyle, 2002b; Lewontin et al., 1984; Lidz et al., 1981; Joseph, 2004b; Pam, 1995.
3. Joseph, 2004b; Mednick & Hutchings, 1977.
4. See Hansen, 1996; Joseph, 2004b.
5. See Boyle, 2002b; Joseph, 2004b; Lewontin et al., 1984; Lidz et al., 1981.
6. For example, the figures from Rosenthal et al., 1971 are: 14/52 index versus 12/67 control, p = .17, Fisher's Exact Test, one-tailed.
7. Rosenthal, 1971a, p. 124.
8. Rosenthal, 1980, p. 4.

hypothesis, which states that index and control adoptee schizophrenia rates are equal (see Chapter 1). Below, I provide some examples of how textbook authors cited statistically non-significant comparisons as evidence in favor of genetics:

Similar to the findings of Heston, the risk of schizophrenia was higher in the adopted-away children of schizophrenic persons.[1]

Three of the forty-six index cases and none of the sixty-seven controls were diagnosed as definitely schizophrenic [statistically non-significant difference[2]]. This high rate in the index group points to a hereditary factor.[3]

In a study of children separated from schizophrenic mothers at an average age of six months, the findings confirmed those of Heston...[4]

Rosenthal's studies strongly support the view that genetic factors are of considerable importance in the transmission of schizophrenia.[5]

Of the 44 index cases, 3 were diagnosed as definitely schizophrenic. None of the controls were considered to be schizophrenic [non-significant difference]. This rate of 7% (3/44) actual incidence of schizophrenia is quite similar to that observed by Heston, 11% (5/47).[6]

Rosenthal et al. reported that adopted-away offspring of a schizophrenic parent had an increased risk compared with adopted-away offspring of well parents.[7]

Rosenthal et al.... found that a significantly higher proportion of the adopted-away offspring of schizophrenic parents from Copenhagen were classified as having schizophrenia or "borderline schizophrenia," than were control adoptees.[8]

Adoption studies of the three main designs...provide evidence that there is an increased risk of schizophrenia in first-degree relatives of probands...[9]

[In Rosenthal's study] genetic interpretations prevailed, since schizophrenia-like disorders were far more common among the offspring of biologically ill than among those of "normal" parents. These and similar findings have provided overwhelming evidence that, in most cases, genetic factors are a principle cause of schizophrenia.[10]

Their initial report found that three of the offspring of schizophrenia parents developed schizophrenia, compared with none of the 47 matched controls [non-significant difference]. When they later extended the study and considered SSD [schizophrenia spectrum disorders], they found that 13 (18.8%) of 69 adoptees of

1. Byerley & Coon, 1995, p. 366.
2. 3/47 versus 0/67, p = .065, n.s. Fisher's Exact Test, one-tailed.
3. Sarason & Sarason, 1984, p. 299.
4. Gelder et al., 1996, p. 268.
5. Murray, 1986, p. 351.
6. Neale & Oltmanns, 1980, p. 195.
7. Levinson & Mowry, 2000, p. 49.
8. Murray & Castle, 2000, p. 599.
9. Zammit et al., 2002, p. 663.
10. Maxmen & Ward, 1995, p. 71.

schizophrenia parents had SSD compared with eight (10.1%) of 79 matched controls [non-significant difference].[1]

And on the basis of his 1968 preliminary findings, for which he claimed no statistically significant results, Rosenthal himself concluded: "The data provide strong evidence indeed that heredity is a salient factor in the etiology of schizophrenic disorders."[2]

Two exceptions include Carson and Sanislow, who wrote that "The main finding of this study had actually *failed* to confirm the hypothesized genetic transmission of a schizophrenia diathesis, according to accepted standards of evaluation,"[3] and Kendler and Diehl's observation that Rosenthal's study "found similar results [to Heston] which, however, fell short of statistical significance, particularly when only parents with a consensus diagnosis of schizophrenia or schizophrenia spectrum were included."[4] Strikingly, of the surveyed texts only Rieder (in his chapter in Bellack's *Disorders of the Schizophrenic Syndrome*) made reference to Haier, Rosenthal, and Wender's 1978 reanalysis, which found that the index/control schizophrenia spectrum consensus diagnosis comparison, using Rosenthal's own criteria, was statistically non-significant (33% vs. 25%).[5]

The Lowing Reanalysis

Patricia Lowing and her colleagues published a 1983 reanalysis of Rosenthal's data using diagnostic criteria from the then recently published DSM-III.[6] Although the majority of post-1983 surveyed texts did not mention this study, those that did claimed that it confirmed Rosenthal's allegedly significant findings. In fact, Lowing and her colleagues confirmed Rosenthal's finding of only one case of chronic schizophrenia in the entire sample, and the index/control comparison remained statistically non-significant even when they counted DSM-III schizotypal personality disorder along with DSM-III chronic schizophrenia. It was only after deciding to expand the spectrum to include schizoid personality — which Kety recognized in 1983 as being genetically unrelated to chronic schizophrenia[7] — that Lowing and colleagues were able to confirm Rosenthal's "original finding concerning the heritability of schizophrenia spectrum disorders."[8]

---

1. Riley et al., 2003, p. 255. 13/69 vs. 8/79, p = .10, n.s. Fisher's Exact Test, one-tailed.
2. Rosenthal, 1970, p. 129.
3. Carson & Sanislow, 1993, p. 311 (emphasis in original).
4. Kendler & Diehl, 1995, p. 945.
5. Haier et al.,1978, Table 3. Index 21/64 (33%), versus control 16/64 (25%), p = .22, Fisher's Exact Test, one-tailed. The textbook reference is for Rieder, 1979.
6. Lowing et al., 1983.
7. See Kety, 1983a.
8. Lowing et al., 1983, p. 1168.

Kendler and Diehl wrote that Lowing et al. performed "a blinded reanalysis using DSM-III criteria, which...found a significant excess of schizophrenia spectrum in adopted-away offspring of schizophrenic parents versus those of control parents."[1] They did not mention that the researchers found only one case of chronic schizophrenia among the 39 index adoptees, or that the "significant excess" was dependent on Lowing's decision to include schizoid cases.

* * *

Many textbooks cited Rosenthal's Danish Adoptees study as providing important evidence in favor of the genetic transmission of schizophrenia, although the facts suggest otherwise. Their authors usually accepted the original investigators' conclusions without subjecting them to critical analysis, and they ignored or misrepresented the results of subsequent reanalyses, which essentially confirmed the *negative* results of the original study.

KETY AND ASSOCIATES' DANISH ADOPTEES' FAMILY STUDIES

As seen in Chapter 3, Kety and associates began with adoptees identified with a schizophrenia spectrum disorder and then obtained information on their biological and adoptive relatives (see Figure 2.1). In their 1968 study, the investigators made blind diagnoses on the basis of psychiatric records. They based their 1975 and 1994 diagnoses on interviews. In all three studies, the investigators found a significantly higher SSD rate among index versus control biological relatives.

Although Kety et al. needed to broaden the definition of schizophrenia in order to find statistically significant results, few textbooks questioned or provided evidence supporting the spectrum's validity. Many simply reported that adoptees were diagnosed with "schizophrenia." For example, "In 33 of the 507 cases, a diagnosis of schizophrenia could be agreed upon by independent judges using an abstracted case history."[2] According to another author, Kety and associates "identified a group of early adoptees who had become schizophrenic."[3] Kety and colleagues designated B1, B2, and B3 as "definite schizophrenia" — an unfortunate and misleading term repeated in several textbooks. When reporting diagnoses among biological relatives, most textbook authors did not mention that Kety et al. gave nearly two-thirds of their SSDs to biological half-siblings, who are second-degree relatives. These textbooks did not mention that this finding runs counter to genetic theory, which predicts that the familial prevalence of genetic disorders will be correlated with the degree of genetic rela-

1. Kendler & Diehl, 1995, p. 945.
2. Rose, 1980, pp. 104-105.
3. Martin, 1981, p. 298.

tionship among biological relatives. As Rosenthal himself once wrote, "To demonstrate that genes have anything to do with schizophrenia.... The frequency of schizophrenia in the relatives of schizophrenics should be positively correlated with the degree of blood relationship to the schizophrenic index cases."[1]

None of the surveyed texts' authors emphasized (and few reported) that the B1 chronic schizophrenia rate among index and control biological relatives was not significantly different in either the 1968 or 1975 study. Although Kety et al. diagnosed five relatives B1 in their 1975 study, four of these diagnoses, contrary to genetic predictions, were made on biological *half-siblings*. But even if we count these half-siblings the same as first-degree relatives, the index/control difference remains statistically non-significant. This is due to the fact that the biological father of control adoptee C9 received a chronic schizophrenia diagnosis in the 1968 study, but had died before he could be interviewed for the 1975 investigation.[2] Although Kety et al. counted several other non-interviewed relative diagnoses in other statistical calculations (e.g., the paternal half-sibling comparison, which I discuss below), they did not count this control relative in their 1975 index/control B1 comparison. Had they decided to count him, their chronic schizophrenia comparison would have been statistically non-significant.[3] None of the textbook authors questioned Kety's failure to count this chronic schizophrenia control first-degree biological relative. Although Kety et al. diagnosed only 2 out of 139 first-degree biological relatives B1 in the 1975 study (one index, one control), we would not discover this by relying on the surveyed textbooks:

> The rate for schizophrenia was greater among the biological relatives of the schizophrenic adoptees than among the relatives of controls, a finding which supports the genetic hypothesis.... The adoption findings reported above were for process [chronic B1] schizophrenia...[4]

> The schizophrenic children had significantly more biological relatives who were schizophrenic than the normal control group.[5]

> Kety and colleagues looked at a group of schizophrenia patients who had been adopted out early in life and found a similar disorder in 12% of their biological parents and in less than 2% of the adoptive parents.[6]

> The results [of the Kety et al. 1975 study] showed that the rate of schizophrenia was much higher in the biological relatives of adoptees with schizophrenia.... The rate of schizophrenia in the biological relatives of adoptees with schizophrenia was about 12%, compared to 1% to 2% in the biological relatives of adoptees without schizophrenia.[7]

---

1. Rosenthal, 1974, p. 589.
2. See Kety et al., 1975, p. 160.
3. 5/173 index versus 1/174 control, p = .11, Fisher's Exact Test, one-tailed.
4. Gelder et al., 1996, p. 268.
5. Gottesfeld, 1979, p. 167.
6. Schuckit, 1986, p. 156.
7. Keefe & Harvey, 1994, p. 83.

Adoption studies have shown that biological relatives of individuals with Schizophrenia have a substantially increased risk for Schizophrenia, whereas adoptive relatives have no increased risk.[1]

Among the biological relatives of index subjects, five were chronic schizophrenics; among the biological relatives of control subjects there were no chronic schizophrenics. Statistically, the difference between the two groups was highly significant.[2]

According to Byerley and Coon, in 1968 Kety et al. "found that the prevalence of schizophrenia was significantly higher in the biological parents.... The Danish adoption studies also found that the biological relatives of schizophrenic persons had elevated rates of 'borderline schizophrenia.'"[3] In fact, in 1968 Kety et al. did not diagnose *any* of their 63 biological index parents with chronic schizophrenia, while diagnosing only *one* with borderline schizophrenia.[4]

In their widely used 1995 textbook, *Essential Psychopathology and its Treatment*, Jerrold Maxmen and Nicholas Ward wrote about Kety's studies as follows:

If schizophrenia were genetically transmitted, the biological parents should have a much higher incidence of schizophrenia than the adoptive parents. On the other hand, if schizophrenia were produced psychologically, just the opposite should occur. The results were striking: Repeatedly, the schizophrenic adoptee's biological parents were schizophrenic, whereas the adoptive parents were "normal."[5]

However, because the Danish adoption agencies likely screened potential adoptive parents for psychopathology, coupled with the likelihood that people who give up children for adoption suffer more psychopathology than average, we should expect to find more psychopathology in the biological parent group for reasons having nothing to do with genetics (see Chapter 10 for further discussion on this point).

Moreover, contrary to Maxmen and Ward's description, Kety and colleagues believed that comparing SSD rates between their adoptees' biological and adoptive relatives was "inappropriate." They wrote in 1976 that "another type of inappropriate comparison that some have made is that between adoptive and biological relatives," thereby justifying their decision to compare diagnoses among their *index versus control* relatives.[6] In a later publication, Kety wrote that conclusions drawn from "improper" comparisons between index biological versus index adoptive relatives are "fallacious."[7] (Popular author Matt Ridley made a similar error in his 2003 book *The Agile Gene*, writing that Kety "found that

---

1. APA, 2000, p. 309.
2. Rosenthal, 1980, p. 4.
3. Byerley & Coon, 1995, p. 366.
4. See Kety et al., 1968, p. 354.
5. Maxmen & Ward, 1995, p. 71.
6. Kety et al., 1976, p. 420.
7. Kety, 1983b, p. 964.

schizophrenia was 10 times as common in the biological relatives of diagnosed schizophrenics who had been adopted as children as it was in their adopting families."[1]) And Maxmen and Ward, like Byerley and Coon, provided a misleading account of the actual diagnostic status of the relatives Kety et al. studied. In their 1975 study, Kety et al. found only one B1 and two B3 diagnoses among the 63 identified biological index parents (a statistically non-significant clustering vs. the adoptive parents, although there were several more index D3 "uncertain borderline" diagnoses).[2]

The Paternal Half-Siblings

Although Kety and colleagues found a significant concentration of SSDs among their index versus control biological relatives, they viewed this result as "compatible with a genetic transmission for schizophrenia, but it is not entirely conclusive."[3] Because of possible factors such as prenatal or perinatal trauma, early mothering experiences, etc., "one cannot, therefore, conclude that the high prevalence of schizophrenia illness found in these biological relatives of schizophrenics is genetic in origin."[4] This statement, unfortunately, did not prevent a generation of textbook authors from concluding that this "high prevalence" did indeed show that schizophrenia is "genetic in origin."

Kety and associates then made their case for the discovery of "compelling evidence" in support of the genetic hypothesis, on the basis of a comparison of their index and control biological paternal half-siblings:

> The largest group of relatives which we have is, understandably, the group of biological paternal half-siblings. Now, a biological paternal half-sibling of an index case has some interesting characteristics. He did not share the same uterus or the neonatal mothering experience, or an increased risk in birth trauma with the index case. The only thing they share is the same father and a certain amount of genetic overlap. Therefore, the distribution of schizophrenic illness in the biological paternal half-siblings is of great interest.[5]

Kety and colleagues counted 16 record- and interview-based spectrum diagnoses among these paternal half-siblings, but found a "highly unbalanced" diagnostic distribution (14 index, 2 control). They concluded, "We regard this as compelling evidence that genetic factors operate significantly in the transmission of schizophrenia."[6]

---

1. Ridley, 2003, pp. 105-106.
2. Kety et al., 1975, p. 158-159. The 1994 study (Kety et al., 1994, p. 448) also failed to find a significantly higher elevation of chronic and/or latent schizophrenia among index biological vs. index adoptive parents.
3. Kety et al., 1975, p. 156.
4. Ibid., p. 156.
5. Ibid., p. 156.
6. Ibid., p. 156.

All textbook authors discussing this claim endorsed Kety and colleagues' conclusion that the significantly higher rate of spectrum diagnoses among their index versus control biological paternal half-sibs provided important evidence supporting a genetic basis for schizophrenia. Aside from the questionable nature of this claim, and the fact that the investigators provided no information about the life circumstances of these half-sibs, *Kety et al. would have found no statistically significant difference if they had counted all schizophrenia spectrum diagnoses.* Although in 1968 and 1975 the investigators counted category C (schizoid and inadequate personality) as a schizophrenia spectrum disorder, they *removed* C diagnoses from their 1975 paternal half-sibling comparison.[1] Strikingly, the index/control difference is not statistically significant when spectrum diagnosis C is included.[2]

Thus, Kety and colleagues excluded category C from their 1975 paternal half-sibling comparison *but not from the spectrum itself.* This means that a comparison alleged to have provided "compelling evidence" in support of genetics was not statistically significant according to Kety and colleagues' 1975 definition of the schizophrenia spectrum — a fact that all surveyed textbook authors failed to mention. The following examples show how they *did* discuss the half-sibling results:

> In investigations of paternal half-siblings of schizophrenia probands, the incidence of schizophrenia was higher than in control cases, ruling out intrauterine contributions to the congenital effects.[3]

> A significant concentration of schizophrenia and uncertain schizophrenia was found in the paternal half-siblings of the schizophrenia index cases with whom they shared no prenatal or postnatal environment.[4]

> Considering the total absence of common environmental factors among the [biological paternal half-sibs], these data are indeed convincing support for a hereditary component in the development of schizophrenia.[5]

> Using the results of both hospital diagnoses and psychiatric interviews, the frequency of hard schizophrenic spectrum in paternal half-siblings of schizophrenic and non-schizophrenic adoptees was determined.... These important data again confirm a genetic hypothesis..."[6]

> A paternal half-sibling study was performed as a part of the Danish adoption investigation and demonstrated that siblings who shared a relationship only through the father had an expected prevalence of schizophrenia even though the offspring had not shared the same uterine environment.[7]

---

1. Joseph, 2004b; Lidz & Blatt, 1983.
2. Kety et al., 1976, p. 418. p = .094, by Kety and colleagues' own calculation.
3. Trimble, 1988, p. 202.
4. Kety & Matthysse, 1988, p. 142.
5. Neale & Oltmanns, 1980, p. 197 (emphasis in original).
6. Plomin et al., 1990, p. 357.
7. Schultz, 1991, pp. 82-83.

In Dr. Kety's work, the biologic paternal half-siblings of schizophrenic adoptees were at greater risk for schizophrenia than the biologic paternal half-siblings of control adoptees.[1]

Also studied were biological paternal half-siblings of ill and well adoptees. The results showed greater risk to biologic paternal half-siblings of schizophrenic adoptees than to biologic paternal half-siblings of control adoptees.[2]

Whereas most surveyed textbook authors failed to mention that Kety and colleagues decided to exclude SSD category C from their 1975 biological paternal half-sibling comparison, several reported the findings of other studies (such as Lowing et al., 1983) which *required* C in order to reach statistical significance.

The Kendler and Gruenberg Reanalysis

Several textbook authors informed their readers that Kety's results were "replicated" in Kendler and Gruenberg's blind 1984 reanalysis, which used DSM-III diagnostic criteria.[3] Even Carson and Sanislow, who took a more critical stance than the others, found Kendler and Gruenberg's data "relatively convincing."[4]

There are, however, some striking results in Kendler and Gruenberg's reanalysis not mentioned in the surveyed textbooks published after 1984. Starting with Kety's original index adoptees, Kendler and Gruenberg found that only 11 of the 17 B1 diagnoses (65%) met DSM-III diagnostic criteria for schizophrenia. Even more striking, they diagnosed only *one* of the original ten B3 adoptees (10%) with schizotypal personality disorder, the DSM-III equivalent of B3. This finding calls into question the entire Danish-American diagnostic process, and speaks to some critics' observation that psychiatric diagnoses are unreliable and often arbitrary.[5] Among the 35 biological relatives of DSM-III schizophrenia index adoptees (first-and second-degree combined), Kendler and Gruenberg found only 2 with the same diagnosis, and the rate among first-degree relatives was 1/10. Kendler and Gruenberg found statistically significant differences only because they, like Kety and colleagues, decided to use a broad definition of schizophrenia.

Kendler and Gruenberg also revealed that several of Kety and colleagues' diagnostic "interviews" never took place, but instead *were fabricated by the Danish investigators*. Kendler and Gruenberg called them "pseudointerviews" and wrote that they could tell the difference between a "real interview with a control adoptee" and a "pseudo interview with an index adoptee."[6] Kety and colleagues'

---

1. Faraone, Tsuang, & Tsuang, 1999, p. 41.
2. Pulver et al., 2002, p. 854.
3. Kendler & Gruenberg, 1984.
4. Carson & Sanislow, 1993, p. 311.
5. Bentall, 2003; Hill, 1986; Kirk & Kutchins, 1992; Kutchins & Kirk, 1997.
6. Kendler & Gruenberg, 1984, p. 556.

use of fabricated interviews, which Lewontin, Rose, and Kamin also discussed in *Not in Their Genes*, published in 1984, was not mentioned in any of the Danish-American publications or in any of the surveyed textbooks.[1] (See Chapter 6 for further discussion of Kendler and Gruenberg's study).

Conclusions

In general, the surveyed textbooks have rubber stamped the original investigators' and contemporary psychiatry's conclusions about the results of schizophrenia adoption research. Their descriptions are frequently inaccurate, and leave the general impression that their authors did not carefully review the original studies, and sometimes did not even read them. Moreover, a clear bias in favor of genetics is evident. In fact, only 4 of the 43 textbooks cited a publication critical of a schizophrenia adoption study's methods and conclusions,[2] and only three provided their own limited critical analysis.[3] While these textbooks occasionally discussed the controversial assumptions of the twin method, only two mentioned the crucial "no selective placement assumption" of adoption studies. As we have seen, a violation of this assumption could lead to a higher index group schizophrenia rate for reasons other than genetics. Because most American, Danish, and Finnish adoptees were placed at a time when eugenic ideas were strong and the status of index adoptees' "tainted" biological relatives was an important factor affecting placement, it is unlikely that index and control adoptees were placed into similar types of rearing environments.[4]

In summary, the authors of psychiatry, psychology, and related textbooks have, in general, provided an inaccurate and misleading description of schizophrenia adoption research. Diane Paul concluded that the genetics textbooks she reviewed "perpetuate a fundamentally inaccurate understanding of the genetics of intelligence."[5] The same is true for textbooks handling the "genetics of schizophrenia" question. As critic Mary Boyle argued in 2004, these textbooks teach "large numbers of people...not to think critically."[6] Clearly, those studying the causes of schizophrenia must be exposed to a wider variety of viewpoints than they currently receive, and inaccurate reporting and bias in favor of genetics must be documented further.

---

1. Lewontin et al., 1984.
2. Adelman, 1987; Carson & Sanislow, 1993; Colemen et al., 1984; Judd & Groves, 1986.
3. Cardno & Murray, 2003; Carson & Sanislow, 1993; Gottesman & Shields, 1982.
4. Joseph, 2004b, 2004c.
5. Paul, 1985, p. 317.
6. Boyle, 2004, p. 81.

CHAPTER 6. IRVING GOTTESMAN'S 1991 SCHIZOPHRENIA GENESIS: A PRIMARY SOURCE FOR MISUNDERSTANDING THE GENETICS OF SCHIZOPHRENIA

Any review of psychiatric and psychological textbook discussions of the genetics of schizophrenia would be incomplete if it failed to discuss what is perhaps the most influential and widely relied upon secondary source on the topic: Irving I. Gottesman's 1991 *Schizophrenia Genesis: The Origins of Madness*. Since the early 1990s this work has served as an important source of information for professionals and students interested in genetic theories of schizophrenia, and many textbooks include Gottesman's calculations (often in tabled form) of the "morbidity risks" of various relatives classified in terms of their degree of genetic relatedness to a person diagnosed with schizophrenia. Unfortunately, this book contributes to the misunderstanding and misinterpretation of a body of research cited in support of genetic influences on schizophrenia.

Gottesman has been a major figure in the schizophrenia genetics field since the mid-1960s, and, along with his colleague James Shields, published a schizophrenia twin study in 1966.[1] He is a highly regarded psychologist, having received the "Award for Distinguished Scientific Contributions" from the American Psychological Association in 2001. In receiving this award, Gottesman was praised for "elucidating the genetic and environmental causes of schizophrenia and criminality by combining the perspective of human genetics, epidemiology, and clinical psychology," and for being a "pioneering researcher, an outstanding mentor, an articulate spokesman for science, and an effective advocate for the mentally ill...[who] has profoundly influenced the field."[2] At the same time, Gottesman has

---

1. Gottesman & Shields, 1966b, 1972.
2. Anonymous, 2001, p. 864.

been a leading voice supporting psychiatric genetic methods and theories as they relate to schizophrenia and other psychiatric diagnoses.

Schizophrenia Genesis, winner of the American Psychological Association's 1992 William James Book Award, was put forward as a relatively accessible, balanced account of the manifestation and causes of schizophrenia. Gottesman included several first-person accounts of people diagnosed with the disorder, quoted liberally from the pages of *Schizophrenia Bulletin*. While every aspect of his book is ripe for critical analysis, I will concentrate on the chapters outlining the evidence supporting Gottesman's advocacy of the "diathesis-stressor" view of schizophrenia, which holds that the disorder is caused by an inherited biological predisposition in combination with environmental conditions or events. In Gottesman's view, schizophrenia is "the same kind of common genetic disorder as coronary heart disease, mental retardation, or diabetes...."[1]

Is Schizophrenia a Genetic Disorder? A Closer Look at Gottesman's "Figure 10" of the Schizophrenia Risk Among Various Types of Relatives

The title of *Schizophrenia Genesis*'s Chapter 5 asks whether schizophrenia is "inherited genetically." Here, Gottesman outlined various kinship correlations that he said were consistent with genetic theories of schizophrenia's causation. To demonstrate this, Gottesman developed his famous Figure 10, which presents this data in the form of an easily understandable figure suggesting that the more closely a person is genetically related to a person diagnosed with schizophrenia, the greater risk that person has of being diagnosed with schizophrenia. According to Gottesman, Figure 10, which is frequently reproduced or discussed in psychiatry and abnormal psychology textbooks, shows the "Grand average risk for developing schizophrenia compiled from the family and twin studies conducted in European populations between 1920 and 1987." The lifetime risks for developing schizophrenia among various types of relatives of people diagnosed with schizophrenia, as Gottesman presented them in his Figure 10, are seen in Figure 6.1.

According to Gottesman, Figure 10 demonstrates that the risk for schizophrenia increases proportionately as the relatives' genetic similarity increases, or as he put it, "the degree of risk correlates highly with the degree of genetic relatedness."[2] However, the percentages presented in Figure 10 (Figure 6.1) also show that the degree of risk correlates roughly with the degree of *environmental* similarity shared by relatives. What I will argue here is that, contrary to mainstream discussions of Gottesman's Figure 10, the risk factors he presented (1) include methodologically unsound and biased research, (2) are not reflective of more recent findings, (3) are inflated by Gottesman's use of the probandwise concor-

---

1. Gottesman, 1991, p. 84.
2. Ibid., p. 96.

dance rate calculation, and, (4) are consistent with a completely environmental etiology of schizophrenia.

Figure 6.1

IRVING GOTTESMAN'S 1991 *FIGURE 10*
OF THE "GRAND AVERAGE RISKS FOR DEVELOPING SCHIZOPHRENIA
COMPILED FROM THE FAMILY AND TWIN STUDIES CONDUCTED IN
EUROPEAN POPULATIONS BETWEEN 1920 AND 1987"

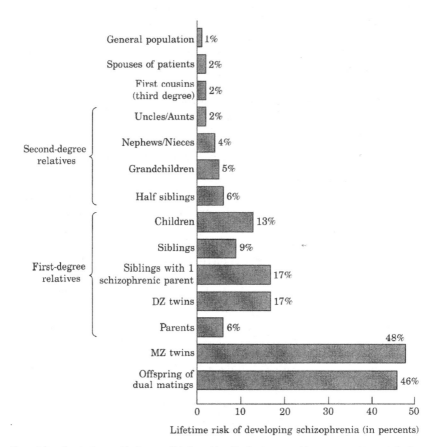

From *Schizophrenia Genesis: The Origins of Madness*, New York, Freeman (Gottesman, 1991, p. 96). Copyright © 1991 by I. I. Gottesman. Reprinted with permission.

Family Data

Gottesman did not name the studies he pooled to create Figure 10, or indicate the weighting of each study in relation to the pooled data. This creates a problem in that many older studies were biased because they were performed non-blinded by investigators strongly devoted to genetic theories, who often advocated eugenic sterilization programs for people diagnosed with schizophrenia, and who used vague and nonstandardized definitions of schizophrenia. A classic example is Franz Kallmann's massive 1938 schizophrenia family study of 1,087 German schizophrenia patients and their 13,851 relatives. In the first chapter of the book describing his study, Kallmann called for directing eugenic measures not only at people diagnosed with schizophrenia, but towards their "heterozygotic taint-carrier" biological relatives as well.[1] For Kallmann, these relatives were "eugenically undesirable" people whose numbers should be "kept at the lowest possible number."[2] Clearly, Kallmann could not have objectively diagnosed the relatives of his "schizophrenia probands," since he already viewed them as being carriers of the "hereditary taint of schizophrenia." Furthermore, Kallmann failed to adequately describe how schizophrenia was defined in his study.

In Chapter 2 of *Schizophrenia Genesis*, Gottesman recognized that "each diagnosis within a schizophrenic's family should be made 'blindly' (completely without knowledge of relatedness to other family members) lest that knowledge contaminate impartial decisions about diagnosis," and that nonblinded diagnoses constitute "poor practice."[3] Yet, since blind diagnostic procedures were not introduced into psychiatric genetics until the 1960s (and only in a limited number of studies), Gottesman's Figure 10 appears to be influenced by data produced by people such as Kallmann, who employed the "poor practice" of nonblinded diagnosis, and made diagnoses on the basis of his vaguely defined notion of what constituted "schizophrenia." As Gottesman recognized in an earlier publication (discussing the pooled data through 1978), these twin and family studies

> were conducted before the heyday of structured interviews, operationalized lists, and grant-funded team approaches permitting blindfolded interviews with probands (index cases) and their relatives. Furthermore, few of the earlier investigators bothered to collect risks in normal or psychiatric controls using the same concept of schizophrenia as in the primary target group of schizophrenia probands.[4]

Gottesman noted that Kallmann's "militant, hereditarian point of view has prevented a full appreciation of his data,"[5] but why should diagnostic data produced by the nonblinded and "militantly hereditarian" Kallmann be accepted?

1. Kallmann, 1938a, p. 3.
2. Ibid., p. 47.
3. Gottesman, 1991, p. 18.
4. Gottesman et al., 1987, pp. 28-29.
5. Gottesman, 1991, p. 107.

Truly, each and every diagnosis Kallmann made in his family and twin studies was "tainted" by Kallmann's views on the genetics of schizophrenia. The impression given in Figure 10 is that (excluding fraternal twins) the average first-degree relative schizophrenia rate in studies conducted between 1920 and 1987 is 11-12%. The pooled data, however, do not include a series of schizophrenia family studies published between 1980 and 1985 from which the pooled first-degree relative schizophrenia rate is only about 4%. Though not without flaws, these studies (seen in Table 6.1) are superior to the older investigations because the researchers made diagnoses blindly, and used structured diagnostic interviews.

Table 6.1

RESULTS OF SCHIZOPHRENIA FAMILY STUDIES PUBLISHED BETWEEN 1920 AND 1987 USING BLIND DIAGNOSES AND STRUCTURED INTERVIEWS

Schizophrenia Diagnoses Among the First-Degree Biological Relatives of People Diagnosed with Schizophrenia					
Authors	Year	Country	First-Degree Biological Relatives	First-Degree Biological Relatives Diagnosed with Schizophrenia	Percentage
Scharfetter & Nüsperli	1980	Switzerland	726	45	6.2%
Tsuang et al.	1980	USA	729	31	4.3%
Pope et al.	1982	USA	199	0	0.0%
Abrams & Taylor	1983	USA	70	2	2.9%
Guze et al.	1983	USA	111	4	3.6%
Baron et al.	1985	USA	376	19	5.1%
Kendler et al.	1985	USA	723	26	3.6%
Frangos et al.	1985	Greece	572	19	3.3%
POOLED			3,536	146	4.1%

Age correction not used.

Gottesman did not include these family studies in Figure 10 (or at least not those performed in the United States) because his results were derived from "family and twin studies conducted in *European* populations between 1920 and 1987 [emphasis added]."[1] Thus, Gottesman omitted at least six American schizophrenia family studies whose pooled first-degree relative schizophrenia rate was only 3.7% (as opposed to the impression of 11-12% from Figure 10), whose results were derived through the "good practice" of blind diagnoses.

Regardless of which family studies one counts or fails to count, however, increased schizophrenia rates that might exist among the spouses, cousins, aunt/uncles, nieces/nephews, grandchildren, half-siblings, full-siblings, parents, and

1. Ibid., p. 96.

children of people diagnosed with schizophrenia can be explained by the more similar physical, familial, and psychological environments these relatives share when compared with randomly selected members of the population (who have a 1% lifetime risk). Most genetic researchers concede this point, as exemplified by Faraone and colleagues correctly pointing out that "'familial' and 'genetic' are not synonymous," because "disorders can run in families for many reasons: genes, cultural transmission, shared environmental adversity, and so forth."[1]

Twin Studies

Given the general consensus that family data can be explained on environmental grounds, the results in Figure 10 more often seen as supporting genetic factors are the schizophrenia concordance rates of identical versus fraternal twins, which Gottesman listed as 48% and 17% respectively. Table 6.2 contains my compilation of the schizophrenia twin studies published between 1920 and 1987, which coincides with the period surveyed by Gottesman.

The concordance rates listed in Table 6.2 are based on pairwise calculations and use the investigators' "strict" definition of schizophrenia in cases where they used differing diagnostic standards in the same study. Gottesman has been a longtime advocate of the "proband" concordance method, which always produces higher concordance rates than the pairwise method. For example, suppose we have 10 pairs of identical twins, where in three pairs both members are diagnosed with schizophrenia, whereas in the remaining seven pairs only one member is diagnosed with schizophrenia. In this case, the pairwise concordance rate would be 3/10 = 30%. Using the proband method, which doubles the numerator, and adds the original numerator to the denominator, the "probandwise" concordance rate is calculated as 6/13 = 46%. Moreover, the pooled pairwise identical twin concordance rate for the methodologically-superior "contemporary" studies (see the discussion below) published between 1963 and 1987 is only 21%. This means that in almost 80% of pairs sharing an identical genetic makeup where one twin is diagnosed with schizophrenia, his or her co-twin is *not* diagnosed. Nevertheless, psychiatric geneticists argue that the contemporary results continue to support the genetic position on the basis of a 4-5 times greater identical versus fraternal concordance rate difference. Of course, this conclusion is based on the acceptance of the equal environment assumption of the twin method, which Gottesman did not address in *Schizophrenia Genesis*.

Classical and Contemporary Studies

Most twin researchers now agree that the schizophrenia twin studies published since 1963 (Tienari and after) are methodologically superior to the older

---

1. Faraone, Tsuang, & Tsuang, 1999, p. 11.

studies published between 1928 and 1961 (Luxenburger through Inouye). Twin researchers made methodological changes following the publication of critiques in the early 1960s by David Rosenthal and Don Jackson, who highlighted several important problems in the studies published to that point.[1] Most "contemporary" studies published since 1963 drew their samples from consecutive hospital admissions or from twin registers. A majority of the older "classical" studies published before 1963 used resident hospital samples, which twin researchers now agree can introduce biases resulting in higher concordance rates. Moreover, the authors of the contemporary studies used superior diagnostic criteria (some had access to national psychiatric registers), and employed more accurate methods of zygosity determination (whether twins are identical or fraternal). The modern view is captured in John Neal and Thomas Oltmanns' 1980 schizophrenia textbook, where they praised the contemporary studies for "using a variety of improved sampling procedures, more accurate means of determining zygosity, and blind diagnostic evaluations."[2]

The general theme of Gottesman's discussion was that, although the contemporary studies are methodologically superior to the classical, both are valid tests of the genetics of schizophrenia. As Gottesman and Shields wrote in their 1982 work *Schizophrenia: The Epigenetic Puzzle,* "We feel quite comfortable in concluding that the twin studies of schizophrenia as a whole represent variations on the same theme and are, in effect, sound replications of the same experiment."[3]

But upon examination of the data in Table 6.2, this does not appear to be the case. My numbers differ from Gottesman's because he included age-correction factors, used the probandwise concordance method, and relied on the original investigators' liberal definition of schizophrenia. In addition, Gottesman did not include the results of Koskenvuo and colleagues' low concordance rate 1984 study, which are missing from virtually all textbooks and reviews of schizophrenia twin research published through 2004.[4] (Koskenvuo et al. actually found a *zero per cent* concordance rate [0/24] for the 24 older pairs in their "Born before 1935" group.[5]) Furthermore, in a table Gottesman listed "Concordance Rates for Schizophrenia in Newer Twin Studies," from which he concluded that the pooled probandwise rates are 48% identical, 17% fraternal.[6] However, he omitted the large NAS-NRC study from this calculation, and failed to include the Koskenvuo et al. data. As we see in Table 6.2, both studies reported low concordance rates.

---

1. Jackson, 1960; Rosenthal, 1960, 1961, 1962a, 1962b. Also see the discussion in Joseph, 2001b, 2004b.
2. Neale & Oltmanns, 1980, p. 192.
3. Gottesman & Shields, 1982, p. 115.
4. For example, see Sullivan & Kendler, 2003.
5. Koskenvuo et al., 1984, p. 327, Table 9.
6. Gottesman, 1991, p. 110.

It is apparent from the data presented in Table 6.2 that there is a large con-cordance rate difference between the "classical" and "contemporary" studies published during the period Gottesman surveyed. Pooled pairwise identical twin concordance in the classical studies is 61%, whereas for the contemporary studies it is only 21%; the classical pooled fraternal rate is 12%, whereas it is only 4% in the contemporary studies. The reduced bias and more careful methods used in the newer investigations indicate that they constitute a more accurate assessment of true identical and fraternal twin schizophrenia concordance rates. Gottesman's 48% probandwise rate more closely matches, and serves as a major source of, the 50% concordance rate estimate found in most psychiatry and psy-chology textbooks. On the other hand, a small minority of biologically-oriented investigators, such as psychiatrist E. Fuller Torrey, have argued that Got-tesman's pooled rates are too high.[1] Elaine Walker and her colleagues concluded in 1991 that a more accurate assessment (again, with the Koskenvuo et al. data missing) finds average pairwise schizophrenia concordance rates of 25% iden-tical and 7% fraternal in the more methodologically sound studies.[2]

Equal Environment Assumption

An important omission from Gottesman's presentation of the schizo-phrenia twin data is his failure to address the critics' main objection to the twin method, which is, of course, that identical twins experience more similar envi-ronments than do fraternal twins. If the equal environment assumption (EEA) is false, Gottesman's pooled concordance rates don't prove anything about genetics and may well reflect nothing more that the greater environmental similarity and psychological bond of identical versus fraternal twins (see Chapter 1). Yet Got-tesman did not mention that drawing conclusions in favor of genetics from iden-tical-fraternal comparisons is dependant upon a controversial and counterintuitive theoretical assumption. He stated, implying that genes are the best explanation, "Undeniably, when twins are reared together, the identical twin of a schizophrenic is much more likely to be schizophrenic than is the fra-ternal twin of a schizophrenic."[3]

The EEA is so counterintuitive that few people, either in or out of science, really believe that fraternal twins' environments are as similar as identicals'. Ironically, twin researchers themselves were the main proponents of the idea that these environments are equal— but only until the advent of the "trait-rel-evant" EEA, as we saw in Chapter 1. Moreover, most schizophrenia twin researchers have concluded that concordance rate differences were at least partly the result of environmental influences,[4] and Gottesman, in the 1960s,

1. Torrey, 1992.
2. Walker et al., 1991, p. 218.
3. Gottesman., 1991, p. 116.
4. Joseph, 2004b., pp. 146-149.

wrote that "a psychological hypothesis such as identification might be used to explain differential concordance rates in MZ and DZ twins..."[1]

The EEA is even more implausible for disorders such as schizophrenia, where psychiatry has long recognized that one member of a closely related pair (whether twin or non-twin) can become psychotic due to the psychological influence of the other member of the pair. *Folie à deux* (shared psychotic disorder) has been defined as "a psychiatric entity characterized by the transference of delusional ideas and/or abnormal behavior from one person to one or more others who have been in close association with the primarily affected person."[2] Family systems therapy pioneer Don Jackson addressed the relationship between *folie à deux* and concordance for schizophrenia among identical twins in his insightful 1960 critique of schizophrenia genetic research.[3] Jackson pointed out that long-standing association and social isolation were common factors linking *folie à deux* and the case histories of concordant identical twins (which were supplied in some of the classical studies). He stressed that, although the twin relationship was not necessarily a positive one, "every twin report I have discussed mentions the strength of attachment between the pair, either in positive terms or in terms of mutual antagonism and jealousy. There are no indifferent cases."[4] Jackson noted that in Kallmann's 1946 twin study, identical twins (average age, 33 years) who had lived apart for five years or more were listed as having a 77.6% concordance rate, while "nonseparated" pairs were listed at 91.5%. He observed that "a separation even past the formative years was apparently very effective in reducing the concordance rate."[5]

We will soon examine other trends in schizophrenia twin research difficult to explain from the genetic perspective, but Gottesman's readers are entitled to know why they should accept that higher identical versus fraternal schizophrenia concordance rates are, as he contends, the result of the more similar *genetic* resemblance of identical versus fraternal twins. A strong advocate of twin research during his career, Gottesman has never written a detailed theoretical defense of the EEA. Although other twin researchers such as Kendler have attempted to do so (see Chapters 1 & 9), their arguments do not hold up to critical examination.[6]

Thus, Gottesman provided little support to his conclusion that genetic factors explain the pooled schizophrenia twin concordance rates he listed in Figure 10. A plausible alternative hypothesis to genetic interpretations of identical-fraternal twin comparisons holds that higher identical versus fraternal schizophrenia concordance rates are completely explained by the more similar

1. Gottesman & Shields, 1966a, p. 55.
2. Gralnick, 1942, p. 232.
3. See Joseph, 2001b.
4. Jackson, 1960, p. 68.
5. Ibid., p. 69.
6. Joseph, 1998, 2004b; Pam et al., 1996.

treatment and environments of identical versus fraternal twins, by identicals twins' stronger emotional bond and "ego fusion," and by bias.

Failure to Include Opposite-Sex Fraternal Concordance Rates

Gottesman also failed to list the pooled concordance rates of *opposite-sex* fraternal twins in his Figure 10. This omission is unfortunate because, due to much lower opposite-sex versus same-sex schizophrenia concordance in several studies reporting such rates, one could draw a different set of conclusions from Figure 10 had Gottesman decided to include these rates.

Although the twin method compares the concordance rates of identical versus same-sex fraternal twins, several studies reported rates for opposite-sex fraternals as well. As Jackson observed long ago, according to genetic theory there should be no difference in same-sex and opposite-sex twin concordance rates in disorders such as schizophrenia, for which the lifetime diagnosis risk does not differ significantly by sex. "On the other hand," wrote Jackson, "if the hypothesis is correct that identical twins are more concordant for schizophrenia because of their 'twinness,' one would expect a higher incidence of concordance for schizophrenia in same-sexed fraternal twins because they are more alike from the identity standpoint than different-sexed fraternal twins."[1] The results are seen in Table 6.3.We see in Table 6.3 that the pooled same-sex fraternal concordance rate in studies published between 1920 and 1987 is 2.7 times greater than the pooled opposite-sex fraternal rate (11.3% vs. 4.7%). As I have discussed elsewhere, to my knowledge neither Gottesman nor any other twin researcher has ever tried to explain this difference on genetic grounds.[2]

Let's take a closer look at the results from Kallmann's 1946 twin study. Although many have criticized this investigation, Gottesman, while recognizing several methodological errors, has since the mid-1960s defended Kallmann's data.[3] Indeed, in *Schizophrenia Genesis* he wrote that although "sadly lacking in details," Kallmann's investigation "is basically sound."[4] However, in addition to finding a large identical-fraternal schizophrenia concordance rate difference, Kallmann found a same-sex fraternal rate of 34/296 (11.5%), but an opposite-sex fraternal rate of only 13/221 (5.9%).[5] Unable to explain this difference in terms of genetics, Kallmann stated that the "morbidity rates for opposite-sexed and same-sexed two-egg [fraternal] twin partners vary only from 10.3 to 17.6%."[6] (These percentages reflect Kallmann's age-adjustment of the raw data.) While

---

1. Jackson, 1960, pp. 64-65.
2. Joseph, 2004b.
3. e.g., Shields et al., 1967.
4. Gottesman, 1991, p. 107.
5. Kallmann, 1946, p. 317.
6. Ibid., p. 321.

stressing that rates varied "only" from 10.3 to 17.6%, Kallmann apparently failed to understand that this difference runs counter to the twin method's assumptions and suggests that identical-fraternal differences could also be explained by non-genetic factors. One could argue that the statistically significant concordance rate difference between his same-sex and opposite-sex fraternal twins invalidates the equal environment assumption, upon which Kallmann based his genetic interpretation of identical versus fraternal concordance rate differences.

Gottesman discussed concordance rate differences between the two types of fraternal twins in a 1966 article co-authored by James Shields. "One significant and consistent difference that emerged from our analyses," they wrote, "was a lower concordance rate for opposite-sex fraternal pairs than same-sex fraternal pairs for studies giving information on this point."[1] In the same article they produced a table presenting Kallmann's same- and opposite-sex fraternal findings, listing the rates as 11% and 6% respectively.[2] Because only one study (by Kringlen) reporting opposite-sex fraternal schizophrenia rates appeared between 1966 and 1987, Gottesman's 1966 and 1991 pooled data were very similar (see the pooled rates in Table 6.3 both with and without Kringlen's data). The difference is that, in Figure 10, Gottesman left out these "significant and consistent" results, which might have led his readers to conclude that concordance rate differences in general can be explained by environmental factors.

Fraternal Twins Versus Siblings

Further evidence that environmental factors influence or completely explain identical-fraternal concordance differences is the comparison between fraternal twins versus the ordinary (non-twin) siblings of people diagnosed with schizophrenia. In Figure 10, for example, Gottesman gave the fraternal twin risk as 17%, but the sibling risk as only 9%. Clearly, genetic theory cannot account for this difference, since both sibling sets have the same genetic relationship to each other. Environmental theories, emphasizing fraternal twins' stronger psychological bond, more similar treatment, and greater physical proximity would expect a greater fraternal twin versus sibling risk, and this is what we find in Figure 10.[3] Unfortunately, Gottesman offered no explanation for these differences in his discussion of the patterns of familial risk.

Dual Mating Studies

The final representation in Gottesman's Figure 10 is a 46% schizophrenia risk among the offspring of two parents diagnosed with schizophrenia, obtained

---

1. Gottesman & Shields, 1966a, p. 76.
2. Ibid., p. 35.
3. Leo & Joseph, 2002.

from "dual mating studies." The percentage Gottesman actually found among the five studies he surveyed, from which he pooled 134 "risk lives," was closer to 33%. He arrived at 46% through the use of an age correction formula.[1] While it is true that many offspring in these types of studies have not passed through the "schizophrenia risk period" (typically 15-45 years of age), age correction can be misleading because it adjusts rates upwards on the basis of offspring *possibly* being diagnosed with schizophrenia in the future. A solution to this problem would be to count only those offspring (or twins in schizophrenia twin studies) over 45 years old. But due to their reluctance to reduce sample sizes and thus decrease statistical power, several researchers have preferred to use age correction formulas. Regarding Gottesman's dual mating calculation, 33% represents reported data; 46% represents an educated guess.

Gottesman did not identify the five studies he used to arrive at a 46% risk, and in his own 1989 dual mating study he listed eight previous investigations studying the "offspring of two schizophrenic parents."[2] Half of these studies were published before 1953, and one was published by Munich School researcher Bruno Schulz in Nazi Germany (more on Schulz, later in this chapter). Moreover, there is no indication that diagnosis in these studies were made blindly. Thus, genetically oriented researchers diagnosed the offspring of two people they viewed as carrying the "hereditary taint of schizophrenia." It is not difficult to imagine that they would tend to see more "schizophrenia" among these offspring than among the offspring of people they knew did not carry a psychiatric diagnosis. In Gottesman's own study based on Danish registers, he found a "morbid risk" of only 10% (1/14, or 7%, before age correction) among the offspring of "reactive psychosis" dual matings.[3]

Another problem with dual mating studies is that it is unusual to find two biological parents diagnosed with schizophrenia who also rear their child. It is important to know at what point in a child's development each parent was diagnosed, as well as the circumstances of the child's upbringing. More than any other type of genetic study, detailed case histories are essential in dual mating studies. Simple numbers or "morbidity rates" will not do. Moreover, like the other risk factors outlined in Figure 10, duel mating results can be completely explained by environmental factors, since these offspring were reared in the family environments of two people diagnosed with schizophrenia. Consistent with emotional deprivation and abuse, Gottesman reported that the offspring of schizophrenia dual matings were at "considerably higher risk for other psychiatric abnormalities."[4] Suppose that a pellagra dual mating study shows high rates of pellagra among the offspring of two parents diagnosed with pellagra.

1. Gottesman, 1991, p. 101.
2. Gottesman & Bertelsen, 1989b, p. 288.
3. Ibid.
4. Gottesman, 1991, p. 101.

This finding would be the result of their parents' diets, not their genes. Thus, as in other types of kinship research, potential genetic and environmental influences are not easily disentangled in dual mating studies.

Conclusions Regarding Gottesman's Figure 10

As we have seen, there are several problems with the schizophrenia lifetime risk percentages compiled by Gottesman in his Figure 10. The most important is that, although it is almost always reproduced or discussed in terms of providing evidence in favor of genetics, the various risk percentages can also be explained by factors such as differential exposure to deviant rearing patterns, modeling the behavior of relatives, hospital or researcher diagnoses made because an individual was known to have a biological relative diagnosed with schizophrenia, common exposure to environmental agents, or the more similar environments experienced by identical versus fraternal twins. Textbook authors relying on Gottesman as a source typically focus more on what Gottesman *concluded* about the data than on plausible alternative explanations. This makes it simple for these authors, who don't have to dig up and analyze primary source material. In doing so, however, they rely on someone who has been devoted to the genetic position for his entire career, and who usually presents data and theories in ways that support the genetic position. Typically, psychiatry and psychology textbook authors fail to inform their readers that there are two sides to the "genetics of schizophrenia" story, and rely on potentially misleading facile graphics exemplified by Gottesman's Figure 10.

GOTTESMAN'S REPORTING OF INDIVIDUAL TWIN STUDIES

Chapter 6 of *Schizophrenia Genesis* is devoted to an overview of individual schizophrenia twin studies published between 1920 and 1987, which are shown in Table 6.2. Unfortunately, there are several errors and omissions in Gottesman's descriptions.

After briefly discussing the early investigations by *Luxenburger* and *Rosanoff et al.* (although failing to mention the significantly different same- versus opposite-sex fraternal concordance rate difference in the latter study), Gottesman turned to *Essen-Möller's* 1941 Swedish investigation. He reported that Essen-Möller found a 64% concordance rate (7/11) among his identical pairs, based on a co-twin diagnosis of "major mental illness with schizophrenic features."[1] However, in a 1970 follow-up, Essen-Möller himself reported that "[f]or schizophrenia, the rate of concordance among those who survived the end of the risk period was 2 out of 7, or 29%."[2] Why did Gottesman report 64% concor-

---

1. Ibid., p. 107.

dance instead of 29%? Perhaps because, as he pointed out, Essen-Möller was a "strict, conservative diagnostician."[1] Yet if we turn to an earlier chapter in *Schizophrenia Genesis*, Gottesman discusses the virtue of being a strict diagnostician:

Table 6.2

PAIRWISE SCHIZOPHRENIA CONCORDANCE RATES IN TWIN STUDIES PUBLISHED BETWEEN 1920 AND 1987

Authors	Year	Country	Identical Pairs	Number Concordant	%	Same-Sex Fraternal Pairs	Number Concordant	%
				"Classical Twin Studies"				
Luxenburger [a]	1928	Germany	17	10	59%	13	0	0%
Rosanoff et al.	1934	USA	41	25	61%	53	7	13%
Essen-Moller [b]	1941/1970	Sweden	7	2	29%	24	2	8%
Kallmann	1946	USA	174	120	69%	296	34	11%
Slater	1953	UK	41	28	68%	61	11	18%
Inouye	1961	Japan	55	20	36%	17	1	6%
				"Contemporary Twin Studies"				
Tienari	1963/1975	Finland	20	3	15%	42	3	7%
Gottesman & Shields	1966b	UK	24	10	42%	33	3	9%
Kringlen [c]	1967	Norway	45	12	27%	69	3	4%
NAS-NRC [d]	1969/1983	USA	164	30	18%	268	9	3%
Fischer [e]	1973	Denmark	25	9	36%	45	8	18%
Koskenvuo et al.	1984	Finland	73	8	11%	225	4	2%
				POOLED RATES				
"Classical"			335	205	61%	464	55	12%
"Contemporary"			351	72	21%	682	30	4%
TOTAL			686	277	40%	1,146	85	7%

Concordance rates based on the authors' narrow ("strict") definition of schizophrenia; age correction factors not included. When two dates are stated, the first indicates the year results were first published, the second indicates the final publication, whose figures are reported in the table.
[a] Based on figures from Gottesman & Shields (1966a). Hospitalized co-twins only.
[b] Identical twin results from Essen-Möller (1970). He did not report fraternal twin results in this 1970 publication. Fraternal twin concordance rate based on 1941 definite cases among co-twins, as reported by Gottesman & Shields (1966a, p. 28).
[c] Based on a strict diagnosis of schizophrenia; hospitalized and registered cases.
[d] National Academy of Sciences-National Research Council. Original report by Pollin et al. (1969); final report by Kendler & Robinette (1983).
Table 1: [e] Final results of an expanded sample originally collected by Harvald & Haugue (1965).

The broader the criteria — say, poor motivation for social participation, a deterioration in personality functioning, and some disordered thought processes — the more cases with generally deviant behavior will be diagnosed as schizophrenia, including "false positive" diagnoses, because such symptoms are *not highly discriminating* for schizophrenia per se [emphasis in original].[2]

---

2. Essen-Möller, 1970, p. 17.
1. Gottesman, 1991, p. 107.

Perhaps Essen-Möller wanted to avoid making "false positive" diagnoses, just as Gottesman described. Moreover, looking back from his 1970 vantage point Essen-Möller concluded that, for his 1941 sample, *the correct rate of concordance for schizophrenia in fact was zero* [emphasis in original]."[1] Readers of *Schizophrenia Genesis*, however, were presented with Gottesman's interpretation of Essen-Möller's results, but not the different results reported by Essen-Möller himself.

Table 6.3

SAME-SEX AND OPPOSITE-SEX FRATERNAL TWIN PAIRWISE SCHIZOPHRENIA CONCORDANCE RATES IN TWIN STUDIES PUBLISHED BETWEEN 1920 AND 1987

Authors	Year	Same-Sex Fraternal Concordance	%	Opposite-Sex Fraternal Concordance	%	Probability*
Rosanoff et al. [a]	1934	5/53	9%	0/48	0%	.036
Kallmann	1946	34/296	11%	13/221	6%	.019
Slater	1953	11/61	18%	2/54	4%	.014
Inouye [b]	1961	2/11	18%	0/6	0%	(ns)
Harvald & Haugue [c]	1965	4/33	12%	2/29	7%	(ns)
Kringlen [d]	1967	3/69	4%	3/64	5%	(ns)
POOLED RATES		59/523	11.3%	20/422	4.7%	
(Pooled Excluding Kringlen)		(56/454)	(12.3%)	(17/358)	(4.7%)	

* Fisher's Exact Test, one-tailed. ns = statistically non-significant at the .05 level.
[a]Based on twins sharing "similar affections" in Rosanoff & colleagues' Table 3 (1934, p. 269).
[b]Based on figures reported by Gottesman & Shields (1966a, p. 50). Includes "schizophrenia and schizophrenia-like disorders," which were the only figures provided.
[c]Preliminary results of the Danish sample. The final results from this sample are found in Fischer's 1973 study, which did not report opposite-sex fraternal twin concordance rates.
[d]Based on Kringlen's strict definition of schizophrenia, hospitalized and registered cases.

Gottesman then moved on to the previously discussed *Kallmann* investigation, and briefly mentioned *Slater*'s 1953 British study. There, he failed to mention several trends in Slater's results that are difficult to explain on genetic grounds. These include a significantly higher same- versus opposite-sex fraternal concordance rate (see Table 6.3), and a significantly higher rate (11.3% vs. 4.6%) for fraternal twin pairs versus the rate among the siblings of twins. As we have seen, both sets have the same genetic — but not familial — relationship to each other. Moreover, Slater reported the concordance rate of *female* fraternal twins as 22.5% (9/40), whereas the rate for the siblings of twins was only 4.6% (26/568). This is a highly statistically significant difference, and the ratio of this concordance rate difference (5:1) is actually greater than Slater's identical-fraternal

2. Ibid., p. 20.
1. Essen-Möller, 1970, p. 315.

twin ratio (4:1). This large fraternal twin/sibling concordance rate difference did not go unnoticed by Slater, who chose not to question the basic assumptions of the twin method and merely wrote that the higher concordance of fraternal twins versus the siblings of twins "is probably attributable to the more thorough investigation of the former."[1]

Gottesman briefly discussed *Inouye's* study, and then arrived at five contemporary studies beginning with *Tienari's* 1963 Finnish investigation. He correctly reported that this study found a 0% identical twin concordance rate, and then claimed that subsequent updates by Tienari showed that this 0% preliminary result was "a fluke."[2] Although it is true, as Gottesman pointed out, that zero percent identical twin concordance was a surprise to hereditarians and environmentalists alike, it turns out that Tienari's final results are not dramatically different from his 1963 "fluke" data. In his final publication, in 1975, Tienari reported pairwise identical twin concordance of only 15%: "The concordance rate for the MZ group is 3:20 (15 per cent) and that for the DZ group is 3:42 (7.5 per cent)."[3]

Readers of *Schizophrenia Genesis*, on the other hand, are provided with different results. According to Gottesman, "Tienari, on follow-up (1975), reported a pairwise concordance rate of 36 percent for MZ twins and 14 percent for DZ twins and is no longer the odd man out."[4] Gottesman and Shields reported similar results in their 1982 book, *Schizophrenia: The Epigenetic Puzzle*. There, they stated a preference for using data from a 1971 update by Tienari, and adjusting them because Tienari allegedly omitted "two organic phenocopies" and diagnosed five co-twins with "schizophrenia related illness."[5] Gottesman and Shields did recognize in 1982 that Tienari reported only 3 of 20 identical pairs as concordant, but in the next amazing sentence they wrote, as if to reveal their intention to boost Tienari's results to levels they were comfortable with, "The possible use of some kind of age correction together with the probandwise method of calculating rates makes the Finnish study no longer such an odd man out in the literature."[6] The difference between 15% and 36% is meaningful because the latter percentage brings Tienari's data into line with the lower range of the older studies. The only problem is that Tienari did not "report a pairwise concordance rate of 36%" for identical twins. Rather, he reported a 15% rate.

In discussing *Kringlen's* 1967 study, Gottesman correctly stated that Kringlen found no differences between his same-sex and opposite sex- fraternal twins. Unfortunately, Gottesman reported data only from Kringlen's study, while failing to mention large differences in the earlier investigations. Thus, he

1. Slater, 1953, p. 86.
2. Gottesman, 1991, p. 109.
3. Tienari, 1975, p. 33.
4. Gottesman, 1991, p. 111.
5. Gottesman & Shields, 1982, p. 103. Also see Tienari, 1971.
6. Gottesman & Shields, 1982, p. 105.

followed his earlier decision in *Schizophrenia: The Epigenetic Puzzle* that Kringlen's results closed the issue, regardless of the previous studies' findings.

Moving through *Fischer's* Danish study, we come to the National Academy of Sciences-National Research Council (NAS-NRC) study based on a twin register of United States armed forces inductees. Although Gottesman reported that the original results were published by *Pollin and colleagues* in 1969, he failed to mention that these investigators found a modest 13% pairwise identical twin concordance rate (11/69).[1] Instead, he declared this study to have been "superseded" by *Kendler and Robinette's* 1981 reanalysis, and highlighted their finding of probandwise rates of identical 31%, fraternal 6%, as well as much higher rates if "any kind of psychiatric diagnosis" is included.[2] Only a careful reader, looking at the bottom of a table listing the newer studies' concordance rates as calculated by Gottesman, would discover that Kendler and Robinette still could find only an 18% pairwise identical twin schizophrenia concordance rate. Moreover, Gottesman omitted this low concordance/large sample study from his calculations for the contemporary studies.[3]

Finally, Gottesman did not mention or was unaware of *Koskenvuo and colleagues'* large 1984 register-based study, which found a modest 11% identical twin concordance rate.

Clearly, Gottesman reported results in ways that would lead his readers to believe that concordance rates were higher than they actually were reported. And as always, the fact that one can draw conclusions in favor of genetics from these results is completely dependent upon the implausible assumption that identical and fraternal twins experience equal environments.

GOTTESMAN'S REPORTING OF SCHIZOPHRENIA ADOPTION RESEARCH

As in the previous chapter, I will refrain from engaging in a detailed critique of schizophrenia adoption research, as this was the subject of Chapter 7 of *The Gene Illusion.* What I will show is that Gottesman's account of these studies is as one-sided as the textbook accounts are.

Gottesman began his evaluation with the astute yet greatly underemphasized observation that "both critics and skeptics" could argue that the family and twin data presented in Figure 10 "can be explained by environmental contagion."[4] For Gottesman, schizophrenia adoption studies are useful to test the hypothesis that deviant communication styles can cause schizophrenia. Yet, adoption study problems such as selective placement, age at transfer, range

---

1. Pollin et al., 1969, p. 600.
2. Gottesman, 1991, p. 113.
3. Ibid., p. 110.
4. Ibid., p. 133.

restriction, researcher bias, and questionable diagnostic procedures are men-tioned only briefly or not at all in *Schizophrenia Genesis*. Furthermore, Gottesman accepted the liberal definition of schizophrenia used by the Danish-American researchers, without which they would have found no statistically significant results (see Chapter 3). However, earlier in his book Gottesman found it "essential that only those who actually suffer from this specific disease be clas-sified, identified, or diagnosed, as schizophrenics..." Moreover, "Disturbed persons who have some other condition must be excluded from further consider-ation to minimize the 'noise' interfering without [*sic*] efforts to detect sometimes weak signals."[1] However, Gottesman did not apply this standard to his evalu-ation of schizophrenia adoption research, whose results depended on counting "weak signal" SSDs as schizophrenia.

Gottesman began with *Heston's* Oregon study. Although he endorsed Heston's methods and conclusions, we saw in Chapter 5 that Heston's diag-nostic process was contaminated by his knowledge the life history, previous hospital diagnoses, and group status of the people he diagnosed. According to Gottesman, "Five of the 47 index adoptees grew up to be schizophrenic, as deter-mined by the consensus of two blindfolded, independent clinicians plus Heston; none of the 50 controls were even considered to be psychotic."[2] However, if the *un*blindfolded Heston confers with his "blindfolded" colleagues, and they jointly arrive at diagnoses influenced by Heston's input, the entire diagnostic process is contaminated.[3]

Moving on to *Rosenthal and colleagues'* Adoptees study, Gottesman, like most genetically-oriented reviewers, failed to highlight the fact that Rosenthal found no statistically significant SSD elevation among the biological offspring of his SSD-diagnosed parents, versus the biological offspring of his control parents. Instead of theorizing about the criteria used for index adoptee diagnoses, or whether age-correction adjustments should have been made, Gottesman should have simply stated that Rosenthal's 1968 and 1971 studies — failing a finding of statistically significant results — found no evidence in support of genetic influ-ences on schizophrenia. However, he implied that Rosenthal's results were con-sistent with genetic theories.

Gottesman noted that "only a handful" of Rosenthal's index adoptees had been placed before their parents had been diagnosed, "thus, the adoption author-ities and the adoptive parents could not engage in any self-fulfilling prophecies about the outcome of the children based upon knowledge about the biological parents' mental illness."[4] However, he failed to comment about what type of "self-fulfilling prophecies" may have been operating in Heston's study, where

---

1. Ibid., p. 17.
2. Ibid., p. 137.
3. Cassou et al., 1980.
4. Gottesman, 1991, p. 140.

100% of the index adoptees were born to previously schizophrenia-diagnosed mental hospital residing biological mothers, in a region and era in which "insanity" was widely viewed as being the result of bad heredity.

We recall from the previous chapter that Rosenthal and colleagues' results were the subject of a 1983 reanalysis using DSM-III diagnostic criteria by Patricia Lowing and her colleagues. Here, Gottesman correctly reported that Lowing et al. found only one case of schizophrenia (2.3%) among the index adoptees they diagnosed, versus zero control diagnoses. This difference is, of course, statistically non-significant. Gottesman could have stopped at this point and concluded that the evidence did not support the genetic position. However, he followed Lowing et al. in claiming that genetic factors were sustained by counting schizotypal and schizoid diagnoses, even though the latter "weak signal" diagnosis, upon which Lowing and colleagues' statistically significant findings were based, was eventually removed from the spectrum by the Danish-American investigators themselves (see Chapter 3).

Gottesman then moved on to the "now classic" Danish-American Adoptees' family study of *Seymour Kety and his colleagues*. He wrote that Kety et al. began their Copenhagen study with "Thirty-three adoptees who, when they grew up, had become schizophrenic..."[1] However, as we saw in Chapter 3, Kety et al. diagnosed only 16 of these with chronic schizophrenia, and they subsequently removed the 10 B2 ("acute schizophrenia") index adoptees. Gottesman also reported that Kety and colleagues found a 6.4% "definite schizophrenia" index biological relative rate in their 1975 study. However, 55% (6/11) of these "definite" cases were B3 "borderline schizophrenia" diagnoses.[2] And we saw in Chapter 5 that the index B1 biological relative chronic schizophrenia rate was not significantly higher than that of the control biological relatives. Moreover, Kety and colleagues gave 7 (64%) of their "definite schizophrenia" diagnoses to second-degree relatives, which runs counter to genetic predictions.

Having reviewed Kety and colleagues' study in detail elsewhere, in *Schizophrenia Genesis* Gottesman chose to focus on Kendler and Gruenberg's 1984 reanalysis of the Kety et al. data, which was based on DSM-III diagnostic criteria (see Chapter 5).[3] According to Gottesman, a major finding of this reanalysis was that "the overall occurrence of schizophrenia or schizophrenia spectrum disorders is still significantly higher in the biological relatives of schizophrenic adoptees, relatives who *did not rear them*, than in the adoptive relatives with whom and by whom they *were* raised. Thus, the original interpretation [by Kety et al.] was sustained and strengthened [emphasis in original]."[4] But as Kendler

---

1. Ibid., p. 143.
2. Kety et al., 1975, p. 154.
3. Kendler & Gruenberg, 1984. For Gottesman's earlier reviews of the Kety et al. studies, see Gottesman & Shields, 1976, 1982.
4. Gottesman, 1991, p. 144.

and Gruenberg reported (and reproduced by Gottesman in a table of his own), only 1 in 10 first-degree, and 1 in 25 second-degree index biological relatives (for a total of 2/35), were diagnosed with chronic schizophrenia. In their own tables, Kendler and Gruenberg selectively left out calculations showing that chronic schizophrenia rates were not statistically significant versus controls, whereas they calculated other comparisons (such as SSD rates) that were statistically significant in the genetic direction.[1] Moreover, Gottesman's claim that Kendler and Gruenberg found significantly more schizophrenia cases among index adoptees' *biological* relatives than their *adoptive* relatives does not appear to be true. Kendler and Gruenberg diagnosed 2 out of 35 index biological relatives with schizophrenia, whereas they diagnosed zero of the 33 adoptive (rearing family) relatives with schizophrenia. This comparison is not statistically significant.[2]

Setting the historical record straight, from 1968 onward Kety and colleagues did not, as Gottesman implied, make any "original interpretation" about diagnostic differences between the biological and adoptive relatives of their index adoptees. (We saw in Chapter 5 that Kety actually called this comparison "inappropriate" and "fallacious.") Rather, they compared diagnoses among index biological relatives versus control biological relatives, and in another comparison between index adoptive relatives versus control adoptive relatives. And Kety and colleagues' 1968 interpretation would have been quite different from Gottesman's, since there was no statistically significant schizophrenia spectrum elevation among the index biological versus index adoptive relatives in the 1968 Kety et al. publication.[3]

Again referring to Kendler and Gruenberg's findings, Gottesman reported a 10.5% (± 9.9%) age-corrected first-degree relative lifetime schizophrenia morbid risk, which he invited his readers to compare to the data in his Figure 10. Like Kendler and Gruenberg, however, he obtained this result on the basis of *one diagnosis* out of 10 first-degree biological relatives — a rate not significantly greater than controls, or versus the general population expectation. Thus, we can draw no valid conclusions about genetic influences (or about "morbid risks") from percentages derived from tiny samples such as this one. The same can be said for Kendler and Gruenberg's claim that the 2/35 schizophrenia rate among all index biological relatives (first- and second-degree combined) is statistically significant versus the control rate. They found statistical significance only because the control biological relative rate was 0/137 (0/91 among the "screened controls"). The fact that this comparison is statistically significant serves mainly as an example of how the results of statistical tests can be mis-

---

1. For example, see Kendler & Gruenberg's (1984) Tables 7 & 8.
2. 2/35 vs. 0/33, p = .26, Fisher's Exact Test, one-tailed. The index biological relative schizophrenia rates can be found in Kendler & Gruenberg, 1984, pp. 560-561 (Tables 7 & 8). The same publication listed the adoptive relative diagnoses on page 562 (Table 10).
3. Kety et al., 1968. Also see Joseph, 2004b, Chapter 6.

leading. Certainly, a large enough grouping of undiagnosed control relatives can transform a very modest index relative rate into a statistically significant finding. However, a comparison left unmentioned by Gottesman, Kendler and other schizophrenia adoption researchers is the index schizophrenia rate *versus general population expectations*. Regardless of possible index-control diagnostic differences, if the index rate is not significantly higher than the population expectation of roughly 1%, researchers can draw no valid conclusions about genetic factors among non-adoptees. Let's explore this point further.

The General Population as an Additional Control Group

Kety has argued that a control group eliminates problems created by the lack of standardized methods and criteria; the control group becoming "the only legitimate basis for comparison."[1] However, if investigators find significantly more schizophrenia among index versus control relatives or adoptees — but not more index schizophrenia versus the general population expectation of 1% — they cannot generalize their results to the non-adoptee population, nor can they conclude in favor of genetics. Although prevalence rates vary according to country and diagnostic criteria, if used conservatively, we can utilize these rates to test whether the schizophrenia rate in a particular population is significantly higher than the general population expectation. Boyle made this point clear:

> If the index biological relatives had a higher than expected prevalence of schizophrenia diagnoses and the control relatives were indistinguishable from the general population *or* the index relatives resembled the general population and the control relatives were exceptionally free from diagnoses, then significant differences, *but carrying very different interpretations*, could appear [emphasis added].[2]

Boyle concluded that "a simple comparison of two groups of biological relatives does not indicate how similar each is to the general population."[3]

No schizophrenia adoption researcher to my knowledge has ever recognized the need to show that index adoptees or biological relatives must have schizophrenia (or even "schizophrenia spectrum disorders") in numbers significantly greater than population expectations, yet it is essential to do so in order to be able to generalize findings to the non-adoptee population. This was never an issue in twin research, where even 30% identical twin concordance demonstrated that an identical twin of a person diagnosed with schizophrenia is about 30 times more likely to be diagnosed than a randomly selected member of the population.

Suppose we conduct an Adoptees' Family study beginning with 300 adoptees who are eventually diagnosed with schizophrenia, and 300 non-diag-

---

1. Kety, 1983a, p. 724.
2. Boyle, 1990, p. 141.
3. Ibid., p. 144.

nosed matched control adoptees. There are 1,000 identified first-degree biological relatives in each group. We then diagnose 17 index biological relatives, but only 2 control biological relatives, with schizophrenia. Although schizophrenia adoption researchers would probably conclude that the significant difference between index and control relatives suggested the operation of genetic factors, they would have overlooked one critical comparison: there is no statistically significant difference between the 17/1,000 index relative diagnoses and the expected general population rate of 10/1,000 (based on a 1% prevalence). Therefore, having a biological relative with schizophrenia did not significantly increase a person's chance of being diagnosed with schizophrenia, compared with the general population expectation. We could conclude that this study produced no evidence supporting genetic factors in schizophrenia, but that it did suggest schizophrenia-producing environmental differences between index and control rearing environments.

According to Rosenthal, in order to "demonstrate that genes have anything to do with schizophrenia," an investigator must show that "the frequency of schizophrenia must be greater in the families of schizophrenics than in the families of nonschizophrenic controls or in the population at large."[1] The last phrase should have read "*and* in the population at large," because a greater index schizophrenia rate versus controls does not absolve the investigator of responsibility for showing that this rate is also significantly greater than the rate expected in the general population.

* * *

While many textbook authors cite schizophrenia adoption research as providing the most important evidence in support of genetic influences on schizophrenia, for Gottesman its main finding has been that when children are removed from theoretically "schizophrenogenic" parents and reared in another family, their risk for developing schizophrenia does not significantly decrease. He wrote that the adoption studies "show that sharing an environment with a schizophrenic parent or another schizophrenic generally does not account for the familiality of cases."[2] Thus, twin research remains for Gottesman the most important evidence in favor of genetics. "The strongest evidence implicating genetic factors in the etiology of schizophrenia," wrote Gottesman and Shields in 1982, "comes from twin studies."[3]

Although Gottesman concluded his *Schizophrenia Genesis* adoption study chapter by writing that the "broad genes-plus-environment hypothesis for explaining the cause of schizophrenia is strengthened by the adoption studies," no such conclusion is warranted from this body of methodologically flawed and environmentally confounded research.[4]

---

1. Rosenthal, 1974, p. 589.
2. Gottesman, 1991, p. 148.
3. Gottesman & Shields, 1982, p. 71.

OTHER ISSUES

A Genetic Predisposition for Schizophrenia?

A running theme of *Schizophrenia Genesis* is that genetically predisposed people develop schizophrenia when exposed to environmental factors triggering the condition. According to Gottesman, schizophrenia is a "heavily genetically influenced disorder rather than a genetically determined disorder."[1] However, his belief that schizophrenia is "heavily genetically influenced" is little more than an acknowledgment that we haven't identified specific environmental triggers. On the other hand, we have the example of PKU, a genetically inherited disorder which leads to mental retardation *only if a dietary intervention is not made when the child is an infant.* Thus PKU, whose genetic influence is certainly "heavier" than schizophrenia's, is prevented with a relatively simple environmental intervention. In the case of schizophrenia, Gottesman and others are able to claim important genetic influences only because effective interventions to prevent or "cure" schizophrenia have not yet been discovered.

In contrast to schizophrenia, Gottesman cited measles as an example of a disease "that deserves to be called environmental":

> Virtually everyone exposed to the virus gets the symptoms if they have not been inoculated against it. Given the universality of the genetic vulnerability to measles and the episodic, acute exposure to the virus, we can easily say that such conditions are "environmentally caused."[2]

But why is measles "environmental" if genetic vulnerability to the virus is "universal"? According to Gottesman, if *some* people are genetically vulnerable to conditions such as schizophrenia, the condition is "heavily influenced" by genetics. Conversely, conditions such as measles, where *everyone* is genetically vulnerable, are "environmentally caused." Thus, if *everyone* is predisposed, then *no one* is predisposed. Suppose that only 50% of human beings contract measles when exposed to the virus. Would it follow that measles is "heavily influenced" by genetics because the population vulnerability dropped from 100% to 50%? One reason that most people would answer no, I would argue, is that we have identified the virus that causes measles.

In discussing the causes of schizophrenia, Gottesman noted that pellagra was once viewed as being heavily influenced by genetics but was later discovered to be caused by a vitamin deficiency. "Once the puzzle pieces were fitted together," he continued, "false ideas about causality related to personal characteristics of the 'hosts,' such as racial and class inferiority, were definitively refuted."[3] Yet he asks us to accept schizophrenia as a "heavily genetically influ-

---

4. Gottesman, 1991, p. 148.
1. Ibid., p. 217.
2. Ibid., p. 218.

enced disorder" even though we possess few of the disorder's "puzzle pieces." And like pellagra and PKU, the discovery of an identifiable environmental puzzle piece for schizophrenia could radically change the way we think about the condition, and could even lead to effective preventative measures and treatments.

According to Gottesman, "In more than 100 years of study, no one has found a single case of schizophrenia caused solely by the conditions of the patient's upbringing."[1] But this would be like saying in 1910 that no one has ever found a single case of pellagra caused solely by environmental factors. As this example shows, it is inappropriate to generalize researchers' failure to find specific environmental factors in schizophrenia to argue against the existence of a factor that solely causes it. Perhaps future researchers will uncover specific environmental risk factors, which all sides of the "genetics of schizophrenia" debate agree must exist.

Besides the obvious fact that no one has ever found a single case of schizophrenia officially caused by *anything*, there is plenty of evidence that disturbed family environments can lead to schizophrenia-like behavior. In Tienari's schizophrenia adoption study, for example, the investigators found that every adoptee eventually diagnosed with schizophrenia was raised in a "severely disturbed" adoptive family environment.[2] And in a 1976 study of blinded Rorschach (ink blot) protocols of normal rearing parents who had adopted children, psychologist Margaret Singer was able to predict with 100% accuracy which parents had raised a child diagnosed with schizophrenia.[3] Additional evidence supporting environmental factors was surveyed by Richard Bentall in his 2003 *Madness Explained.*[4]

Morten and Karl

In a discussion of psychological and environmental stressors associated with schizophrenia, Gottesman retold the story (culled from Danish records) of twins Morten and Karl as described in Gottesman and Bertelsen's award-winning 1989 study of the offspring of Danish identical twins discordant for schizophrenia (one twin diagnosed with schizophrenia, the other not diagnosed).[5] Although Karl lived to the age of 88 and was never diagnosed with schizophrenia, his identical co-twin Morten was hospitalized at age 23, diagnosed with paranoid schizophrenia. Morten began to exhibit schizophrenia-like

---

3. Ibid., p. 62.
1. Ibid., p. 18.
2. Tienari et al., 1987.
3. Wynne et al., 1976.
4. Bentall, 2003. Also see Read et al., 2004.
5. Gottesman & Bertelsen, 1989a. These authors were awarded the 1991 Kurt Schneider Prize for this publication.

symptoms after he fell from a height of four meters and hit his head on a stone cornice. "With time," Gottesman wrote, "he became more reserved and increasingly psychotic, with auditory hallucinations and persecutory and bizarre somatic delusions."[1] According to Gottesman, "the head injury to Morten appears to have been enough to release his predisposition to schizophrenia," and he and Bertelsen decided to retain diagnoses among the offspring of this discordant pair.

One might argue, however, that Gottesman and Bertelsen should have removed Morten and Karl and their offspring from their study due to the possibility that Morten's "schizophrenia" was the result of organic brain damage caused by his fall, or, as geneticists like to say, his brain damage could have resulted in a "nongenetic phenocopy" of schizophrenia. This seemed to be Gottesman's view in an earlier chapter of *Schizophrenia Genesis*:

> Schizophrenia has been diagnosed after, for instance, *a head injury* or the flu, but these trauma-induced, schizophrenia-like psychoses *are no longer accepted as true schizophrenia*; the majority are now seen as parts of the broader category of organic brain syndromes associated with physical disorders. An ambiguous "no-man's land" still exists for those few cases of schizophrenia triggered by a trauma that is just one more risk-increasing factor in the combined liability [emphasis added].[2]

Nevertheless, Gottesman and Bertelsen diagnosed Morten with schizophrenia. However, without the offspring produced by Morton and Karl, they would not have found statistically significant results for an "important contrast" they viewed as supporting genetic influences on schizophrenia.[3]

Politicians and Scientists

The accumulated evidence supporting genetic and unspecified environmental influences on schizophrenia is, according to Gottesman, so solid that "Resistance to such a balanced conclusion, when it appears, must be based on ideological reasons." He went on to dismiss Lewontin, Rose, and Kamin's 1984 book *Not in Our Genes* as a "politicized tract" by "well-respected scientists working on the periphery of their expertise."[4] This enabled him to ignore Lewontin and colleagues' important argument out of hand by attacking their alleged motives and questioning their expertise. Elsewhere, Gottesman accused

---

1. Gottesman, 1991, p. 163.
2. Ibid., p. 93.
3. Gottesman & Bertelsen, 1989a, p. 870. See Joseph, 2004b, Chapter 6 for more details on Gottesman & Bertelsen's 1989 study of the offspring of discordant identical twin pairs.
4. Gottesman, 1991, pp. 216-217. Elsewhere, Gottesman commented that Lewontin et al., although "scientists...are speaking as ideologues in that particular book [*Not in Our Genes*] and in many of their other writings, all of which I find generally offensive..." (quoted in Healy, 1998, p. 396).

them of "very quickly get[ting] emotional in talking and writing about [nature/nurture issues] and I find that is very dangerous."[1] Moreover, Gottesman did not consider it necessary to answer other critics of genetic theories because, by his own definition, any challenge to his theory must be "ideological." Thus, he has refused to appear on television with Peter Breggin, "a very ideological, psychiatrically trained physician," because "I don't want to give him any credibility by allowing him to appear with a scientist."[2]

Like most other genetic researchers, Gottesman desires to protect "science" from political intrusion, as if his own beliefs and political opinions have no relation to his research or to his evaluation of other people's research. However, the separation of "science" and "politics" is a myth, and, as is superbly documented in Allan Chase's *The Legacy of Malthus* and Edwin Black's *War Against the Weak*, nowhere is this seen more clearly than in human genetic research.[3] Harvard biologist Ruth Hubbard has written that although most scientists believe that "science is immune from political and societal pressures," this view "has been proved wrong time and again." Moreover, she argued that "scientists, as a group, tend to provide results that support the basic values of their society," which is "particularly true when scientists are studying people."[4]

With an understanding that people often fear "extremist" positions, Gottesman, though for many decades a strident and at times lonely advocate for the importance of genetics, placed himself in the "balanced" moderate position on the genes/environment question in opposition to both environmental determinism *and* genetic determinism. Positioning oneself in the "sensible middle position" is common in ideological debates and political campaigns, but has little to do with impartial conclusions based on sound scientific research. As critic Adam Hedgecoe observed, "A strategy of caution deflects criticism that researchers are over-enthusiastic and guilty of genetic hype, and allows readers to see the authors as reasonable and objective in their assessments."[5] And according to Mary Boyle,

> The vulnerability-stress hypothesis — widely interpreted as implying biological or genetic vulnerability — has proved to be an extraordinarily useful and effective mechanism for managing the potential threat to biological models of "schizophrenia"...The usefulness of the hypothesis lies partly in its lack of specificity — since the nature of the claimed vulnerability has never been discovered, anything can count as an instance of it. Its usefulness also lies in its seeming reasonableness (who could deny that biological and psychological or social factors interact?) and its inclusiveness (it encompasses both the biological and social — surely better than focusing on only one?) while at the same time it firmly maintains the primacy of biology, not least through word order, and potentially de-empha-

1. Quoted in Healy, 1998, p. 397.
2. Ibid., pp. 398-400.
3. Chase, 1980; Black, 2003.
4. Hubbard & Wald, 1993, p. 7.
5. Hedgecoe, 2001, p. 893.

sizes the environment by making it look as if the "stress" part of the vulnerability-stress model consists of ordinary stresses which most of us would cope with, but which overwhelm only "vulnerable" people. We are thus excused from examining too closely either the events themselves or their meaning to the "vulnerable" person.[1]

Although Gottesman might believe that his political and social views do not influence his interpretation of the scientific evidence, he nonetheless devoted an entire chapter of *Schizophrenia Genesis* to a discussion of "Schizophrenia, Society, and Social Policy," which consisted largely of Gottesman's views on genetic counseling, on how society should deal with "schizophrenics," and of whether society should be concerned with, or wish to control, the reproduction of people diagnosed with schizophrenia. As a pair of sociologists pointed out, "scientists' and clinicians' emphasis on the separate nature of the social [versus the scientific] realm, in which their knowledges and technologies apply, does not square with their social concerns about this area."[2]

In a 1971 article he co-authored with L. Erlenmeyer-Kimling, entitled "Foundations for Informed Eugenics," Gottesman discussed the "new eugenic" view of society and its "human gene pool," which lamentably no one is "minding."[3] The new eugenics, according to the authors, advocates ideas such as "artificial selection" and "rational elitism," but eschews old eugenic "excesses" such as compulsory sterilization and "genocide in Nazi Germany." Gottesman was concerned with the "differential reproduction" not only of people suffering from presumed hereditary disorders, but even among those whose "disorders have no appreciable genetic loading," since "the reproductive behavior of such individuals and the characteristics of their offspring will influence human ecology by increasing social costs." Gottesman urged "would-be eugenists to abandon their fixation on IQ as *the* trait to be maximized in our species in favor of an Index of Social Value (ISV)":

> The ISV would take into account the fact that a garbage hauler and a brain surgeon make essential contributions to the smooth functioning of society, but that society would break down more rapidly if garbage haulers went on a prolonged strike than if brain surgeons did. Once the ISV has been devised by optimizing the traits that enter it, the eugenist then has the equally challenging task of deciding the optimal degree of phenotypic diversity required to fill the various ecological niches.[4]

The fact that scientists' deeply held opinions about human beings and society influence their research and conclusions is not deplorable — it is inevitable, and often positive. Instead of perpetuating the myth that science is above or separate from politics, why not simply recognize that scientists' worldviews,

1. Boyle, 2002a.
2. Cunningham-Burley & Kerr, 1999, p. 660.
3. Gottesman & Erlenmeyer-Kimling, 1971.
4. Ibid., p. 87.

beliefs, and economic interests influence what they decide to study, the way they interpret data, and the funding they receive. And to be fair, the same can be said of their critics both in and out of the scientific community. If anthropologists can recognize that their own "ethnocentrism" might influence how they view cultures other than their own, why can't genetic researchers at least acknowledge that they tend to view human beings from their own "genocentric" point of view? Neither anthropological nor genetic research is invalidated by their investigators' biases, but the *recognition* of bias does encourage a more critical perspective of their research and conclusions.

The issue isn't that Lewontin and colleagues are influenced by ideology and that Gottesman isn't, but rather that their "ideologies" *differ*. A recognition that scientists' conclusions, including those by genetic researchers, can be influenced by their beliefs and political perspectives (although this process works in both directions) simply acknowledges the true state of affairs. Confusion can occur, however, when genetic researchers deny to themselves or others that their findings and conclusions are influenced by their own genocentrism and the belief systems that flow from it.

Gottesman on the Relationship Between Psychiatric Genetics and German National Socialism

Gottesman's role as a historian of psychiatric genetics must also be addressed. As seen in Chapter 2 of *The Gene Illusion*, several leading contemporary psychiatric geneticists have written in ways that obscure their discipline's collaboration with German National Socialism and the crimes of the Nazi regime. Their general line is that psychiatric genetics was founded by Ernst Rüdin and others, who carried out important scientific work out of their Munich institute, while at the same time Hitler's regime enacted eugenic sterilization laws which set the stage for the its genocidal policies in the name of eugenics and "racial purification." According to these accounts, the Nazis *misused* psychiatric genetic findings for evil purposes. For example, in their 1999 psychiatric genetics textbook, Faraone, Tsuang, and Tsuang wrote that "Adolph Hitler and his Nazi regime began a systematic program first to sterilize and then to kill 'genetically defective' people.... Contemporary researchers in psychiatric genetics are especially disturbed to learn that the Nazis used [German psychiatric genetic] research to justify their eugenics policies regarding the mentally ill..."[1] And in *Schizophrenia Genesis*, Gottesman wrote positively of the work of Rüdin, Hans Luxenburger, Bruno Schulz and other leaders of the "now-famous Munich school of psychiatric genetics."[2] Elsewhere in the book he referred to "thoroughly the scientist" Rüdin,[3] to Rüdin's collaborator Bruno Schulz as a "star

---

1. Faraone, Tsuang, & Tsuang, 1999, pp. 223-224.
2. Gottesman, 1991, p. 14.

member of Rüdin's Munich school,"[1] and to Kallmann as merely a "refugee from Hitler's Germany."[2]

Gottesman discussed the Nazis' crimes in the name of genetics and eugenics separately from his account of the work of Rüdin and his associates, writing, for example, that "'social biology' can be perverted to evil ends and become 'political (pseudo) biology,' as it was by Adolf Hitler and the Nazis in implementing their insane policies of murdering psychiatric patients, genocide, and the Holocaust."[3] Thus, Gottesman falsely implied that the founders of psychiatric genetics had nothing to do with the sterilization laws or Nazi atrocities. In fact, Rüdin and Luxenburger were militant supporters of the German "Racial Hygiene" movement and the forced sterilization of hundreds of thousands of people, for which they helped lay the groundwork long before the Nazis came to power.[4] In addition, Rüdin, Schulz, Luxenburger, and Kallmann supported the 1933 sterilization law while arguing that it didn't go far enough in protecting society from the reproduction of "eugenically undesirable" people (see below).[5]

Unfortunately, subsequent psychiatric genetic commentators continue to perpetuate myths about the origins of their field, such as Peter McGuffin falsely claiming in 2002 that "Luxemburger [*sic*] and Schultz [*sic*] opposed" compulsory sterilization.[6] He claimed elsewhere that Schulz and other German psychiatric geneticists "were appalled at what they saw" after the Nazis took power,[7] and that Luxenburger and Schulz opposed eugenic policies "on both moral and scientific

---

3. Ibid., p. 13.
1. Ibid., p. 94.
2. Ibid., p. 95.
3. Ibid., p. 185.
4. Joseph, 2004b.
5. Ibid., Chapter 2.
6. McGuffin, 2002, pp. 2-3. It is worth pointing out, if only because it would be unthinkable in almost any other field, that Gottesman's former student Peter McGuffin (2002), currently and for many years one of the world's leading psychiatric geneticists, misspelled the names of three important founders of his discipline. (The correct spellings are "Rüdin," "Luxenburger," and "Schulz," not "Rudin," "Luxemburger," and "Schultz"). And in another publication, Cardno & McGuffin (1999, pp. 344-345) referred to Swedish Munich School student and subsequent schizophrenia twin researcher as "Hans Essen-Möller" (actually *Erik* Essen-Möller), and to Franz Kallmann as "Franz Kallman." Although I am well aware of the difficulties presented by language barriers, clearly McGuffin and his colleagues are insufficiently acquainted with the original psychiatric genetic authors and documents. They may be prolific researchers, but they are poor historians indeed, and usually rely on other poor historians to support their claims. Another poor historian is twin researcher Kerry Jang, in his 2005 book *The Behavioral Genetics of Psychopathology*. There, he three times referred to prominent genetic proponent and *The Bell Curve* co-author Richard Herrnstein as "Bernstein" (Jang, 2005, pp. 10, 175, 196; Herrnstein & Murray, 1994), and made the remarkable claim (p. 10) that in the 1970s, "behavior genetics was not associated with either side in [the nature-nurture] debate."
7. Farmer & McGuffin, 1999, p. 482.

grounds."[1] Regarding Rüdin, accounts of this type take at face value his postwar conviction by the Allies as a mere "fellow traveler" of the Nazis, ignoring the evidence implicating Rüdin as a major scientific mastermind of Nazi atrocities.[2]

In Chapter 10 of *Schizophrenia Genesis*, Gottesman denounced Nazi crimes while at the same time implying that the founders of psychiatric genetics had no role in these crimes. Following Robert Jay Lifton in *The Nazi Doctors*, Gottesman described the "six steps culminating in the Final Solution." Let's review each of these steps as described by Gottesman, followed by documentation showing that, contrary to the impression he gives in *Schizophrenia Genesis*, "Munich School" leaders such as Rüdin were there at every step. Although some of the information and quotes I cite below are taken from publications appearing after 1991, much was available before that time.

Steps One and Two

Gottesman:

> The new Nazi government, in a hurry to heal the physical and economic health of Germany in 1933 after Hitler's inauguration, passed laws compelling eugenic sterilization. Judicial safeguards, cynically cosmetic, required a review by a Hereditary Health Court of everyone, institutionalized or not, known by a physician to have a 'hereditary' condition.... At least 350,000 sterilizations were carried out under the legislation...[3]

In 1935, "the sterilization laws were renewed and extended to marriage control via the 'Law for the Protection of the Genetic Well-being of the German People,'" also known as the Nuremberg Laws, which were aimed at Jews and others.

> The last half of the nineteenth century saw a growing, misguided fear that a dysgenic tide would flood society with both insane and retarded persons who had been abandoned by their families and who would multiply 'like rabbits.' Repressive legislation and the atrocities of the Third Reich were fed by such unfounded fears.[4]

Rüdin:

Rüdin, who early in his career had, according to a historian, "constantly proposed a 'merciless extinction' of patients with dementia praecox [schizophrenia],"[5] co-authored the official German publication in support of the 1933 sterilization law.[6] In 1934, Rüdin recognized that "only through the political

---

1. Cardno & McGuffin, 1999, p. 344.
2. Joseph, 2004b.
3. Gottesman, 1991, p. 207.
4. Ibid., pp. 195-196.
5. Peters, 1999, p. 89.

work of Adolf Hitler, and only through him has our more than thirty-year-old dream become reality: to be able to put race hygiene into action."[1] In 1935, Rüdin proposed extending the sterilization law to include "valueless individuals."[2] In 1939, he spoke favorably of the Nuremberg Laws and of Hitler's views on racial hygiene.[3] Also in 1939, Hitler awarded Rüdin the Goethe medal for art and science, which was accompanied by a telegram from Nazi Interior Minister Wilhelm Frick, which read, "To the indefatigable champion of racial hygiene and meritorious pioneer of the racial-hygienic measures of the Third Reich I send...my heartiest congratulations."[4] In 1942, Rüdin wrote favorably of the 1935 Nazi "Law for the Protection of the Genetic Health of the German people," which prohibited "marriages undesirable for the national community."[5] Furthermore, according to Rüdin, "Already on September 15, 1935 the decree of the Nuremberg Laws of German Citizenship and the Law for The Protection of Blood were put into effect, bringing about the progressive reduction of Jewish influence and especially the hindrance of further intrusions of Jewish blood into the German gene pool."[6]

Gottesman discussed the "misguided fear" that a "eugenic tide would flood society," a fear that helped set the stage for the "atrocities of the Third Reich." In a 1933 article entitled "Eugenic Sterilization: An Urgent Need," Rüdin (whose colleagues nicknamed the "Reichsführer for Sterilization"[7]) wrote, "Probably our greatest eugenic anxiety is caused by the vast army of psychopaths..."[8] Furthermore, "As concerns *mental defectives*, there is, of course, no necessity for accessory methods of preventing procreation in those low grades which require permanent segregation. The public however is insufficiently aware of the results of allowing feeble-minded males the liberty to procreate. The danger to the community of the unsegregated feeble-minded woman is more evident. Most dangerous are the middle and high grades living at large who, despite the fact that their defect is not easily recognizable, *should nevertheless be prevented from procreation* [emphasis in original]."[9]

Luxenburger:

---

6. Gütt et al., 1934.
1. Quoted in Weingart, 1989, p. 270. Nevertheless, McGuffin (2002, p. 2) claimed that "the situation became less favourable" for Rüdin's Munich institute "as the Nazis...came to power in 1933."
2. Quoted in Müller-Hill, 1998a, p. 33.
3. Rüdin, 1939.
4. Quoted in Weinreich, 1946, pp. 32-33.
5. Rüdin, 1942, p. 321 (my translation).
6. Ibid., p. 321 (my translation).
7. Seidelman, W. 2001. *Science and Inhumanity: The Kaiser-Wilhelm/Max Planck Society.* Retrieved online 6/11/05: http://www.doew.at/thema/planck/planckl.html.
8. Rüdin, 1933, p. 102.
9. Ibid., p. 102.

In 1931, two years before Hitler took power, Hans Luxenburger wrote that the procreation of the "unfit" could lead to a situation where "the fit will be overtaken and annihilated" by a "huge army of dull inferiors." He believed that "it is not only the responsibility of eugenics but of all of society to make sure that this horrible vision will never become a reality in our culture."[1] Also in 1931, Luxenburger advocated eugenic sterilization while recognizing it as only a half-measure "because a radical eradication of degenerate hereditary properties is still impossible today.[2]

Luxenburger believed that the 1933 Nazi sterilization law did not go far enough. For example, in his chapter in Rüdin's 1934 edited volume *Erblehre und Rassenhygiene im Völkischen Staat* [Genetics and Racial Hygiene in the Völkish State] he wrote, "We should demand an extension of the definition of sickness according to the Law for the Prevention of Genetically Diseased Offspring [the 1933 sterilization law] to apply to both identical twins, including the twin who is not ill."[3]

Luxenburger continued to support eugenic measures carried out by the Nazis. The following English-language summary is found in the 1938 edition of the American Psychological Association's *Psychological Abstracts*.

> Luxenburger, H. Zur frage der Erbberatung in den Familien Schizophener. (The question of eugenic advice in families of schizophrenics.) *Med. Klin.*, 1936, Part 2, 1136. — The principles of eugenic advice in 21 possible relationships of an applicant with schizophrenics [sic], psychopaths, etc., are discussed and brought together in an outline. — *P. L. Krieger* (Leipzig/Munich).[4]

And in the 1939 edition of *Psychological Abstracts*, we find the following:

> *Luxenburger, H. Psychiatrische Erblehre.* (A psychiatric study of heredity.) Munich: Lehmann, 1938. pp. 134...The author attempts to utilize the accepted scientific findings of genetics as a basis for eugenic measures as well as to indicate further problems for research. Personality and its pathological deviations are treated in detail as an introduction to the discussion of modern genetics, the genetic cycles of psychopathology, the hereditary organic nervous disorders, the hereditary background of non-hereditary psychoses, and the hereditary psychopathology of personality. — *H. Luxenburger* (Munich).[5]

Schulz:

In 1934, Bruno Schulz published an article in the SS-controlled journal *Volk und Rasse* (People and Race) in support of the regime's eugenic sterilization law.[6] Feeling that the law left too many people unaffected, he wrote that it "is without

---

1. Luxenburger, 1931, p. 124 (my translation).
2. Quoted in Burleigh, 1994, p. 41.
3. Luxenburger, 1934, p. 306 (my translation).
4. American Psychological Association, 1938, p. 27.
5. American Psychological Association, 1939, p. 649.
6. Schulz, 1934.

question that the reproduction of another large number of persons who do not fall under the law are undesirable in terms of racial hygiene."[1] In discussing the scope of the law, Schulz wrote that "children of someone who later becomes schizophrenic have to be considered as qualitatively equal to the children of obvious schizophrenics."[2] According to a 1934 report to the Rockefeller Foundation by Danish eugenicist Tage Kemp, Schulz was "doing a great deal of statistical work concerning mental diseases of practical value for the sterilization law and the eugenical legislation in Germany."[3] (In 1998, however, Gottesman wrote that Schulz "was somebody who just kept his nose to the grindstone, didn't get involved in politics and was extremely talented mathematically."[4])

Kallmann:

While still active in Germany, Kallmann argued that the sterilization law did not go far enough in preventing the spread of the "hereditary taint" of schizophrenia. In 1935, he wrote, "It is desirable to extend prevention of reproduction to relatives of schizophrenics who stand out because of minor anomalies, and, above all, to define each of them as being undesirable from the eugenic point of view at the beginning of their reproductive years."[5] According to German geneticist Benno Müller-Hill, about 20% of the German population would have been sterilized under Kallmann's proposal.[6]

Upon arriving in the United States, Kallmann published a 1938 article in *Eugenical News* calling for "negative eugenic measures" against people carrying the "schizophrenic taint."[7] Furthermore, Kallmann wrote,

> From a eugenic point of view, it is particularly disastrous that these patients continue to crowd mental hospitals all over the world, but also afford, to society as a whole, an unceasing source of maladjusted cranks, asocial eccentrics and the lowest types of criminal offenders. Even the fanciful believer in the predominance of individual liberty will admit that mankind would be much happier without those numerous adventurers, fanatics and pseudo-saviors of the world who are found again and again to come from the schizophrenic genotype.

Step Three

Gottesman:

> "Mercy killing" of the mentally ill was discussed in the inner circle of Hitler's advisors as early as 1935.... At first, only infants and institutionalized children under age three were designated for medical killing, after review, *of course*, and unanimous

---

1. Ibid., pp. 138-139 (my translation).
2. Ibid., p. 142 (my translation).
3. Kemp, quoted in Black, 2003, p. 419.
4. Quoted in Healy, 1998, p. 393.
5. Quoted in Müller-Hill, 1998a, p. 11.
6. Müller-Hill, 1998b.
7. All quotations in this paragraph are taken from Kallmann, 1938b.

approval by a panel of three physicians. An overdosing with sedatives was the primary method of killing.... At least 5,000 institutionalized retarded children were killed in the program [emphasis in original].[1]

Rüdin:

Regarding the murder of children, in 1942 Rüdin, in his own words, emphasized "the value of eliminating young children of clearly inferior quality."[2] In the words of Rüdin's biographer, Matthias Weber, "Rüdin considered the broadening of the criteria for killing handicapped newborns to be a scientific issue 'of importance to the war effort.'"[3]

According to a 1998 account by German researchers V. Roelcke, G. Hohendorf, and M. Rotzoll, who were granted access to archival materials:

> Ernst Rüdin, director of the Deutsche Forschungsanstalt für Psychiatrie (DFA) in Munich, was one of the leading psychiatrists in Nazi Germany and inaugurator of the "Munich School of Psychiatric Genetics."...According to the prevailing historical research, neither Rüdin nor any of his co-scientists at the DFA were actively involved in the systematic killing of patients. In contrast to this, a reevaluation of the historical sources available clearly shows that due to the fragmentary character of the evidence, any exculpation of particular individuals or institutions is premature to date. Furthermore, new documents...prove that Rüdin had a genuine interest in research which on the one hand made profitable use of the killings, and on the other hand was aimed at formulating scientific criteria for the systematic selection and "euthanasia" of those supposedly unworthy to live. Julius Deussen, since 1939 a member of the DFA, was also a close co-worker of Carl Schneider at the University of Heidelberg. He coordinated the research on children carried through in the context of the "euthanasia" programme between 1943 and 1945. This research sought to systematically correlate clinical and laboratory findings with the histopathological data of the victims' brains. From the beginning, it included the killing of the patients. Central elements of the research programme had been formulated by Deussen already in Munich. Rüdin supported the activities of Deussen in Heidelberg and repeatedly pointed out that they were of importance for the population policy of the Nazi regime.[4]

Step Four

Gottesman:

> The fourth step is of most relevance to the schizophrenia story and took place in the fall of 1939 with Hitler's decree...that 'patients considered incurable' could be granted a mercy death.... The actual killing was kept secret. The code name 'T4' was used for the program to kill the much larger numbers of persons with schizophrenia, epilepsy, senile dementia.... Hitler...selected carbon monoxide as 'the most humane' method of killing. It is estimated that 100,000 psychiatric patients were murdered in this program.[5]

1. Gottesman, 1991, pp. 208-209.
2. Quoted in Weber, 1996, p. 329.
3. Weber, 2000, p. 255.
4. Roelcke et al., 1998, p. 474. See also Koenig, 2000; Abbott, 2000.

Rüdin:

In a memorandum on the T4 "euthanasia" murder program, Rüdin and his collaborators suggested ways that the killings could be justified:

> Even the euthanasia measures will meet with general understanding and approval, as it becomes established and more generally known that, in each and every case of mental disease, all possible measures were taken either to cure the patients or to improve their state sufficiently to enable them to return to work which is economically worthwhile, either in their original professions or in some other occupation.[1]

In a 2004 account, German psychiatric geneticists Thomas Schulze, Heiner Fangerau, and Peter Propping wrote,

> When the National Socialists came to power in Germany in 1933, it was the members of the Munich school who guided psychiatric genetics, hand in hand with eugenic theories, along the pernicious path from sterilisation of psychiatric patients to the killing in organized euthanasia programs. Although psychiatrists and geneticists, in general, did not plan these atrocities, they helped to prepare the intellectual ground making them possible. As regards the prominent figure Rüdin, however, modern research hints at a more pronounced involvement. According to Roelcke, Hohendorf, and Rotzoll...[Rüdin] was one of the main scientific actors and supporters of the process of selecting 'unworthy' patients for sterilisation and euthanasia.[2]

Steps Five and Six

Gottesman:

> From April 1940, Jews in institutions, as well as Jews and other 'non-Aryans' in concentration camps within Germany, were moved to killing centers (code name 14f13) and exterminated.... A new quicker poison gas had been developed, hydrogen cyanide, for the sixth step, the Final Solution to purifying the Aryan race.... An estimated 4 million Jews were murdered in the camps, a further 2 million in the course of the war, and a further estimated total of 4 million non-Jewish, non-Aryan civilians.[3]

Rüdin:

Rüdin wrote the following lines in the 1942/1943 edition of *Archiv für Rassen- und Gesellschaftsbiologie* (Archive for Racial and Social Biology), in celebration of ten years of Nazi rule:

> The results of our science had earlier attracted much attention (both support and opposition) in national and international circles. Nevertheless, it will always remain the undying, historic achievement of Adolf Hitler and his followers that they dared to take the first trail-blazing and decisive steps toward such brilliant race-hygienic achievement in and for the German people. In so doing, they went beyond

5. Gottesman, 1991, p. 209.
1. Quoted in Müller-Hill, 1998a, p. 46.
2. Schulze et al., 2004, p. 254.
3. Gottesman, 1991, pp. 209-210.

the boundaries of purely scientific knowledge. He and his followers were concerned with putting into practice the theories and advances of Nordic race-conceptions...the fight against parasitic alien races such as the Jews and Gypsies...and preventing the breeding of those with hereditary diseases and those of inferior stock.[1]

* * *

The crimes of Rüdin and his colleagues have been documented in books such as Benno Müller-Hill's *Murderous Science*, Robert Proctor's *Racial Hygiene: Medicine Under the Nazis*, and Lifton's *The Nazi Doctors*. Gottesman cited these books in *Schizophrenia Genesis*, yet he failed to mention facts and quotations relating to Rüdin's criminal activities as documented by these authors.

The main purpose of Rüdin's "Munich School," where the field of psychiatric genetics was born, was to conduct research for the purpose of providing alleged scientific evidence in support of compulsory eugenic (called "racial hygiene" in Germany) measures against those considered to have hereditary diseases. Leaders of the Munich School such as Rüdin and Luxenburger were passionate supporters of racial hygiene and sterilization. They had little interest in "treating" people's suffering. Rather, they branded people as "taint carriers" in need of sterilization in order to protect the "purity" of the German gene pool. In the understated words of Rüdin's biographer, "As did other psychiatrists in Germany, Rüdin tended to be interested less in understanding and alleviating individual mental illness than in all too willingly making psychiatry's knowledge of genetics and its eugenic plans available to the current [Nazi] rulers in order to prevent the supposed social threat."[2]

Given the information available when he wrote *Schizophrenia Genesis*, Gottesman's depiction of the "thoroughly scientific" investigators of the Munich School, and the "murderous, genocidal" Nazi regime, as though they were separate unrelated elements, is unfortunate. In subsequent writings on Rüdin and the Munich School, he continued this theme.[3]

On the basis of the historicla facts I have outlined, available in libraries and bookstores throughout the world in the English language, it is unfortunate that leading contemporary psychiatric geneticists such as Gottesman, Faraone, Tsuang, McGuffin, Farmer, and Kendler, who have written about the history of their discipline, have not come clean about the role that the founders of psychiatric genetics played in supporting massive forced eugenic sterilization programs and the murder of mental patients, and in helping lay the scientific groundwork for Nazi genocide. Why not simply state that, although these founders were guilty of aiding unspeakable crimes against humanity and used their research to justify these crimes, they also pioneered research and statistical

---

1. Quoted in Müller-Hill, 1998a, p. 67.
2. Weber, 2000, p. 257.
3. See Joseph, 2004b, Chapter 2. Gottesman's later writings include Gottesman & Bertelsen, 1996; McGuffin et al., 1994; Torrey et al., 1994.

methods which, despite their misuse and despite the motives of their creators, are still viewed as useful today in psychiatric genetics for assessing the role of genetic influences on psychiatric disorders. Most people would understand this, as have several contemporary psychiatric genetic investigators who perform their work while recognizing, at times in published statements, that the founders of their movement committed terrible crimes.[1]

An example is psychiatric geneticist Myron Baron, who in 1998 wrote of the "past crimes of our discipline."[2] He accused Rüdin of "having played a central role in inspiring, condoning and promoting forcible sterilization and castration of schizophrenics...[whose] sterilization program was a precursor to the notorious 'euthanasia' program, which the Nazis implemented with characteristic efficiency and brutality."[3] Furthermore, "the Law for the Prevention of Genetically Diseased Offspring, formulated and championed by Rüdin and his colleagues, was nothing short of a euphemism for mass sterilization and a prelude to mass murder."[4] Baron also pointed out that the information Rüdin and others had previously collected on the families of people diagnosed with schizophrenia was probably used in the Nazi era to target people for sterilization and murder: "It is highly likely that Rüdin's own research subjects — thousands of patients and family members were enrolled in his programs — were among those who fell pray to the evil he helped inculcate."[5] This raises the additional question of whether Luxenburger supplied the names of the twins he studied to the Nazi Hereditary Health Courts.

Thus, as German psychiatric geneticist Peter Propping described it in a paper he presented in accepting the 2004 Lifetime Achievement Award from the International Society of Psychiatric Genetics, the "sinister history" of his field has become "a permanent Damocles sword for psychiatric genetics."[6]

* * *

Apart from these crimes, and from the dubious "genetic disorders" it has created on the basis of massively flawed and biased research, what positive contributions to the human condition has the field of psychiatric genetics made in its roughly 100 years of existence? On balance, its influence has been overwhelmingly negative. Although contemporary psychiatric geneticists frequently justify their work on the grounds that finding genes may lead to the development of drugs tailored to fit a person's particular genetic profile ("pharmacogenomics"),[7] we will see in Chapter 11 that these postulated genes remain undiscovered.

---

1. For example, see Baron, 1998; Gershon, 1997; Lerer & Segman, 1997.
2. Baron, 1998, p. 96.
3. Ibid., p. 96.
4. Ibid., p. 96.
5. Ibid., p. 97
6. Propping, 2005, p. 3.
7. McInnis & Potash, 2004; Dinwiddie et al., 2004.

Other psychiatric geneticists argue that their field has "propelled our understanding of mental disorders...."[1] However, one could argue the exact opposite: that unsubstantiated genetic theories have helped *obscure* the true causes of mental disorders. Swiss ex-psychoanalyst Alice Miller made a similar argument in the 1980s in relation to Sigmund Freud's theory of infantile sexuality ("drive theory").[2] For Miller, Freud's theory played a major role, for three generations, in blinding society to the damage caused by childhood trauma and abuse. Psychiatric genetic theories play a similar role in blinding society to the destructive psychological impact of these and other environmental events. And, despite all its supposed scientific underpinnings, psychiatric genetics is comparable to Freudian drive theory in the sense that its theories are based mainly on unsupported assumptions, conjecture, and dogma.

CONCLUSIONS

Although it was put forward as a popular exposition of schizophrenia's causes, *Schizophrenia Genesis* has played an important role, on the basis of some questionable facts and interpretations, in strengthening theories emphasizing the primacy of genetic influences on schizophrenia. And the fact that this book has served as a major source for textbook authors since the early 1990s only compounds the problem, in that Gottesman's account continues to be retold again and again.

That *Schizophrenia Genesis* is a leading source of information on the genetics of schizophrenia exemplifies historian of science Thomas Kuhn's observation, in his classic *The Structure of Scientific Revolutions*, that textbooks and popular works can be "systematically misleading" when recording "the stable outcome of past [scientific] revolutions and thus display the bases of the current normal-scientific tradition."[3] And even more today than in 1991, the diathesis-stressor model of psychiatric disorders, with the accompanying biological/psychopharmacology treatment protocol, is the dominant paradigm in psychiatry. A new paradigm emphasizing psychological, familial, political, socioeconomic, and other environmental factors, with an accompanying de-emphasis of diagnoses and medical and pharmacological interventions, must begin by exposing the lack of scientifically acceptable evidence supporting genetic theories of human psychological distress.

---

1. Schulze et al., 2004, p. 246.
2. Miller, 1984, 1997.
3. Kuhn, 1996, p. 137.

Chapter 7. Autism and Genetics: Much Ado About Very Little

Never in the recent history of psychiatry have so many definitive claims been made in support of genetics, in the face of so little evidence, as in the case of autism. And yet, although there is no such thing as a "genetic epidemic,"[1] autism rates appear to be skyrocketing. Mark Blaxill of the Safe Minds organization has documented the rising incidence of autism, which he views as "a matter of urgent public concern."[2] Furthermore, he observed that "Causal theories that emphasize genetic inheritance carry greater weight if disease frequency is unchanged over time, whereas rising incidence demands environmental explanations."[3] In this chapter I will show that there is little evidence that autism is caused by genetic factors, which carries the implication that future research should focus on potential environmental factors.

The autism prevalence rate in the US is usually given as about 5 per 10,000, although recent reports indicate that this rate is increasing rapidly, and some put the figure as high as 40 per 10,000.[4] Boys are diagnosed 4-5 times more often than girls.[5] The disorder, first described by psychiatrist Leo Kanner in 1943,[6] is diagnosed on the basis of three domains of behavioral characteristics: (1) "marked communication abnormalities such as language delay, echolalia, and deficits in the pragmatic use of language," (2) "significant social deficits ranging from an absence of interest in interacting with others to a limited ability to

---

1. Silverman and Herbert, 2003.
2. Blaxill, 2004, p. 536.
3. Ibid., p. 537.
4. The 40/10,000 figure was found in a study by Bertrand et al., 2001. For other reports increasing autism rates, see Croen et al., 2002; Gillberg et al., 1991; Seligman, 2005.
5. Sadock & Sadock, 2003.
6. Kanner, 1943.

engage in reciprocal social interactions," and (3) "an excess of stereotyped, ritualistic, and repetitive behaviors...."[1]

A distinguishing feature of autism is that, as opposed to most psychiatric disorders, there is some evidence suggesting that it is caused by biological factors affecting the brain's development. The evidence includes early onset, the apparent lifelong course, and that several physical diseases are associated with autism. Moreover, there is some evidence linking autism to childhood vaccinations containing the preservative Thimerosal, which contains mercury and which could cause damage in developing brains.[2] But we must, as always, guard against conflating "biological" with "genetic." Many diseases are caused by nongenetic biological factors such as bacteria, chemicals, viruses, and so on. As Gottesman has correctly pointed out, "Everything that is genetic is biological, but not all things biological are genetic."[3]

Textbook chapters and review articles tend to present the evidence in support of genetics as overwhelming, often claiming that autism "is more heritable" than most other psychiatric disorders. Some examples follow: "The importance of hereditary factors in the etiology of autism is now well recognized";[4] it is a "fact that autistic disorder is a complex genetic disorder";[5] "The last ten years of research on autism clearly implicate a genetic component in the etiology of this disorder";[6] "Autism is influenced by complex, yet strong genetic factors;"[7] "Autism is under a high degree of genetic control...";[8] and "there is robust evidence that autism is a highly heritable condition...."[9] Finally, psychiatrist David Skuse regarded autism as "a quintessentially highly heritable condition."[10] It is rarely mentioned that, because reviewers draw these conclusions on the basis of family and twin data, environmental interpretations based on the more similar environments experienced by family members versus the general population, and by identical twins compared with fraternal twins, call into question genetic inferences drawn from this body of research.

---

1. Piven, 2002, p. 43.
2. Blaxill et al., 2004; Goldman et al., 2001; Hobson, 2003; Kirby, 2005. A Centers for Disease Control study published in late 2003 (Verstraeten et al., 2003) found no association between Thimerosal-containing vaccines and neurodevelopmental disorders. In response, the Safe Minds organization (Safe Minds, 2003) argued that this study contained "bias and conscious manipulation of samples," and that the data "provide support for a causal relationship between Thimerosal exposure and childhood developmental disorders."
3. Gottesman & Hanson, 2005, p. 265.
4. Piven, 2002, p. 45.
5. Pericak-Vance, 2003, p. 268.
6. Smalley et al., 1988, p. 959.
7. Cook, 1998, p. 113.
8. Bailey et al., 1995, p. 63.
9. Thapar & Scourfield, 2003, p. 151.
10. Skuse, 2001, p. 395.

Interestingly, although the results of *schizophrenia* family studies are usually accompanied by the caveat that they are explainable on environmental grounds, autism family results are frequently cited as evidence in support of genetic causation. The transformation of merely the "best" evidence into "conclusive" evidence has a long history in psychiatry and psychiatric genetics. Let's take the example of schizophrenia. In the absence of other types of studies, family pedigrees were once seen as conclusive proof of genetics. When researchers performed systematic family studies, they became conclusive proof. With the advent of schizophrenia twin research, family pedigrees and systematic family studies were belatedly recognized as potentially confounded by environmental factors, and identical-fraternal schizophrenia concordance rate differences were seen as conclusive. By the 1960s, psychiatric genetic researchers such as Kety, Rosenthal, and Wender began to view identical-fraternal twin concordance differences as potentially confounded by environmental factors, putting forward their adoption studies as conclusively demonstrating the genetic basis of schizophrenia.

In the case of autism, however, no adoption studies have been performed. Whereas some ADHD and schizophrenia researchers have stressed the need for adoption studies to confirm the role of genetic influences, to my knowledge no prominent autism researcher or reviewer has written that adoption studies are necessary to confirm genetic influences on autism. This is revealing, given that autism twin studies are few in number and contain small samples. Moreover, despite worldwide efforts to do so (including at least 10 genome scans), no autism genes have been discovered (see Chapter 11).[1] The following observation by genetic researchers Michelle LaBuda and colleagues is unfortunately all-too-rare in the "genetics of autism" literature: "Although the twin data are consistent with some kind of genetic contribution to the etiology of autism, compared with other psychiatric disorders there is a paucity of corroborating evidence."[2]

The autism rate among the siblings of children diagnosed with autism is usually reported as 2-4%.[3] Although many times greater than the general population expectation, we can attribute this elevated rate to any number of possible environmental factors shared by family members. While most genetic researchers understand this, in the autism literature the elevated rate among siblings is usually discussed in the context of genetic theories. For example, autism family researcher Patrick Bolton and his colleagues concluded that their "findings suggest that the autism phenotype extends beyond autism as traditionally diagnosed; that etiology involved several genes, [and] that autism is genetically heterogeneous...."[4] And molecular genetic researchers Jun Li and col-

1. Rutter, 2005.
2. LaBuda et al., 1993, p. 52.
3. Sadock & Sadock, 2003.
4. Bolton et al., 1994, p. 877.

leagues wrote, "Siblingship reoccurrence risk is about 3%, 50-100 times greater than the population prevalence, demonstrating a substantial contribution to autism by hereditary factors...."[1]

Because twin studies are seen as providing conclusive evidence in favor of genetics, it is necessary to examine them closely. I will cover autism molecular genetic research in Chapter 11.

AUTISM TWIN STUDIES

In Chapter 1, we saw that genetic inferences drawn from identical-fraternal comparisons are confounded by the greater environmental similarity of identical versus fraternal twins. This position requires some clarification regarding autism, which is diagnosed at an early age and for which unequal social or psychological environments might not be a factor influencing concordance rates. Identical twins, however, experience more similar postnatal and prenatal physical environments, which could lead to their being more similarly exposed to environmental agents, such as chemicals or viruses, that could cause autism.

Unequal Prenatal Environments

Although autism twin researchers rarely mention this, identical twins, in addition to experiencing more similar postnatal environments, experience more similar *prenatal* environments. For example, about 60% of identical twins share the same chorion and placenta.[2] These monochorionic (MC) twins are contrasted with *all* fraternal twins, and with about 40% of identical twins, who are dichorionic (DC) and develop *separate* placentas and chorions. Although it is unlikely that viral, chemical, or infectious agents cause psychiatric disorders such as schizophrenia, in autism, for which biological explanations are plausible, such an agent may play a role. And just as dissimilar postnatal environments confound the results of psychiatric twin studies, the dissimilar prenatal environments of identical and fraternal twins (based on their chorionic status) suggests that the twin method cannot disentangle possible genetic and non-genetic biological factors influencing abnormal prenatal development in twins. In order to rule out twins' chorionic status as a major factor in twin concordance for autism, comparisons between MC and DC twins (independent of zygosity) must be performed.

---

1. Li et al., 2002, p. 24.
2. Bulmer, 1970.

In a passage highly relevant to this discussion as it relates to twin studies of autism, schizophrenia investigators James O. Davis and his colleagues discussed the differing prenatal environments of MC and DC twins:

> One conspicuous difference between MC and DC placentation involves fetal blood circulation. Most (85% -100%) MC twins exchange blood through shared vascular communication, whereas DC twin pairs (whether MZ or DZ) very rarely exchange blood.... This is relevant to the viral hypothesis [of schizophrenia] because the shared vascular communication would encourage mutual infection when an infectious agent crosses the shared placenta of an MC twin pair. On the other hand, infections and other toxic insults could breach the placenta of only one twin in a DC pair, leaving the cotwin unaffected.[1]

"MC twins," they concluded, "are likely to be mutually affected by such insults as bloodborne infections, while DC twins — with their separate fetal circulations — cannot share infections through an exchange of blood."[2] According to another group of investigators,

> The importance of the human placenta with regard to the prenatal environment, especially with twins, is obvious. It is the physical and physiological link between mother and child, and it exhibits variations with regard to membrane type, size, shape, and circulation which may be important in themselves or may affect the nutrition of the embryo or the transport of drugs, toxins, and other agents which can influence brain development.[3]

British commentator D. I. W. Phillips published a 1993 article in *The Lancet*, writing, "A greater concordance for disease in monozygous than dizygous twins cannot provide proof that the disease has genetic determinants.... The results of twin studies may be especially misleading in disorders in which the prenatal environment is thought to play a part in their aetiology."[4]

And there is evidence suggesting that prenatal factors play a role in causing autism. In a 2003 study, for example, Holmes, Blaxill, and Haley found that the mothers of autistic children had a significantly higher exposure to mercury when pregnant, versus control group mothers.[5] These mothers were far more likely to have received Thimerosal-containing Rho D immunoglobulin injections while pregnant. Moreover, there are remarkable similarities between the symptoms of autism and the symptoms of mercury poisoning, including a higher rate among males versus females.[6] This has led a group of researchers to theorize that autism is "a novel form of mercury poisoning."[7]

---

1. Davis et al., 1995, p. 359.
2. Ibid., p. 359.
3. Melnick et al., 1978, p. 426.
4. Phillips, 1993, p. 1009.
5. Holmes et al., 2003.
6. Bernard et al., 2001.
7. Ibid., p. 462.

The Four Studies

Four autism twin studies have been published to date. Each of these investigations studied pairs of twins reared together in the same family. There have
been no studies of reared-apart twins where one or both members of the pair
have been diagnosed with autism. Because autism is a relatively rare disorder,
sample sizes in twin studies have been low. The results of these studies are
shown in Table 7.1.

Table 7.1

PAIRWISE AUTISM TWIN CONCORDANCE RATES

Authors	Year	Identical	%	Fraternal	%
Folstein & Rutter	1977	4/11	36%	0/10	0%
Ritvo et al.	1985	22/23	96%	4/17	24%
Steffenburg et al.	1989	10/11	91%	0/10	0%
Bailey et al. [a]	1995	12/17 [b]	69%	1/12 [c]	9%
POOLED RATES		48/62	77%	5/49	10%

[a] This study investigated twins previously reported in Folstein & Rutter's 1977 investigation, plus additional pairs. In
order to avoid double counting, only these additional pairs are listed.
[b] Bailey et al. did not clearly state this result. Although they reported a 69% pairwise concordance rate, the figure derived
from 12/17 is 70.6%.
[c] Although in a table Bailey and colleagues listed 0/11 fraternal twin pairs as concordant for autism, towards the end of
their publication they reported that one concordant fraternal pair "was ascertained after the close of the current study"
(p. 73). The investigators decided to count this pair in another comparison in their 1995 publication (p. 73).

These results show a large difference in identical and fraternal twin concordance rates. Moreover, in half the studies the fraternal rate is 0%. All of the
investigators interpreted their identical-fraternal concordance differences as
supporting genetic theories, while at the same time failing to mention that such
conclusions are dependent upon the twin method's equal environment
assumption.

Folstein and Rutter, 1977

British investigators Susan Folstein and Michael Rutter performed the
first systematic autism twin study, which they published in 1977.[1] Their sample
consisted of 11 identical and 10 same-sex fraternal pairs, of which 4 identical
pairs were judged concordant for autism (36%). Conversely, none (0%) of the
fraternal pairs was judged concordant (4/11 identical vs. 0/10 fraternal). The
probability (p-value) of the identical-fraternal concordance rate comparison in
this tiny sample was listed as .055.

---

1. Folstein & Rutter, 1977a, 1977b.

Folstein and Rutter (and most subsequent reviewers) concluded that the higher identical concordance rate "strongly suggests the importance of hereditary influences in the aetiology of autism." Although they wrote that "some caution is needed before drawing too sweeping conclusions" about genetics from their small sample, they added to their claims in favor of genetics by noting that concordance rates rose to 82% versus 10% for co-twins having cognitive disorders.

Folstein and Rutter also found that among their *discordant* pairs, biological hazards in the birth process were associated with the development of autism. Among these hazards they listed severe hemolytic disease, a delay in breathing of at least 5 minutes after birth, neonatal convulsions, multiple congenital anomalies, and a second birth delayed by at least 30 minutes following the birth of the first twin. Strikingly, using the investigators' broad definition of biological hazard, in 12 of the 17 discordant pairs the autistic twin "probably or possibly suffered a brain injury," whereas the non-autistic twin had not.[1] (In 4 of the discordant pairs, neither twin had experienced a perinatal hazard, and in 2 pairs both had experienced a hazard.) These findings led Folstein and Rutter to conclude that "some form of biological impairment, usually in the perinatal period, strongly predisposed to the development of autism."[2]

Given their tiny sample, the borderline p-value, and the possible role of perinatal factors, Folstein and Rutter might have concluded that, while their study raised some interesting questions, no important conclusions could be drawn from the data. However, in the two original publications describing their study (one of which appeared in the prestigious scientific journal *Nature*), they repeatedly claimed that their results strongly supported a role for genetics. Indeed, in their brief *Nature* article they concluded in favor of important genetic influences no less than five times, making claims about "genetic determination" and the "importance of genetic factors in the aetiology of autism."[3] Although their conclusions in support of genetics were unwarranted on several grounds, to their credit Folstein and Rutter took care to control for some of the biases plaguing earlier psychiatric twin studies. Moreover, they included lengthy case summaries of the twin pairs, enabling reviewers to better understand the children's symptoms and the circumstances surrounding their birth and development.

Ritvo and Colleagues, 1985

The second and largest autism twin study was published by University of California at Los Angeles researcher Edward Ritvo and his colleagues in 1985.

---

1. Folstein & Rutter, 1977a, p. 728.
2. Folstein & Rutter, 1977b, p. 305.
3. Folstein & Rutter, 1977a, pp. 727-728.

These investigators studied 40 pairs (23 identical, 17 fraternal), and found concordance rates of 96% identical and 23.5% fraternal. The identical twin figure
stood in contrast to Folstein and Rutter's 36% rate, and the 23.5% fraternal rate
is clearly distinguished from the 0% rates in two other studies (see Table 7.1).
The investigators concluded that their "results to date are compatible with autosomal recessive inheritance, which predicts 100% concordance in monozygotic
pairs and 25% concordance in dizygotic pairs."[1]

Ritvo and colleagues' 23.5% fraternal rate, however, is consistent with
environmental influences in light of reports that the autism rate among the
ordinary siblings of people diagnosed with autism is only around 2-4%. Let us
recall that fraternal twins and ordinary siblings have the same average genetic
relationship to each other (50%), but fraternal twins experience more similar
pre- and postnatal environments than sibling pairs. Thus the 23.5% Ritvo et al.
fraternal twin autism rate, which represents a 6-11-fold increase over the
expected rate among ordinary siblings, suggests that the more similar environments shared by fraternal twins (which could involve aspects of their status as
twins) accounts for this elevation. As a commentator observed, the fraternal
twin/sibling difference "indicates a substantial twin-specific environmental etiology for autism."[2] Other explanations are of course plausible, yet Ritvo and colleagues, who were prepared to speculate about genetic factors on the basis of
their identical-fraternal differences, failed to address the issues raised by their
high fraternal twin concordance rate.

Rather than address how Ritvo's 23.5% fraternal twin concordance rate is
consistent with genetic theories, or with environmental theories, Folstein,
Rutter, and the authors of the two subsequent twin studies attempted to discredit, dismiss, or ignore this finding. In their 1991 review article, Folstein and
Piven wrote of the "number of methodological problems" of Ritvo and colleagues' study.[3] According to Steffenburg et al., the investigation was not an
"acceptable twin study of infantile autism," because "cases were recruited largely
from a pool of replies to a newsletter announcement of the National Society for
Autistic Children," which could "lead to an over-inclusion of concordant and
monozygotic cases." They added that it was improper to use opposite-sex pairs,
"which is not appropriate given the usually high boy: girl ratio in autism."[4]
Finally, in their 1995 twin study, Bailey, Gottesman, Rutter and others simply
wrote Ritvo et al. out of the history of autism twin research, discussing the Folstein and Rutter and Steffenburg studies as the "two previous epidemiological

---

1. Ritvo et al., 1985a, p. 75.
2. Sturt, 1985, p. 1521.
3. Folstein & Piven, 1991, p. 768.
4. Steffenburg et al., 1989, pp. 405-406.

studies of autistic twins....”[1] Amazingly, Bailey et al. neither mentioned nor cited Ritvo's study.

Of course, Ritvo and colleagues' study contained several methodological problems, and even they admitted in a subsequent response to a critic, “we ourselves refrain (and we advise other to refrain) from conducting inapplicable analyses and drawing unwarranted conclusions from our acknowledged nonsystematically ascertained pairs of twins.”[2] However, enormous methodological problems have plagued the twin method since its origins in the 1920s.[3] Some examples from the schizophrenia twin study literature include Rosanoff and colleagues' failure to use a zygosity determination procedure (other than their hunches based on observations),[4] the frequent failure to define schizophrenia,[5] Inouye's admission that the “twin subjects of the present study were not collected by the so-called systematic study method...,”[6] and the failures of Luxenburger, Rosanoff, Essen-Möller, Kallmann, Slater, Inouye, Tienari, and Kringlen to make blind diagnoses.[7] Moreover, despite having recruited their volunteer subjects from media appeals, which biased the sample in favor of similarity,[8] the famous Minnesota reared-apart IQ and personality twin studies of Bouchard and colleagues are widely (albeit incorrectly) cited as providing important evidence in favor of genetic influences on many traits. In fact, most reared-together twin studies are biased because female twins and identical twins each typically constitute two-thirds of the samples, which is far greater than expectations derived from the general twin population (about 50% and 30% respectively).[9] Couldn't we therefore conclude that most twin studies are unacceptable due to their use of non-representative samples?

Generally speaking, twin researchers make great efforts to legitimize the most slipshod and haphazard of studies and rarely, if ever, argue that another twin study is invalid. Why then is Ritvo's investigation singled out and discredited, or simply ignored, when many twin studies in psychiatry could also be disregarded with a comparable level of critical appraisal? The most plausible explanation is that Ritvo's fraternal twin concordance rate simply does not fit the argument of those eager to claim autism as a genetic disorder.

---

1. Bailey et al., 1995, p. 63.
2. Ritvo et al., 1985b, p. 1521.
3. Also see Joseph, 2004b.
4. Rosanoff et al., 1934, p. 25.
5. Sullivan & Kendler, 2003.
6. Inouye, 1961, p. 524.
7. Essen-Möller, 1941; Inouye, 1961; Kallmann, 1946; Luxenburger, 1928; Rosanoff et al., 1934; Slater, 1953. Tienari, 1963, Kringlen, 1967.
8. See Joseph, 2001c, 2004b, Chapter 4.
9. See Lykken et al. 1978, 1987 for documentation of the “Rule of Two-Thirds” in twin research.

Steffenburg and Colleagues, 1989

A third study of autism in twins was performed by Suzanne Steffenburg and colleagues with a sample obtained in the Scandinavian countries. In addition to their comments on the Ritvo et al. study, Steffenburg and colleagues correctly observed that "firm convictions cannot be drawn" from Folstein and Rutter's small sample of twins.[1]

The investigators obtained their sample by contacting medical officers in Denmark, Finland, Iceland, Norway, and Sweden. They requested records for same-sex twin pairs under the age of 25, where at least one twin had been diagnosed with or was suspected of having autism. There is, however, a bias in favor of identical twin concordance in obtaining twins in this way, because concordant identical pairs are more likely to come to the attention of the referring doctors than are discordant pairs. Referral biases of this type are well known in schizophrenia twin research, where sampling from consecutive hospital admissions or twin registers was initiated in order to reduce bias. Bailey and colleagues described the "tendency for the referral of concordant pairs [which] may have been exacerbated by the Scandinavian group accepting cases from five countries," and found the higher Scandinavian versus British concordance rate plausibly explained by "referral bias" in the former study.[2] Genetic researchers Thapar and Scourfield observed that "twin studies of autism are unusual within the field of child and adolescent psychiatry in having been based on clinical cases rather than on non-referred populations, as has been the rule with most other diagnoses."[3] More forcefully, Kendler has written that, as opposed to the preferred practice of using population-based samples, "Basing twin studies on treated samples may yield results that are biased and cannot therefore be extrapolated to the total population of affected individuals."[4]

Moreover, Steffenburg et al. made nonblinded diagnoses (although they performed a blind inter-rater reliability check on the case reports). They also stated that "Zygosity tests were performed in 18 (82%) of the 22 sets of twins.... In the remaining four pairs zygosity was determined on the basis of a combination of placental evidence and physical appearance."[5] But the researchers did not clearly state what types of "zygosity tests" they used, and in four cases they used a zygosity determination method judged decades ago by twin researchers as not meeting acceptable standards of accuracy.

Given the methodological problems, biases, and dubious assumptions in twin research in general, however, this study *does not* stand out as a noteworthy example of biased research. I mention these issues only to point out that Stef-

1. Steffenburg et al., 1989, p. 405.
2. Bailey et al., 1995, p. 72.
3. Thapar & Scourfield, 2003, p. 148.
4. Kendler, 1993, p. 907.
5. Steffenburg et al., 1989, p. 408.

fenburg and colleagues' study — like Ritvo and colleagues' and most other twin studies — is subject to several potentially invalidating methodological problems and biases. Thus, it was improper for Steffenburg and others to single out Ritvo's study as being qualitatively more biased than other twin studies of autism.

From their tiny sample Steffenburg et al. found concordance rates of 91% (10/11) identical and 0% (0/10) fraternal, concluding that the "striking difference in concordance between MZ and DZ pairs suggests the importance of hereditary influences in the aetiology of autistic disorder."[1] Like Folstein and Rutter, Steffenburg et al. found an association between autism and perinatal hazards. Thus, "In all pairs concordant for AD [autistic disorder], both twins (but not all the members of the set of identical triplets) experienced some reduction of pre-, peri and/or neonatal optimality."[2] Moreover, "In 7/10 DZ pairs discordant for AD, and in the only discordant MZ pair, the AD twin had more perinatal stress (face presentation, foot presentation, hyperbilirubinaemia or asphyxia). In not a single one of the pairs discordant for AD was there a reversed relationship in this respect."[3] Like Folstein and Rutter, the investigators concluded that in addition to genetic factors, "perinatal stress is involved in some cases."[4] Yet, they might have concluded that autism associated with perinatal stress confounded their results (not to mention that they, like Folstein and Rutter, failed to discuss or justify the validity of the equal environment assumption).

Bailey and Colleagues, 1995

The most recent autism twin study was published in 1995 by Bailey, Le Couteur, Gottesman, Bolton, Simonoff, Yuzda, and Rutter, and carried the provocative title, "Autism as a Strongly Genetic Disorder: Evidence from a British Twin Study." This study consisted of a follow-up of the twins in Folstein and Rutter's 1977 investigation, in addition to several newly added pairs. Of the 47 pairs they reported (27 identical, 20 same-sex fraternal), 19 were ascertained from Folstein and Rutter's sample (10 identical, 9 same-sex fraternal), while 28 new pairs were added (17 identical, 11 same-sex fraternal). The investigators listed overall identical twin concordance as 60% (69% in the new sample), and the overall fraternal rate as 0%. After calculating higher concordance by expanding the dependent variable to include other cognitive deficits, Bailey et al. concluded, "These findings suggest that the genetic liability is for the development of specific cognitive and social abnormalities, with autism as the most severe phenotype."[5]

---

1. Ibid., p. 410.
2. Ibid., p. 410.
3. Ibid., p. 411.
4. Ibid., p. 411.
5. Bailey et al., 1995, p. 68.

The investigators estimated heritability for autism at about .92, and claimed that "the liability to autism appears to be largely genetically determined."[1] Psychiatric textbooks published after 1995 frequently cite this heritability figure. The unfortunate impression they convey is that autism is a genetic disorder for which little can be done to prevent. However, we have already seen that in some cases even a proven genetic disorder, such as PKU, can be prevented once the environmental trigger is identified.

The investigators also claimed that, because autism is a rare disorder, "even modest concordance rates in MZ twins indicate the action of substantial genetic influences."[2] Not necessarily. For example, even 100% identical twin concordance does not, in and of itself, prove anything about genetics. The reason is that identical twins share very similar environments (and much more similar than fraternals, as most studies show), and therefore both members of the identical twin pair will be exposed to potentially autism-producing toxins, viruses, chemicals, etc. Taking the world population as a whole, it is very rare to find someone who speaks Lithuanian. This does not mean, however, that 100% identical twin concordance for speaking Lithuanian "indicates the action of substantial genetic influences" on speaking Lithuanian. Given this obvious example, it is unfortunate that, lacking supporting evidence or theories, Folstein could assert in a 2001 review article that the "concordance of autism in MZ pairs cannot be accounted for by shared prenatal or perinatal difficulties...."[3]

Rather than discuss the possibility that higher autism concordance rates among identical twins versus fraternal twins are strongly influenced by the former's more similar postnatal and prenatal environments — regardless of whether this shows up in perinatal hazard counts — Bailey and colleagues emphasized alleged genetic factors. Attempting to discount the previous studies' findings of an association between perinatal hazards and autism, they claimed that "no concordant DZ pairs were ascertained in the previous studies...."[4] Bailey and colleagues calculated a pooled 3.2% fraternal twin rate across three studies, which was "virtually identical to the rate of autism in the siblings of autistic singletons."[5] They arrived at this 3.2% fraternal rate, which corresponded neatly with the 2-4% sibling rate, because they counted a concordant fraternal pair obtained "immediately after the close of the current study."[6] Although Bailey et al. used this concordant fraternal pair to calculate their pooled 3.2% fraternal rate, they did not include it in their concordance table, which listed a 0% rate. Thus, Bailey and colleagues' fraternal concordance rate was actually 9%. More importantly, their 3.2% pooled fraternal rate was based on their decision to

---

1. Ibid., p. 72.
2. Bailey et al., 1995, p. 72.
3. Folstein & Rosen-Sheidley, 2001, p. 945.
4. Bailey et al., 1995, p. 73.
5. Ibid., p. 73.
6. Ibid., p. 73.

ignore the Ritvo et al. 1985 results. As seen in Table 7.1, the actual pooled fraternal twin concordance rate is 10.2%, a 3-4 fold increase compared with the sibling rate and unexplainable on genetic grounds.[1]

In a companion autism family study, Bolton and colleagues wrote that "the findings from British twin studies have shown that the rate amongst DZ co-twins, who are exposed to a raised rate of OCs [obstetric complications] is no greater than the rate of autism found in siblings, as might be expected if OCs cause autism."[2] These investigators emphasized British studies, while ignoring the pooled and American figures consistent with the idea that "obstetric complications cause autism." Again ignoring Ritvo's finding, Gottesman more recently has written that autism fraternal concordance rates are "typically close to zero and ranging up to 10% with large standard errors."[3] Unfortunately, students and professionals alike are taught that autism is "highly heritable" on the basis of researchers' uncritical acceptance of implausible assumptions, combined with questionable methods of counting relatives performed by Bailey, Gottesman, Rutter, and others. Genetics popularizers such as Matt Ridley then cite Bailey et al. in popular works to the effect that the autism "concordance rate for fraternal twins is 0 percent," thereby transmitting false information to a wider audience.[4]

Bailey and colleagues also addressed the previous twin studies' finding of an association between autism and perinatal hazards. They believed that the effects of these hazards had been "overestimated" in the previous studies, and that "late obstetric hazards may only be consequences of earlier abnormal development."[5] Due to their belief that autism carries an important genetic component, circularly deduced from the results of twin studies, they wrote that these obstetric hazards "appear to be the consequences of genetically influenced abnormal development...."[6] Thus, they assumed that abnormal prenatal development was genetically influenced, when several non-genetic biological factors could also explain such development. Moreover, the interaction of non-genetic biological factors (e.g., exposure to or ingestion of toxic substances by the mother) with twins' chorionic status could lead to higher identical versus fraternal autism concordance for autism for reasons unrelated to genetics.

---

1. Unfortunately, subsequent reviewers (e.g., Gerlai & Gerlai, 2003) cite Bailey and colleagues' incorrect figures, which then, as we saw in the schizophrenia genetics literature, take on a life of their own.
2. Bolton et al., 1994, p. 878.
3. Gottesman & Hanson, 2005, p. 275.
4. Ridley, 2003, p. 103. Other examples of reviewers failing to mention Ritvo's study and his fraternal twin concordance rate include Lamb et al., 2000; Muhle et al., 2004; Veenstra-VanderWeele et al., 2004.
5. Bailey et al., 1995, pp. 65-66.
6. Ibid., p. 63.

ARE TWINS MORE SUSCEPTIBLE TO AUTISM THAN SINGLETONS?

In order to be able to generalize findings from twin studies to the non-twin population, researchers must demonstrate that twins and non-twins alike are diagnosed with the condition in question at comparable rates (see Figure 1.1). If this is not established, these findings are applicable only to the population of twins. Gottesman and Shields wrote in 1982 that genetic interpretations of Folstein and Rutter's 1977 results "must be treated with caution.... Before any firm conclusions about genetics can be drawn from twin data, it is necessary to rule out the possibility that the twinning process itself might contribute to the development of the trait under study."[1]

Two studies, by David Greenberg and colleagues in the United States, and by Betancur et al. in France, found a significant excess of twins among their samples of autistic sibling pairs.[2] Specifically, Greenberg et al. found a 7-fold increase of twins versus the expected population rate, whereas they found no elevation among a comparison group of sibling pairs diagnosed with insulin-dependent diabetes mellitus. Although the investigators did not challenge genetic theories of autism, they considered the possibility that identical-fraternal concordance rate differences reported in autism twin research are "due to differences in the twinning processes for MZ twins versus DZ twins or to the intrauterine environment."[3] Betancur et al. found a "remarkably high [14-fold increase] proportion of MZ pairs among affected sib pairs," while finding no increase among fraternal pairs.[4]

In response, Hallmayer and colleagues presented data from their own sample and two other studies showing "only a slight-to-moderate increase in the risk for multiples, compared to singletons, to be diagnosed with autism."[5] These investigators did acknowledge that, if the twinning process is an important risk factor for autism, "this would have major consequences for the interpretation of twin studies."[6] In a subsequent investigation published in 2005, a group of researchers studied a sample of 802 twins and concluded, "As has been suggested for autism, twin status may incur increased liability to subthreshold autistic symptomatology, particularly in males."[7]

Given that the question of elevated autism rates among twins remains open, conclusions in favor of genetic influences on autism among the population of non-twins (i.e., among most children) are premature on this basis alone. Even

---

1. Gottesman & Shields, 1982, p. 162.
2. Greenberg et al., 2001; Betancur et al., 2002.
3. Greenberg et al., 2001, p. 1065.
4. Betancur et al., 2002, p. 1382.
5. Hallmayer et al., 2002, p. 945.
6. Ibid., p. 941.
7. Ho et al., 2005.

Gottesman recognized in 2005 that the Greenberg and Betancur studies suggest that "the twin method may not be completely valid for this trait [autism]."[1]

CONCLUSIONS REGARDING THE AUTISM TWIN DATA

The results of the four twin studies reported here are the bedrock of contemporary claims of important genetic influences on autism. However, these results are confounded by the more similar environments — both postnatal and prenatal — experienced by identical versus fraternal twins. Amazingly, not a word is mentioned in any of these four studies about these differing environments, although in 2001, Rutter and other behavior geneticists recognized that, in general, the equal environment assumption "is likely often to be violated" in twin studies.[2] And with the exception of Ritvo and colleagues, all autism twin researchers concluded that pre-or perinatal influences could cause autism. Even Bailey and colleagues, clearly more determined than the others to explain everything in terms of genetics, found that autism was associated with increased head circumference. Moreover, they interpreted the results of the previous studies as suggesting that "some cases of autism might have been environmentally determined."[3] Subsequent population-based research has supported an association between autism and birth complications.[4]

The 3-4 times higher pooled rate among fraternal twins versus the concordance rate among ordinary siblings, unexplainable on genetic grounds, is consistent with the evidence in most of the studies that pre- and perinatal factors can cause autism. Moreover, higher rates among identical versus fraternal pairs can be explained by a number of possible environmental factors, not the least of which is that the majority of identical twins share the same chorion and placenta, whereas fraternal twins do not. In light of the multiple possible etiological factors which autism twin research has been unable to disentangle, the investigators' rush to judgment in support of genetics is clearly the result of their bias in favor of genetic interpretations.

Let's review non-genetic factors that could account for high identical twin concordance for autism, as well as the observed identical-fraternal concordance rate difference:

- Identical twins share the same chorion and placenta about 60% of the time, whereas fraternal twins never share the same chorion or placenta (prenatal equal environment assumption is false).

1. Gottesman & Hanson, 2005, p. 275.
2. Rutter et al., 2001, p. 304.
3. Bailey et al., 1995, p. 69.
4. Glasson et al., 2004; Hultman & Sparen, 2004; Zwaigenbaum et al., 2002.

• Identical twins share a more similar postnatal environment and are treated more similarly than fraternals (postnatal equal environment assumption is false).

• Twins may be more susceptible to autism than singletons.

• Most twin studies found that autism is associated with perinatal hazards.

• There was ascertainment bias in favor of identical twin concordance.

• The genetic bias of the investigators likely influenced their conclusions.

• The results from small samples might not reflect true concordance rates.

Gottesman concluded in 1976 that "biological, probably congenital (but not genetic) etiological agents" cause autism.[1] That a handful of twins found in a few subsequent studies caused him to change his mind is more an indication of Gottesman's devotion to the twin method than of convincing evidence in support of genetic influences on autism.

AUTISM AND POLIOMYELITIS

The problem of inferring genetic influences from the results of identical-fraternal comparisons on disorders possibly caused by environmental factors, such as viruses or chemicals, becomes clearer by comparing autism twin results to those of a disease known to be caused by a virus. In this section, I discuss *polio* twin research to illustrate how large concordance rate differences can be explained by non-genetic factors, and how similar findings in autism twin research also could result from such factors.

A 1951 twin study of poliomyelitis (polio) by C. Nash Herndon and Royal G. Jennings, entitled "A Twin-Family Study of Susceptibility to Poliomyelitis," could be renamed as "*The Fallacy of the Twin Method.*" Among their North Carolina twin sample, Herndon and Jennings found polio concordance rates of 36% identical (5/14) and 6% fraternal (2/33). They concluded,

> Although these data present evidence of the existence of a measurable genetic influence on susceptibility to the paralytic form of poliomyelitis, they do not allow us to reach any conclusions concerning the number or kind of genes conditioning such susceptibility.[2]

The fallacy here is that identical twins were more concordant because they shared a more similar environment and spent more time together than fraternal twins, and were therefore more similarly exposed to the polio virus. This study proved nothing about genetic influences on polio.

---

1. Hanson & Gottesman, 1976, p. 226.
2. Herndon & Jennings, 1951, p. 44.

Substituting the word "autism" for "the paralytic form of poliomyelitis," we find many statements comparable to Herndon and Jennings' about the role of genetics in autism, similarly deduced from the results of twin studies. Folstein and Rutter, for example, concluded that their results

> indicate the importance of a genetic factor which probably concerns a cognitive deficit involving language.... However, uncertainty remains on both the mode of inheritance and exactly what it is which is inherited.[1]

And according to Ritvo and his colleagues,

> the assumption that pathogenic genes are present poses the fascinating task of determining where on the gene map they reside, precisely what pathogenic ciphers they transmit, and where we can deduce their presence from clinical clues.[2]

Finally, Bailey and colleagues wrote,

> The concordance findings from the various twin studies suggest that autism is a very strongly genetic neuropsychiatric disorder. Whether autism unassociated with medical conditions is a single or heterogeneous genetic disorder is unknown....[3]

Autism twin researchers appear to have overlooked potential environmental confounds in their studies, and unfortunately have learned little from experiences such as the 1951 polio twin study.

Although Herndon and Jennings understood that it had been shown "beyond question" that "acute poliomyelitis occurs only with invasion of the host by the specific virus...,"[4] they performed their study because "it is equally clear that only a small proportion of the individuals exposed to the virus develop clinical signs of the disease."[5] Contemporary psychiatric geneticists, it should be noted, use a similar argument in support of an important genetic predisposition for the conditions they study. Herndon and Jennings concluded that the 36% identical twin polio concordance rate (falling well short of 100%) "would seem to indicate that environmental factors are of major importance in determining the reaction to exposure even in persons of identical genetic endowment."[6] But if a virus was known to cause the disease, why bother speculating about whether twin data "seem to indicate" the importance of environmental factors?

The authors of a 1942 family pedigree study of people diagnosed with the paralytic form of polio, John Addair and Laurence Snyder, concluded in favor of "an autosomal recessive gene for susceptibility to paralytic poliomyelitis."[7] Like

---

1. Folstein & Rutter, 1977b, p. 310.
2. Ritvo et al., 1985a, p. 77.
3. Bailey at al., 1995, p. 73.
4. Herndon & Jennings, 1951, p. 17.
5. Ibid., p. 17.
6. Ibid., p. 45.
7. Addair & Snyder, 1942, p. 307.

Herndon and Jennings they were aware that the illness was caused by a virus, but argued that genetic factors were also important. Moreover, they interpreted family data as "indicat[ing] that the major determinant for crippling lies in the host rather than the parasitic factors."[1] These polio genetic researchers were merely echoing the earlier teachings of eugenicist Charles Davenport, who, using syphilitic paresis, delirium tremens, and tuberculosis as examples, wrote in 1911 that, "in general, the causes of disease as given in the pathologies *are not the real causes*. They are due to inciting conditions acting on a susceptible protoplasm. The real cause of death of any person is his inability to cope with the disease germ, or other untoward conditions [emphasis added]."[2]

Some may accuse me of using discarded studies and ideas in an attempt to discredit current research, but contemporary psychiatric genetic theories are similar because, although they recognize a role for environmental factors, the major "determinant" for susceptibility to autism and other psychiatric disorders is seen as *lying in the host*.[3]

Psychiatric geneticists argue that studying genetics remains an important task even after the discovery of significant (known or unknown) environmental factors. But why have researchers stopped conducting *polio* family and twin studies? Furthermore, why aren't they searching for "polio susceptibility genes"? Had they followed the example of most contemporary psychiatric genetic researchers, Salk and Sabin would have searched for polio genes instead of developing their vaccines.

CONCLUSION

Behind the ubiquitous claims of autism as a "highly heritable strongly genetic" disorder lie a handful of small sample-size twin studies. There are no studies of twins reared apart. There are no adoption studies. There are no gene discoveries. The investigators' conclusions are, as we have seen, influenced by their genetic bias. And, as is always the case in psychiatric twin studies, the results are explainable on non-genetic grounds. We saw how in addition to the implausible traditional postnatal equal environment assumption of the twin method, the important fact that monochorionic and dichorionic twins experience differing prenatal environments is never mentioned in autism twin study publica-

---

1. Ibid., p. 307.
2. Davenport, 1911, pp. 253-254.
3. Interestingly, some contemporary authors still cite polio twin research in support of genetics. In 2004, for example, psychologist Daniel Hanson could write, "The truth that polio is a viral disease is only a partial truth. Polio is also a genetic disease — or so suggests Herndorn [sic] and Jennings's (1951) twin data with an MZ...concordance of 36% and a DZ...concordance of 6%" (Hanson, 2004, p. 207).

tions. Perhaps a comparison of MC identical versus DC identical twin concordance rates would reveal a significantly higher concordance rate in the former group. This finding might suggest that nongenetic biological or chemical factors (e.g., mercury, viruses) are the cause of autism. Phillips called for modified twin studies in which identical MC pairs, as well as identical DC pairs with significantly different birth weights, would be excluded. He concluded that twin studies "may have misled us into believing in a genetic origin of many diseases."[1]

Genetically oriented researchers and authors frequently cite autism as an example of what they see as the fallacy of purely environmental explanations of psychiatric disorders, often citing discredited decades-old "refrigerator mother" theories of autism. But even if autism were found to be caused by faulty genes, this would do little to strengthen genetic arguments about behavior in general. That true genetic disorders exist, such as Huntington's Disease, does not mean that variations in human psychological traits in general have a genetic component, just as the fact that brain tumors exist does not mean that behavioral disorders in general are caused by brain diseases. But, it just so happens that there is little scientifically acceptable evidence in support of autism as a genetic disorder.

---

1. Phillips, 1993, p. 1008.

CHAPTER 8. THE 1942 "EUTHANASIA" DEBATE IN THE *AMERICAN JOURNAL OF PSYCHIATRY*

Some professions have episodes in their history that they would just as soon forget. One such episode in the history of psychiatry concerns a 1942 *American Journal of Psychiatry* debate on whether or not "feebleminded" children and adults should be killed.[1] The participants were the noted neurologist Foster Kennedy (1884-1952), who was in favor of killing, child psychiatrist Leo Kanner (1894-1981), who was against killing, and the anonymous authors of an editorial siding with Kennedy's position.[2]

As Edwin Black documented in his 2003 *War Against the Weak*, American, British, and German eugenicists openly discussed using "lethal chambers" to kill "defectives" in the decades prior to 1942.[3] Although many eugenicists were opposed to murder, some saw the lethal chamber as a possible eugenic intervention. For example, in the 1918 edition of *Applied Eugenics* Popenoe and Johnson wrote, "From the historical point of view the first method that presents itself is execution…. Its value in keeping up the standard of the race should not be underestimated."[4]

The Kennedy/Kanner exchange took place in 1941/42, when the popularity of eugenic ideas among the American intelligentsia was at a high point. Moreover, Nazi Germany had been exterminating mental patients and "defectives" for two years under the guise of performing "euthanasia." Both Kennedy and Kanner were well-known investigators whose opinions held a certain degree of influence. I will present their arguments in detail for the purpose of

---

1. This chapter is based on an article published in *History of Psychiatry* (Joseph, 2005a).
2. All citations in this chapter attributed to Kennedy, Kanner, or to the editorial writers are taken from Kennedy, 1942, Kanner, 1942, or Anonymous, 1942.
3. Black, 2003, Chapter 13.
4. Quoted in Black, 2003, p. 251.

shedding light on the debate, as well as highlighting the common threads of both Kennedy's and Kanner's arguments. Following this, we will look at the incredible final words of the anonymous editorial writers. The ideas expressed in the debate may come as a shock to those who view Nazi Germany and mid-20th century America as political opposites. As we will see, during this period the thinking of influential psychiatrists in both countries was disturbingly similar.

Foster Kennedy Calls for Killing

Kennedy's article, entitled "The Problem of Social Control of the Congenital Defective: Education, Sterilization, Euthanasia," was based on a May, 1941 address to a meeting of the American Psychiatric Association. He began by stating that "it is not easy to know how to start to talk on such a subject as this, for this subject has to do with the whole of life and death." For Kennedy, "feeblemindedness" was a great problem facing society. While recognizing that people diagnosed with schizophrenia and manic-depression "so largely fill our mental hospitals," the elimination of these people would be costly to society because future generations might only produce a population of "mediocrities, capable of pushing but not leaping; and it's the leap that counts."

"On the other hand," wrote Kennedy, "we have too many feebleminded people among us, something like 60,000." He viewed these people as "hopelessly unfit." Kennedy was opposed to "euthanasia" for those who fell ill from various diseases, on the grounds that they could recover. "But I *am*," he wrote, "in favor of euthanasia for those hopeless ones who should never have been born — Nature's mistakes." Like Hitler and his psychiatric collaborators in Germany, for Kennedy "euthanasia" meant killing people who did not wish to be killed, and was simply a euphemism for the murder of adults and children.

According to Kennedy's plan, "defective children" reaching the age of five, with the consent of their guardians, should have their case reviewed by "a competent medical board." If, after several months and at least three examinations, the board finds that the "defective has no future or hope of one,"

> then I believe it is a merciful and kindly thing to relieve that defective — often tortured and convulsed, grotesque and absurd, useless and foolish and entirely undesirable — the agony of living.

How kind and merciful, Kennedy must have believed, to relieve these "defectives" from the "agony of living." Kennedy's use of the word "undesirable" is taken directly from standard psychiatric genetic and Nazi literature of the era, which saw frequent calls for sterilizing the "eugenically" or "racial hygienically undesirable" (the German phrase is "rassenhygienisch unerwünscht").

Anticipating objections that these "creatures have immortal souls," Kennedy answered "that to release the soul from its misshapen body which only defeats in this world the soul's powers and gifts is surely to exchange, on that soul's behalf, bondage for freedom." Similar to the way Hitler began the German killing program, Kennedy discussed parents of "defective children" from all over the United States who had appealed to him with "sad pleas" to assist that "their unhappy offspring be mercifully released from life." Sadly, in Kennedy's mind, these pleas remained unanswered, due to the "laws and social mores" of the nation.

Kennedy was opposed to euthanasia for "normal adults" who had become ill (he didn't discuss the question of whether this was requested by the patient). For him, "to legalize such euthanasia may put a weapon in the hands of wicked men, or, worse, a tool in the hands of the foolish." As an eminent and educated man, Kennedy believed he could determine who is normal and in need of protection, versus who is abnormal and in need of extermination:

> So the place for euthanasia, I believe, is for the completely hopeless defective: nature's mistake; something we hustle out of sight, which should never have been seen at all. These should be relieved the burden of living, because for them the burden of living at no time can produce any good thing at all.... For us to allow them to continue such a living is sheer sentimentality, and cruel too; we deny them as much solace as we give our stricken horse. Here we may most kindly kill, and have no fear of error.

Thus Kennedy was able to publish his call for killing in the pages of American psychiatry's leading publication. As in Germany, Kennedy's 1942 proposal carried the added attraction of ridding society of the burden of caring for "defectives" at a time when great resources were needed for the war effort. Kennedy, however, did not make this argument, most likely because the US had not yet entered World War II at the time his speech was delivered. Nor did he stress the eugenic desire to eliminate "defectives" in order to breed for the master race. Nevertheless, he believed that society's way of dealing with "tortured and convulsed, grotesque and absurd, useless and foolish and entirely undesirable" people should be to "hustle them out of sight" and kill them.

LEO KANNER'S "EXONERATION OF THE FEEBLEMINDED"

Given that Kennedy was calling for the murder of thousands of children and adults, one might have expected child psychiatrist Leo Kanner to have replied with as much outrage as one could muster in the pages of a learned professional journal. But this is not quite how Kanner responded.

Kanner began by questioning the validity of the "feeblemindedness" concept so widely used during the first four decades of the 20th century. He wrote that "'feeblemindedness' and 'mental deficiency,' in spite of existing gradations, are terms used very much in the manner of clichés, somewhat reminiscent of the designations 'insanity' and 'lunacy' as they were applied in the days of yore." He then highlighted the two main points of his article: (1) that an IQ score does not necessarily correlate with a person's worth to society, and (2) that people considered "feebleminded" or "mentally deficient" often play an important role in the United States' economic system. As an example of this second point, Kanner spoke of "the garbage collector's assistant who has served our neighborhood for many years." This was a "sober, conscientious, and industrious fellow," who is "deservedly respected by his employer, his co-workers and his spare time companions." Still, "with an IQ of 65, he is rated by us psychiatrists as feebleminded or mentally deficient." In contrast, Kanner spoke of a "handsome, dashing, reckless blade who has driven his parents frantic with alcoholism, debts and amorous adventures, has made his wife miserable, has deserted her and their offspring, [and] has not done a single thing that can be considered socially useful...." However, "with an IQ that nearly hits the ceiling, he receives from us the honor of being considered 'mentally superior.'" Although Kanner recognized that there was "nothing new or original in making this contrast," he sought to demonstrate that people with low IQs sometimes can contribute more to society than people with high IQs.

Kanner distinguished between two groups of people. The first group "consists of individuals so markedly deficient in their cognitive emotional and constructively conative potentialities that they would stand out as defectives in any type of existing human community." The second group is made up of individuals "whose limitations are definitely related to the standards of the culture which surrounds them," who tend to function well in less complex societies, and could "make successful peasants, hunters, fishermen, tribal dancers." For Kanner, the shortcomings of this second group only become apparent in more intellectually demanding cultures such as that found in the United States. "The members of the second group," he continued, "are not truly and absolutely feebleminded or mentally deficient. Their principle shortcoming is a greater or lesser degree of inability to comply with the intellectual requirements of the community." Kanner therefore believed that it was wrong "to label these individuals as mentally deficient, together with the idiots and imbeciles."

Kanner then addressed the argument that the feebleminded are a drag on society and "hamper the progress of civilization." He pointed to examples in history where atrocities were committed by people who were not mentally deficient. As a contemporary example he mentioned Hitler (to whom he referred by his birth name "Schicklgruber"), "whose IQ is probably not below normal, [and] has in a few years brought infinitely more disaster and suffering to this world than have all the innumerable mental defectives of all countries and generations combined." While recognizing that "the absence of vice is not necessarily in itself a virtue," Kanner discussed ways in which the "mentally deficient" contribute to society:

> Sewage disposal, ditch digging, potato peeling, scrubbing of floors and other such occupations are as indispensable and essential to our way of living as science,

literature and art. Cotton picking is an integral part of our textile industries. Oyster shucking is an important part of our seafood supply. Garbage collection is an essential part of our public hygiene measures. For all practical purposes, the garbage collector is as much of a public hygienist as is the laboratory bacteriologist. All such performances, often referred to snobbishly as "the dirty work," are indeed real and necessary contributions to our culture, without which our culture would collapse within less than a month.

He added that people performing these tasks free "the time and energies of others for tasks which involve planning and creative activities."

Although Kanner agreed with Kennedy that "idiots and imbeciles cannot be trained in any kind of social usefulness," he disagreed with Kennedy's conclusion that, in Kanner's words, "we are justified in passing the black bottle among them" through the procedure some "dignify with the term euthanasia." Kanner linked such ideas to reports of Nazi atrocities, and asked, "Shall we psychiatrists take our cue from the Nazi Gestapo?" Still, Kanner agreed with Kennedy and others that "sterilization is often a desirable procedure" for "persons intellectually or emotionally unfit to rear children." However, he objected to sterilization performed "solely on the basis of the IQ."

For Kanner, in this debate among the American educated elite, an important point was the elite's need for low IQ people to do their dirty work: "Do we really wish to deprive ourselves," he asked, "of people whom we desperately need for a variety of essential occupations?" He spoke of a "disaster" greater than the "present world-wide holocaust" that would occur "if we decided to annihilate the intellectually inadequate today." He urged his fellow psychiatrists to "leave the cotton pickers, oyster shuckers and bundle wrappers alone, regardless of their IQ, so long as they are industrious and good natured!" (He did not mention what might happen to those not "industrious and good natured.") And heaven forbid that Kanner, Kennedy and their fellow elitists would have to pick their own cotton or shuck their own oysters.

Kanner, who never mentioned Kennedy by name, ended with an appeal to extend "the democratic ideal to the feebleminded," rather than follow "carping critics and whining would-be protectors of future generations."

THE FINAL WORD: AN ANONYMOUS EDITORIAL COMMENT

The editors of *The American Journal of Psychiatry*, not content to leave the issue to Kanner and Kennedy, decided to weigh in on the question themselves in the form of an unsigned editorial opinion in the same issue, entitled "Euthanasia."

The editorial writers began by noting that although "recognized authorities" such as Kennedy and Kanner "might appear to represent quite contradictory standpoints," a "careful perusal of the texts" reveals that "the differences narrow down to a single point." Kennedy "proposes a method of disposal which he believes would bring relief to all concerned," whereas Kanner "prefers to let the situation remain as it is." They also added, correctly, that both writers supported forced sterilization. This was followed by a list of six objections to killing found among the populace. The first two dealt with religious and secular aversions to the taking of human life. Their focus then turned to the parents of those slated to be killed:

> A third variety of reaction results from an accusing sense of obligation on the part of the parents towards the defective creature they have caused to be born. The extreme devotion and care bestowed upon the defective child, even with sacrifice of advantages for its normal brothers and sisters, is a matter of common observation. This position is understandable, but to the impersonal observer may appear to partake of the morbid. Disposal by euthanasia of their idiot offspring would perhaps unbearably magnify the parents' sense of guilt.

Thus, psychiatrists were informed that parents' "morbid" devotion stood in the way of the "disposal by euthanasia of their idiot offspring." But the parents were guilty of much more, having "caused" their "defective creature" to be born in the first place.

A fourth problem faced by euthanasia advocates was the parents' dread of their neighbors' opinions. The fifth was the parents' "instinct and love," which leads them to resist the idea that their child should be killed. But the writers added a twist: Truly devoted parents could show their devotion more by allowing their child's "merciful passage from life" than by "insisting that a crippled vegetative existence be continued at all costs...." The sixth factor was people's general tendency to reject "any new drastic procedure." Clearly, the authors were by now writing in favor of, as Kanner put it, "passing the black bottle" among the "defectives." But they wanted to appear reasonable, reminding their readers that it is only "to the lowest grade of defectives for whom alone euthanasia has been proposed."

The remaining portion of the anonymous editorial elaborated upon the theme that, due to their pathologically misplaced sentimentality, the parents of condemned children are an important focus of psychiatric attention:

> It is submitted that the state of mind of the parents of an idiot may as fairly become a subject of psychiatric concern as the interrelationships in the families of psychotic patients, and the unwholesome reactions stand as much in need of correction in one case as in the other.

Whereas scientists "presumably have reached their convictions by more or less impersonal routes," the person "who has the misfortune to be the parent of a low-grade defective is actuated by strongly personal motives which he may or may not be capable of setting out clearly in his own consciousness." In other words, parents' failure to realize that their child "should be relieved of the burden of living" stems from their "personal motives" for wanting to keep their child alive. Ironically, according to the authors, many of these parents would be

178

nonetheless relieved if "natural causes" could write a "lethal finis to the painful chapter."

The next question was "whether the attitude of the parent to the defective child can be regarded as morbid, and if so, whether anything can and should be done about it." Naturally, their answer was yes. They found it difficult to understand how parents could feel "normal affection... for a creature incapable of the slightest response." Psychiatrists and others attempting to relieve these parents of the "unhappy obsession of obligation or guilt...would seem to be [practicing] good mental hygiene." The authors acknowledged that, despite Kennedy's "strong arguments in support of his position," the proponents of euthanasia were in the minority. Kanner, they wrote, presented no argument against killing other than his belief that parents would oppose it. In fact, Kanner's main argument was that lower IQ people perform important roles in society and that their extermination, in addition to being immoral, might force himself, Kennedy, and other members of the intellectual elite to haul their own garbage, scrub their own floors, and shuck their own oysters.

Returning to psychiatry's task of convincing parents of the necessity of releasing their "creatures" from the "burden of living," the editorial writers argued that the parents' feelings are:

> precisely the psychiatric problem this over-lengthy discussion has been trying to get at, namely, the "fondness" of the parents of an idiot and their "want" that he should be kept alive. It is this parental state of mind that we believe deserves study — to the extent to which it exists, in fact and not merely as a generalization of opinion, what underlying factors such as those set forth above are discoverable, whether it can be assessed as healthy or morbid, and whether in the latter case it is modifiable by exposure to mental hygiene principles.

They recognized that "enabling legislation will be required" if "euthanasia is to become at some distant day an available procedure." The authors thereby recognized a difference between fascist Germany and the democratic United States: whereas Hitler began to exterminate the "feebleminded" after stage-managed appeals by parents followed by secret orders to kill, in the US "enabling legislation" would be necessary. The passing of "euthanasia" laws would have to overcome public opinion against extermination, and the psychopathology of the "parents of the candidates for the contemplated procedure" was seen as an important focus of psychiatric attention. Thus, the "whole question must center" on psychiatrists' "evaluation and melioration of this parental attitude."

Finally, the editorial writers wrote that in addition to killing, "the story of sterilization will doubtless be repeated on an extended scale." This rounded out their vision for America: killing "defectives" and greatly expanding compulsory eugenic sterilization. Their model, of course, was Hitler and Rüdin's German racial hygiene state, with its killing program and over 300,000 forced sterilizations.

CONCLUSION

Reviewing these articles has been personally disturbing to me, and it remains almost beyond belief that they could have appeared in the most prominent psychiatric journal in the United States at the time. But as historians have noted, eugenic sterilization was legally sanctioned in the US long before the Nazi sterilization law of 1933. The logical procession from sterilization (killing presumed genes) to "euthanasia" (killing presumed gene *carriers*) occurred much more slowly in the US, but accelerated in the early 1940s under German influence. The procession from sterilization to killing is "logical" because, once it has been established that the state should actively participate in preventing the reproduction of "genetically undesirable" people through compulsory sterilization, it eventually seems more "efficient" to wipe out the alleged gene carriers themselves. In a chilling and prophetic 1923 statement, Swedish Member of Parliament and sterilization opponent Carl Lindhagen asked, "Why shall we only deprive these persons, of no use to society or even for themselves, the ability of reproduction? Is it not even kinder to take their lives? This kind of dubious reasoning will be the outcome of the methods proposed today."[1]

The steps taken in Germany were: (1) the belief that mental traits and disorders are largely genetically determined and cannot be altered; (2) the production of alleged scientific evidence confirming of this belief; (3) the establishment of the state's duty to forcibly prevent the reproduction of the "carriers of genetic defects"; and (4) the physical destruction of people presumed to carry genetic defects. We have seen in this chapter that the United States was also moving in the direction of the final stage in the early 1940s. Due to the post World War II revelations of Nazi atrocities, "euthanasia" ceased to be debated in US academic journals. However, today we are witnessing the reestablishment, albeit on the basis of false ideas derived from misinterpreted and biased research, of the idea that psychological trait differences and psychopathology are largely determined by the genes. Whether this leads to the third stage, as it did in Germany, the United States, and several Scandinavian countries, or to the final stage, as it did in Germany and could have done in the US, remains to be seen. The main focus of current opposition to genetic determinism and its consequences is to prevent the full acceptance of the first stage belief that our futures are determined at birth by heredity, and the second stage position that research shows this to be true. Thus, thoroughly exposing unsound research — and the beliefs that drive it — is an important task facing us today.

---

1. Quoted in Broberg & Tydén, 1996, p. 104.

CHAPTER 9. THE TWIN METHOD'S ACHILLES HEEL: A CRITICAL REVIEW OF THE EQUAL ENVIRONMENT ASSUMPTION TEST LITERATURE

Throughout this book and in previous publications I have elaborated upon the views of three generations of critics, who have argued that genetic conclusions derived from the twin method are confounded by environmental factors. Although it is now widely accepted that identical twins experience much more similar environments than fraternals, the twin method continues to be used in psychiatric research because, as we saw in Chapter 1, twin researchers decided to change the definition of the equal environment assumption (EEA) when the original 40-year-old definition turned out to be false.[1]

Over the past forty years or so, twin researchers have produced a body of evidence in an attempt to validate their frequently-criticized research method. Paradoxically, the starting point of most "EEA test" studies is a recognition that identical twins experience more similar environments than fraternals. It is then argued that the greater environmental similarity of identical twins does not contribute to their greater behavioral similarity, or to higher identical versus fraternal concordance for various psychiatric disorders. We saw in Chapter 1 that this is an astonishing claim when we realize that, by extrapolation, it must follow that *no-one's* behavior is influenced by their family, social, or cultural environments! Had twin researchers thought through this implication of their argument, they might have abandoned the twin method long ago.

We also saw in Chapter 1 that, since 1983, psychiatric geneticist Kenneth Kendler has argued that the EEA test literature has consistently supported the validity of the assumption as defined in the trait-relevant sense, meaning that to invalidate the EEA, *critics* must demonstrate that identical and fraternal twins'

---

1. Joseph, 2004b, Chapter 3.

environments differ in aspects relevant to the trait under study. We saw, however, that placing the burden of proof on critics violates a basic tenet of science. Moreover, even if critics were able to show that the environments of identical and fraternal twins differ on trait-relevant dimensions, it is likely that twin researchers would continue to uphold the EEA on the basis of their claim that twins "create their own" trait-relevant environments.

Kendler's main focus is defending the EEA as it pertains to psychiatric and behavioral twin studies (as opposed to cognitive ability), which is the focus of this chapter. Although he usually does not cite studies of reared-apart twins in defense of the EEA, others sometimes do. In a 2005 publication using twin method data to argue that people's political views are shaped by their genetic heritage, a group of political scientists wrote, "The most powerful refutation," of criticism of the EEA "comes in recent studies utilizing MZ and DZ twins raised apart. These studies uniformly validate MZ and DZ differences found in earlier studies of twins reared together."[1] However, these studies provide no such validation of the twin method (see Chapter 1).[2]

Kendler has written that "the EEA has been tested in five different ways."[3] A review of these studies is necessary because Kendler presents his readers with a large body of empirical evidence, which, he claims, supports the ideas he puts forth. I review these studies individually (all of which deal with twins reared together), while at the same time I highlight common themes. Following this, I briefly describe several additional EEA test studies performed in the past few years. Studies in italics indicate those Kendler cited in support of the EEA in a 1995 article co-authored by J. Hettema and M. C. Neale.[4] This article serves as our reference point.

But first, a word of caution. I review the EEA test literature only because mainstream psychiatry accepts it as validating the twin method. As I argued in Chapter One, and argue at the conclusion of this chapter, however, the widely recognized greater environmental similarity of identical versus fraternal twins *invalidates the twin method on its face.* Thus, the twin method is confounded by environmental factors regardless of what EEA-test researchers claim. What they actually must demonstrate — without qualification — is that identical twins and fraternal twins experience equal environments. Readers who agree could skim or skip the bulk of this chapter and move on to the conclusions.

---

1. Alford et al., 2005, p. 155.
2. See also Joseph, 2004b, Chapter 4.
3. Hettema et al.,1995, p. 327.
4. Ibid.

THE "FIRST METHOD"

The first method of testing the EEA, according to Kendler, "involves direct observation of twins in a social situation in which the behavior of other individuals is divided into those that are self-initiated and those that occur in response to behavior of the twins."[1] Kendler cited one study, by the Canadian psychologist Hugh Lytton, which attempted to carry out such observations.[2]

Lytton, whose sample consisted of 2 ½ -year-old male twin pairs (17 identical, 29 fraternal), attempted to determine whether the acknowledged differential treatment of identical and fraternal twins is *initiated* by parents, or is a *response* to their twins' behavior. Lytton made a distinction between "child-initiated parental responses" and "parent-initiated actions." He defined the latter as parent-actions not preceded by a child-action within the previous 10 seconds. Thus, according to Lytton's criteria, a mother waiting eleven seconds before reacting to her child's behavior is taking a "parent-initiated action," whereas a child may defy her parent on Tuesday morning in response to a spanking she received Monday night, and Lytton would classify this as a "child-initiated action." As EEA critics Alvin Pam and colleagues commented, "We find such an inference dubious since it is based on the supposition that one can discern which behavior is 'imposed' or 'elicited' — but any family therapist will insist that child-parent encounters are interactional."[3]

For parent-initiated actions, Lytton found that identical twins were treated significantly more alike in only one of the eight categories, and non-significant results were recorded for the remaining seven. He concluded that "parents respond to, rather than create, differences between the twins..." and that the EEA is therefore valid.[4]

Lytton's rating system contained a clear bias. Observations were made in the families' homes, and the raters observed parent/child interaction and scored behavior counts for the family members. Lytton indicated that there were two raters. Thus, two non-blinded raters are sitting in the homes of families for whom in many cases the zygosity of the twins was apparent. It is difficult to believe that their biases did not affect the subjectively rated behavior counts they were making as they observed these families. Moreover, the presence of a rater in one's home will influence the way that parents treat and respond to their twins. (We can make a similar point about so-called television "reality programs.") As Lytton acknowledged in an earlier publication, "It is obvious that the introduction of an observer in the home must affect relationships to some extent and produce some distortion of the 'normal interaction'...the presence of

---

1. Ibid., p. 327.
2. Lytton, 1977.
3. Pam et al., 1996, p. 352.
4. Lytton, 1977, p. 459.

an observer...increases the positive 'desirable' responses of parents."[1] He also wrote that parents "played and romped with their children perhaps more than usual for our benefit," that parental behavior such as kissing "may have been staged," and that "slapping also occurred in our presence...probably with reduced frequency."[2] And although Lytton claimed that ratings were made in a "naturalistic setting," the investigation required that "at least one parent should, for the most part, remain in the living room area."[3] This restriction could have influenced the interactions (and the raters' behavior counts) between parents and twins since, among other reasons, parents were not permitted to follow their twins into another room and interact with them there.

The reliability of the raters' scores, which were derived through a compli-cated scoring system requiring 50-60 hours of training over several weeks, pre-sented another problem.[4] "It must be clear," Lytton recognized, "that hoping to obtain the usual psychometric test standards of reliability in a free-flowing interaction situation is straining after the impossible."[5] Not surprisingly, therefore, Lytton reported an "interobserver agreement for the observed behavior" of only 63.9%, which he recognized was "lower than is usually demanded by psychometric investigations."[6]

One rater's recorded observation of an identical twin pair reveals a description which, although apparently unnoticed by Lytton and Kendler, runs counter to genetic theory. As a newborn, one twin suffered from a respiratory infection and was maintained in a hospital incubator for several weeks after birth, whereas the other twin came home with his mother from the hospital. The interviewer's notes read as follows:

> Mother doesn't think of them as twins, J. has been behind. Their personalities warrant their being treated differently. The differences that mother makes are those they demand, or that events produce. J. is 10 times worse than D. in climbing on cupboards and tables and is usually spanked. Mother often has to spank J. for things that D. does not have to be spanked for. D. is more sensitive, responds to a look or being sent to his room. Mother spends about half an hour holding and cud-dling D. and about 15 minutes with J. or as much time as he'll allow.[7]

The phrase "Their personalities warrant their being treated differently" reflected Lytton's belief, which neatly fit his defense of the EEA, that children's personalities dictate the treatment they receive. Although Lytton quoted this passage as an example of a mother who treats her identical twins differently because their personalities are different, as opposed to the treatment she ini-

---

1. Lytton, 1973, p. 8.
2. Ibid., p. 8.
3. Lytton & Zwirner, 1975, p. 771.
4. Lytton, 1973.
5. Ibid., p. 7.
6. Lytton & Zwirner, 1975, p. 772.
7. Lytton, 1977, p. 458.

tiates, he overlooked the fact that differences in her twins' personalities were apparently caused by an *environmental* event (the hospital stay). From the genetic perspective, however, their behavior should not have been so different. It seems that an environmental event played an important role in shaping the differing behaviors of these genetically identical individuals.

Lytton's study provides no support to the EEA for reasons that include a small and very young twin sample, rater bias, inter-rater unreliability, and the unsupported claim that one can determine the direction of interactions between parents and children. Finally, how can the experience of twins J and D be reconciled with Kendler's assertion that "monozygotic twins would develop similar phenotypes regardless of the similarity of their social environment"?[1] Clearly, their story does not support this theory.

THE "SECOND METHOD"

The second method takes as its starting point twin researchers' current consensus position that identical twins experience more similar environments than fraternal twins. However, according to Kendler,

> If these environmental experiences influence twin similarity, controlling for zygosity, the degree of similarity in childhood environment or the frequency of contact as adults should predict twin similarity. However, the vast majority of studies using this method have found no relationship between these measures of environmental similarity and twin similarity for [personality and various psychiatric conditions].[2]

Kendler cited several investigations in support of this idea. We begin with *Morris-Yates and associates*, who studied 343 Australian adult same-sex twin pairs.[3] Twins filled out a mailed childhood experience questionnaire and completed two personality tests. It has been shown in numerous studies, however, that retrospective questionnaires have serious reliability problems.[4] Yet, Morris-Yates and colleagues' conclusions depended on this type of questionnaire. The investigators found that although questionnaire responses showed that identical twins experience much more similar environments than fraternals, there is "no significant association between intra-pair similarities in either neuroticism, anxiety or depression and the extent of similar treatment either identical or DZ twins had imposed on them."[5] They arrived at this conclusion after dividing

---

1. Kendler, 1983, p. 1414.
2. Kendler, 1993, p. 906.
3. Morris-Yates et al., 1990.
4. Bradburn et al., 1987; Halverson, 1988; Hardt & Rutter, 2004; Holmberg & Holmes, 1994; Reuband, 1994; Robbins, 1963; Yarrow et al., 1970.
5. Morris-Yates et al., 1990, p. 324.

their twins' environments into two categories similar to Lytton's: (1) parentally determined environment imposed on the twins, and (2) environmental experience elicited by the twins themselves. The researchers provided no information on how they distinguished between these two alleged types of childhood environments. Furthermore, they did not disclose which questionnaire responses loaded on which factors.

The failure to provide sufficient data is common in the EEA test literature and, as Leon Kamin showed, scores on the extreme ends of a questionnaire scale can often suggest the existence of environmental effects.[1] Pooled or "blunted" figures can conceal correlations between these extreme scores and the trait in question. For example, Kamin cited a study that reported twins' responses to whether they were dressed alike as "yes" or "no," whose authors determined that there was no significant correlation between the answer to this question and IQ. However, the twins actually had answered a *four*-point scale, and having access to the raw data, Kamin found a significant intrapair IQ correlation difference between twins who answered "almost always" and those who answered "seldom." In other words, he found an environmental effect when none was apparent from the data provided in the published report:Many studies I review in this chapter fail to provide adequate details on the frequency of responses to various questions put to twins or to their parents. For this and other reasons, their authors' claims to have found no significant environmental effects on behavioral correlations must be viewed with caution. Moreover, as pointed out by British psychologists Sarah Norgate and Ken Richardson, "any strategy that searches for the absence of an association (or 'proving a negative') is acutely prone to Type II errors."[2]

According to Morris-Yates et al., "On the basis of several indicators of environmental similarity, the equal environments assumption appears to be invalid." However, since in their view childhood environmental differences are the product of *genetic* resemblance, they concluded that "the equal environments assumption is valid"[3] Circular reasoning of this type is common in twin research, as Morris-Yates and colleagues' belief that genetic endowment determines childhood environments was based largely on the results of previous twin studies. Like other twin researchers, they implicitly *assumed* the validity of the EEA in order to defend it.

According to twin researchers *Martin and colleagues*, "social attitudes" are genetically transmitted.[4] They reached this counterintuitive conclusion after analyzing mailed questionnaire data on the social and political views of more than 4,000 Australian and British twin pairs. Due to the higher observed correlations among identical versus fraternal twins, the investigators concluded that 14

---

1. See Kamin, in Eysenck vs. Kamin, 1981.
2. Richardson & Norgate, in press.
3. Morris-Yates et al., 1990, p. 325
4. Martin et al., 1986.

of the 50 items "showed significant evidence of both genetic and social compo-nents of twin resemblance."[1] These fourteen items included: "Sabbath obser-vance," "Hippies," "Divine law," "Socialism," "Moral training," "Legalized abortion," "Student pranks," "Royalty," "Nudist camps," "Church authority," "Caning," "Mixed marriage," "Casual living," and "Bible truth."

Although Martin and associates concluded that humans are genetically predisposed for their attitudes towards hippies, student pranks, nudist camps, and so on, their results could be interpreted as evidence *against* the EEA and the twin method, because attitudes obviously shaped by social interaction and various cultural influences show statistically significant identical versus fra-ternal correlational differences. But rarely does this enter the mind of geneti-cally-oriented twin researchers carrying out these types of studies, who tend to focus on evidence they perceive as validating genetic conclusions.

Next, we come to *four studies by Kendler and colleagues* based on female twin pairs from the Virginia Twin Registry.[2] These studies were derived from the same sample and used the same methodology in assessing correlations between similarities of childhood experiences and adult concordance for the conditions under review (major depression, phobia, generalized anxiety disorder, and alco-holism). The investigators claimed a genetic component for all of these condi-tions, but my purpose here is to examine the relationship between twins' environments and concordance for the psychiatric disorders under study.

In each study the researchers asked twins how often as children they (1) "shared the same room," (2) "had the same playmates," (3) "were dressed alike," and (4) "were in the same classes in school."[3] To assess closeness as adults, they asked twins an unspecified number of questions to determine "how frequently they were in current contact with their co-twins, with response options ranging from 'living together' to 'once a year or less.'"[4] According to Kendler et al., "con-trolling for zygosity, twin similarity for depression was not influenced by the similarity of environment in childhood...or adulthood...."[5]

The questions researchers asked the twins reflect a misunderstanding of what constitutes "the environment." It would have been better to have asked the types of questions posed by Einer Kringlen in his 1967 schizophrenia twin study, where twins were assessed for items such as "identity confusion in childhood," being "brought up as a unit," and "global twin closeness" (see Table 1.1). These items shed more light on twin closeness and identification than simply deter-mining how much time twins spent together, or whether they were dressed alike.[6] Perhaps Kendler and associates, making these types of assessments, would have recorded far different results than the ones they obtained.

We now come to a group of studies assessing the relationship between adult twin contact and concordance for behavioral and psychological traits.

1. Ibid., p. 4366.
2. Kendler, Heath, et al., 1992; Kendler, Neale, et al., 1992a, 1992b, 1992c.
3. Kendler, Neale, et al., 1992a, p. 259.
4. Ibid., p. 259.
5. Ibid., p. 261.
6. Kringlen, 1967, p. 115.

Heath and colleagues looked into the alcohol consumption levels of 1,984 Australian female twin pairs. In addition to questions about their marital status, alcohol consumption, and so on, they asked twins about the amount of social contact they had with their adult co-twin. The researchers concluded,

> There was no consistent evidence for increased twin resemblance in sisters who were living together or had frequent social contact, compared with those who had less frequent social contact. In three twin groups — young MZ women, young DZ women and older DZ women — there was no significant correlation between absolute intrapair differences in alcohol consumption and amount of social contact.[1]

In 1986, *Kendler and colleagues* assessed the relationship between the frequency of identical twin contact and anxiety and depression, reporting that "the number of significant correlations observed (4/70) did not exceed chance expectations."[2] It is also important to know the mean intraclass correlational differences between groups, but they did not provide this information.

Kendler then cited two studies which *did* report a significant association between adult twins' shared experience and behavioral similarity. The first, by *Kaprio and associates*, was based on 550 Finnish twin pairs (ages 18-25) rated for alcohol use, "extraversion," and "neuroticism." The researchers concluded that "conventional comparisons of MZ/dizygotic (DZ) twins are confounded by MZ/DZ differences in social contact, and, for some behaviors, social contact does correlate with co-twin resemblance."[3]

Clifford et al. surveyed 572 British twin pairs for alcohol use, anxiety, and depression, and correlated their scores with the cohabitation status of the adult twin pairs. They concluded that their findings did not support the EEA:

> Much controversy has surrounded the basic assumption of the classical twin method: that the effect of shared environment is the same for (like-sexed) identical and fraternal twins. From our analysis this assumption would appear not to be satisfied for these measures...[4]

However, *Lykken and associates* maintained that the evidence suggests that "similarity leads to contact, rather than the other way around."[5] "The most plausible explanation," they concluded, is that identicals "enjoy each other's company because they are so similar in personality, interests and attitudes."[6]

---

1. Heath et al., 1989, p. 44.
2. Kendler et al., 1986, p. 216.
3. Kaprio et al., 1990, p. 274.
4. Clifford et al., 1984, p. 76.
5. Lykken et al.,1990, p. 547.
6. Ibid., p. 560.

THE "THIRD METHOD"

Kendler described the third method of testing the equal environment assumption as follows:

> Parents of MZ twins may treat their twin offspring more similarly than parents of DZ twins. In a study of academically talented adolescent twins, parents of MZ twins indeed reported treating their twins more similarly than parents of DZ twins, but the similarity of parental treatment was unrelated to twin similarity for cognitive abilities, personality, or vocational interests.[1]

Kendler was referring to *Loehlin and Nichols's 1976 twin study*, which remains one of the most frequently cited investigations in the behavior genetics literature.[2] Indeed, in a 2005 response to my criticism of genetic research, a journalist supporter of behavior genetics based his entire argument in defense of the EEA on this 30-year-old study.[3]

A critical examination of Loehlin and Nichols' study's relevance to the EEA is found in *The Gene Illusion*, where the main points were as follows:

• Loehlin and Nichols obtained their data from mailed questionnaires and tests, which meant that they had no control over who was filling them out, or how they were filled out.

• The investigators' conclusion that the similarity of parental treatment did not correlate with twin resemblance on various measures depended on parents' responses to retrospective questionnaires. However, research has shown that retrospective accounts are of questionable reliability, and are influenced by the views of respondents at the time of questioning.[4] Moreover, parents' accounts are influenced by societal expectations of how "good parents" *should* act, as well as their own idealized recollection of how they parented their children. This was reflected in Loehlin and Nichols's Parent Questionnaire, where parents reported that, regardless of their twins' zygosity or sex, they treated their twins alike almost 100% of the time! And Loehlin and Nichols themselves recognized the "fallibility of parents' retrospective accounts of the early history of their children."[5] Thus, their biased Parent Questionnaire responses cannot be used to assess the relationship between twins' treatment and their personality or IQ resemblance. Retrospective questionnaires are, in fact, an unacceptable substitute for a carefully designed longitudinal study.[6]

---

1. Kendler, 1993, pp. 906-907.
2. Loehlin & Nichols, 1976.
3. See Joseph, 2005b; Miele, 2005.
4. Bradburn et al., 1987; Halverson, 1988; Hardt & Rutter, 2004; Holmberg & Holmes, 1994; Reuband, 1994; Robbins, 1963; Yarrow et al., 1970.
5. Loehlin & Nichols, 1976, p. 47.
6. Yarrow et al., 1970.

• Although behavior geneticists frequently cite Loehlin and Nichols's findings in support of the EEA, they rarely mention these investigators' own cautionary statements in their 1976 study.[1] For example, Loehlin and Nichols understood that their twin sample was biased because it (1) contained many more identical twins than would be expected by chance, (2) contained many more female pairs than would be expected by chance, (3) consisted only of twins willing to respond to a mailed questionnaire, and (4) consisted of high-achieving Merit Scholarship test takers. They also cautioned that their sample may have been biased for similarity of experience[2] and that, in general, interpretations of their data "do not altogether exclude a completely environmentalist position."[3]

• The investigation could serve as a case study of how flawed research can achieve near mythological status with the passage of time, the absence of critical examination, and the failure to mention the original investigators' cautions about interpreting their data. Although Loehlin and Nichols believed that "fallible data interpreted with some caution are clearly better than no data at all...,"[4] subsequent behavior geneticists' caution-free and uncritical acceptance of their conclusions have led to far *worse* results than if they had reported no data at all.

Loehlin and Nichols's Parent Questionnaire results showing that parents treat their twins the same nearly 100% of the time — regardless of zygosity — place Kendler and his EEA theory in a difficult position because, according his theory that the greater genetic similarity of identical twins causes them to be treated more similarly, parental treatment of identical and fraternal twins *must* be different. If, as Loehlin and Nichols found, identical and fraternal twins correlate at .50 and .28 respectively on personality measures, it must follow, according to Kendler's theory, that identical twins were treated far more alike than fraternals. But this is not what Loehlin and Nichols found.

Like others before and after, Loehlin and Nichols denied that the greater similarity of treatment on the basis of physical appearance constitutes an environmental confound in identical-fraternal twin comparisons: "Most probably, identical twins are treated more alike because they look and act more alike."[5] Contrary to the authors' opinion, this statement calls the EEA into question because the more similar appearance-related treatment of twins, which is an environmental effect, suggests that the twin method is unable to disentangle potential genetic and environmental influences on psychological trait variation.

---

1. Loehlin & Nichols, 1976, pp. 8-9.
2. Ibid., p. 87.
3. Ibid., p. 94.
4. Ibid., p. 47.
5. Ibid., p. 87.

Kendler cited Loehlin and Nichols's study in support of the EEA because, in his view, it found that "the similarity of parental treatment did not systematically relate to twin similarity for cognitive abilities, personality or vocational interests."[1] However, the study's numerous problems invalidate claims that it supports the equal environment assumption.

THE "FOURTH METHOD"

The fourth (or "reverse-zygosity") method assesses twins whose zygosity status has been misidentified by their parents or by the twins themselves. These misidentified twins are then compared to correctly identified twins in order to test the effects of "true" versus "perceived" zygosity. Kendler cited several studies in support of the idea that true zygosity is the best predictor of twin correlations on behavioral and cognitive ability measures.[2]

The first study to test the reverse-zygosity hypothesis was published by *Sandra Scarr in 1968*. While acknowledging that identical twins are treated more similarly than fraternal twins,

> Not all parents of twins are correct about their twins' zygosity, however, and these parents offer a critical test of environmental bias in twin studies. By examining the cases of parents who are *wrong* about their twins' zygosity, it is possible to separate parental reactions to similarities and differences based on *genetic relatedness* from parental behaviors which arise from their *belief* that their twins should or should not be similar.[3]

Scarr interviewed the mothers of 52 twin pairs (identical = 23, fraternal = 29) and administered them the Adjective Check List and the Vineland Social Maturity Scale. In eleven cases (identical = 4, fraternal = 7), there was a discrepancy between the twins' true zygosity (as determined by blood-grouping) and the mothers' perceived zygosity of their twins. Genetic theory holds that behavioral resemblance is not influenced by the twin pair's (or their parents') perception of their true zygosity. The environmental position holds that such beliefs might affect behavioral resemblance. A "true zygosity effect" denotes correlations attributed to genes. A "perceived zygosity effect" denotes correlations due to a belief, and is therefore an environmental effect. Scarr concluded that "the trends are clear" that "genetic relatedness of the twins determines the simi-

---

1. Kendler et al., 1994, p. 580.
2. It is difficult to determine the originator of the reverse-zygosity hypothesis, but the idea was put forward by Seymour Kety as early as 1959, when he proposed "a comparison of the concordance rates in monozygotic twins whose zygosity had been mistakenly evaluated by the twins themselves and by their parents and associates." See Kety, 1959, p. 1594.
3. Scarr, 1968, p. 38. (Emphasis in original)

larity of parental treatment," even though "the numbers are too small to yield statistical significance."[1] Statistically speaking, however, there can be no "trends" without significant results.

Scarr provided a table showing mean difference scores on selected Adjective Check List and Vineland Social Maturity scales for the four classifications of twins (true identical, misidentified identical, true fraternal, misidentified fraternal). However, she included the results from only 3 of the 26 Adjective Check List personality scales, which she justified on the grounds that only these three scales showed significantly greater identical versus fraternal twin intraclass correlations. Scarr should have reported all scales because, from either a genetic *or* an environmental perspective, we would expect differences between identical and fraternal scores on all of the scales.

Scarr's results showed that misidentified fraternal twins' mean difference scores were not significantly different from correctly identified identicals on two of the three scales, which is consistent with environmental expectations. Genetic theory predicts that misclassified fraternal twins would be as different as correctly classified fraternals, when they clearly were not. On the Vineland Social Maturity measure, misidentified identical pairs were as different as correctly identified fraternals, which is also consistent with environmental expectations. Although other comparisons were consistent with genetic expectations, Scarr wrote, "venturing out on the slim branch of small numbers.... [the results] suggest that beliefs about zygosity also have an effect on MZ pairs," although these differences are "not as potent as the critics charge."[2] We now turn to Scarr's most substantial attempt to study the effects of true and perceived zygosity on twin correlations.

Scarr and Carter-Saltzman's 1979 study, entitled "Twin Method: Defense of a Critical Assumption," utilized a sample of 400 same-sex, 10- to 16-year-old twin pairs recruited from the Philadelphia metropolitan area.[3] They found that 41 pairs misperceived their zygosity status (20 true fraternals thought they were identical, and 21 true identicals thought they were fraternal).

As proponents of the "twins create their environment theory" (see Chapter 1), Scarr and Carter-Saltzman recognized the "overwhelming" evidence that identical twins experience a greater environmental similarity than fraternal twins.[4] However,

> If genetic similarity were the sole determinant of behavioral likeness, then DZ twins who believe themselves to be MZs will be no more alike than other DZs, and MZs who mistake themselves for DZs will be no more different than other MZs. If, however, beliefs about zygosity determine the extent to which cotwins are behav-

---

1. Ibid., p. 38.
2. Ibid., p. 40.
3. Scarr & Carter-Saltzman, 1979.
4. Ibid., p. 528.

iorally similar, then DZ twins who believe they are MZs will be as similar as true MZs. Likewise, MZs who believe they are DZs will be as different as true DZs.[1]

They divided their sample of 400 twin pairs into four groups: identicals who correctly believed themselves to be identical; identicals who incorrectly believed they were fraternal; fraternals who correctly believed themselves to be fraternal; and fraternals who incorrectly believed they were identical. "This study," they wrote, "tests the hypothesis that actual zygosity, not the twins' (correct or erroneous) beliefs about it, determines the degree of behavioral similarity between the cotwins."[2]

In a table provided by the investigators, fraternal twins who believed they were identical had a lower mean difference score than true identicals on the two personality measures.[3] Additionally, the mean difference score of identicals who mistakenly believed they were fraternals (0.85) does not appear to differ significantly from the true fraternals (0.93). These results are consistent with environmental expectations that fraternals who believed themselves to be identical would have a mean difference score similar to true identicals, and that identicals who believed themselves to be fraternal would have difference scores comparable to true fraternals. As in 1968, Scarr found that perceived zygosity effects do influence twin concordance rates.[4]

The investigators ended with an endorsement of the EEA and the twin method:

> The critical assumption of equal environmental variance for MZ and DZ twins is tenable. Although MZ twins generally experience more similar environments, this fact seems to result from their genetic similarities and not to be a cause of exaggerated phenotypic resemblance.[5]

Their critical *error*, and that of twin researchers ever since, is that the *reason* identical twins experience more similar environments than fraternals has no bearing on the validity of the twin method. Contrary to the views of most contemporary twin researchers, the only relevant question is *whether*, as opposed to *why*, identical and fraternal environments differ.[6]

Munsinger and Douglass gave language ability tests to 37 identical and 37 same-sex fraternal 3- to 17-year-old pairs, along with several of their siblings.[7] Following Scarr, they identified eight identical pairs whose parents misclassified

---

1. Ibid., p. 529.
2. Ibid., p. 529.
3. Ibid., p. 533.
4. Using blood group analysis, Scarr and Carter-Saltzman found that fraternal twins who believed they were identical were more similar genetically than other fraternal pairs. They concluded that "Beliefs about zygosity were highly related to the genetic similarity of the DZ twins" (Scarr & Carter-Saltzman, 1979, p. 532).
5. Scarr & Carter-Saltzman, 1979, p. 541.
6. Joseph, 2004b.
7. Munsinger & Douglass, 1976.

as fraternal, and 12 fraternal pairs whose parents misclassified as identical. They presented a table showing the twins' intraclass correlations for the language ability measures according to mistaken or true zygosity. Although some comparisons suggested true zygosity effects, it is not clear if any of the correlational differences were significantly different. True identicals who were thought to be fraternals did not correlate significantly higher than true fraternals, which is consistent with environmental expectations. Munsinger and Douglass concluded that perceived zygosity effects had a small effect on children's language skills. However, their sample of misidentified twins was small, and perceived zygosity status was based on the *parents'* beliefs, not those of the twins themselves. Considering that some twins were over 13 years old, it is surprising that their own beliefs were not also determined.

Adam Matheny reported IQ scores for 101 identical and 71 fraternal pairs in the Louisville Twin Study, given when the twins were three-years-old. The parents of 18 identical pairs mistakenly classified their twins as fraternal, and 7 fraternal pairs were misclassified as identical. After analyzing the data, Matheny concluded that "parental error in assigning zygosity was not systematically related to difference in IQ."[1] Although Kendler cited Matheny's study in support of the EEA for twin studies of psychiatric disorders, one wonders how the results of three-year-old children's testing for IQ could provide much evidence in defense of his thesis.

ADHD twin researchers *Goodman and Stevenson* studied 102 identical pairs, 25 of whom their parents misidentified as fraternal. The authors referred to these latter pairs as "unrecognized" identical twins. Of the 111 fraternal pairs, 12 were misidentified as identical by their parents. Measures were administered to the twins' mothers, fathers, and teachers. Twins were separated into three groups: recognized identicals (whose parents were correct about their zygosity), unrecognized identicals, and same-sex correctly recognized fraternals. In addition to the researchers' goal of assessing genetic influences on attentiveness and hyperactivity, they also assessed the relationship between the effects of zygosity and expectancy on identical twins' correlations for the two behaviors. By expectancy effects, the investigators accounted for the possibility that parents' and teachers' behavioral ratings are influenced by how they *expect* identical and fraternal pairs to act. For the attentiveness scales, Goodman and Stevenson found an "apparent absence of expectancy bias," but on the three hyperactivity (HA) scores, "the intraclass correlations...provide evidence both for a substantial heritability and for substantial expectancy effects."[2]

Another study Kendler cited was given as "Kendler et al., 1995," but there is no listing of any 1995 study by Kendler et al. in the survey under review.[3]

---

1. Matheny, 1979, p. 159.
2. Goodman & Stevenson, 1989b, p. 696.
3. Hettema et al., 1995.

Kendler may have been referring to a 1994 study by *Kendler, Neale, et al.*[1] I review this investigation along with a 1993 study by *Kendler and his colleagues* because of the similarities between them.[2] In the 1993 investigation, Kendler et al. assessed 1,030 pairs of twins for five psychiatric diagnoses: major depression, generalized anxiety disorder, phobia, bulimia, and alcoholism. The investigators' 1994 study used a smaller sample from the same pool (844 pairs), and found 69 true identical pairs whose mothers had misidentified them as fraternal. In the 1993 study, the researchers identified 64 misclassified pairs on the basis of twin self-reports. In both investigations, they found no evidence of significant perceived zygosity effects. However, neither study provided concordance rates for twins in the correctly and incorrectly perceived zygosity groups. Instead, the investigators presented path analysis diagrams. Path analysis was used in this instance as a test of the "impact of perceived zygosity on twin resemblance for psychiatric disorders."[3] The problems with using this technique as a method of testing the EEA have been discussed elsewhere,[4] and Kendler's model is based on circular reasoning, since he *assumes* genetic influences in order to measure genetic influences. According to Kendler, "Path analysis, which assumes a polygenic-multifactorial model, is a method that has considerable power at discriminating genetic from cultural transmission."[5] And elsewhere he wrote that, according to his model, "Residual common environment is, by definition, assumed to be perfectly correlated in all twin pairs."[6] In Kendler and colleagues' model-fitting procedures, they assume that environmental exposure is "random and is therefore uncorrelated among relatives."[7]

Another theme in Kendler's publications is that researchers can use computer-based statistical procedures to test the EEA's validity. Thus, Kendler has recommended that "psychiatric twin researchers would be well advised to continue to test the EEA rather than to assume its validity."[8] It is not necessary to engage in a detailed discussion of the technical aspects of the test and how it is performed, because the question is more theoretical and empirical. The EEA is not true or false for any particular set of psychological or psychiatric data; rather, it is a statement about the nature of interactions between twins and their relationship to each other and to their social and physical environments, as well as their treatment similarity. The validity of the EEA in psychiatric twin research can be determined only by empirical, theoretical, and sociological data pertaining to the nature of twinship in general, and not for a particular study.

---

1. Kendler et al., 1994.
2. Kendler, Neale, et al., 1993a.
3. Ibid., p. 24.
4. Pam et al., 1996.
5. Kendler, 1987, p. 709.
6. Kendler, Neale, et al., 1993, p. 24.
7. Kendler & Eaves, 1986, p. 279.
8. Kendler et al., 1994, p. 588.

The fifth method examines whether twins' physical resemblance is corre-lated with their personality resemblance or concordance for various psychiatric disorders. According to Kendler,

> Resemblance in twins may be influenced by the similarity with which they are treated by their social environment, which is a result of their degree of physical resemblance. If this is the case, controlling for zygosity, physical similarity of twin pairs should be correlated with trait similarity. Three studies have examined this question...and none suggested that twin resemblance was substantially influenced by physical similarity.[1]

In fact, we could interpret every twin study finding a significantly higher identical versus fraternal concordance rate as evidence supporting the idea that the "physical similarity of twin pairs should be correlated with trait similarity," since identical twins are more similar in appearance than fraternal twins. But Kendler and colleagues apparently do not see this. And, we have already seen that Loehlin and Nichols conceded the likelihood that "identical twins are treated more alike because they look and act more alike."

Kendler cited an unpublished 1982 study by *Kendler and Robinette*, who reviewed the charts of 164 identical twin pairs in the National Academy of Sci-ences — National Research Council twin sample.[2] This study was supposed to have found "no correlation between degree of physical similarity and concor-dance rates for schizophrenia"[3]:

> When the schizophrenic monozygotic twin-pairs are divided on the basis of their degree of physical similarity (based on hair and eye color, height, and weight at induction), concordance for schizophrenia is no higher in those who were versus those who were not very physically similar.[4]

Kendler and Robinette, who did not personally investigate these twins, ranked them on the basis of information they obtained from US Armed Forces induction charts. In fact, several twin researchers have shown that identical twins' eye and hair color are exactly the same over 90% of the time.[5] In 1975, Cohen and colleagues asked mothers to describe the physical resemblance of their identical twins: "To what extent are the twins similar at this time for the following features?" For hair and eye color, 99.5% of the mothers answered that they were "exactly similar," and 99.5% rated their twins' height as "exactly similar" or "somewhat similar."[6] Newman and associates found that the mean

---

1. Kendler, 1993, p. 906.
2. Kendler, 1983.
3. Ibid., p. 1415.
4. Kendler & Robinette, 1983, p. 1557.
5. Cohen et al., 1975; Newman et al., 1937.
6. Cohen et al., 1975, p. 1374.

identical twin difference in height was a mere 0.7 inches.[1] Yet Kendler and Robinette would have us believe that they could make a meaningful distinction between identical pairs on the basis of eye color, hair color, and height.

A more meaningful demonstration of the effects of physical resemblance on concordance for schizophrenia would compare concordance rates among same- and opposite-sex *fraternal twins*. The fact that same-sex fraternal twins were significantly more concordant than opposite-sex fraternals in three major schizophrenia twin studies, and that the pooled schizophrenia same-sex fraternal rate is more than double the pooled opposite-sex rate (see Table 6.4), suggests that physical appearance does have an important effect on twin concordance rates.[2] Unfortunately, Kendler rarely mentions these results in his numerous review articles and textbook chapters on the genetics of schizophrenia.

The 1995 article by *Kendler and colleagues* used as the reference point of this chapter reports the authors' own Virginia Twin Registry-based study on the relationship between the similarity of physical appearance and concordance for depression, generalized anxiety disorder, phobia, alcoholism, and bulimia.[3] Physical similarity was determined by blinded ratings of color photographs of 882 adult twin pairs, and was rated on a seven-point scale. The researchers found no evidence for a significant effect of physical similarity on twin concordance for all diagnoses except bulimia, where a relationship was found. According to the authors, "Correlations between physical similarity score and diagnosis are not reported because they are not significant." In their model-fitting calculations, the investigators reduced the seven-point scale to three-point and two-point measurement scales, further obscuring possible environmental effects by combining ratings into larger groups. The verdict on this study is similar to that of most other Virginia Twin Register studies published by the Kendler group: The researchers provided insufficient information from which to make an independent evaluation of the data and of the authors' conclusions.

Plomin and associates' 1976 study examined the relationship between twins' physical resemblance and personality correlations.[4] They studied ninety-five pairs aged 2 to 6 by matching parental ratings of twins' personalities with their degree of physical resemblance. The researchers found no significant relationship between the degree of physical resemblance and personality correlations, leading them to conclude that "similarity of appearance does not seem to lead to similarity of personality, and hence does not represent a serious source of bias in twin studies of personality."[5] The physical similarity rating was based on

1. Newman et al., 1937.
2. Kallmann, 1946; Rosanoff et al., 1934; Slater, 1953. Also see discussion in Joseph, 2004b, Chapter 6.
3. Hettema et al., 1995.
4. Plomin et al., 1976.
5. Ibid., p. 50.

a four-question survey asking mothers to rate the "confusability" of their twins, but the actual questions were not reported. In any event, the fact that some parents were better than others at telling apart their 2- to 6-year-old identical twins is trivial in comparison to the general striking physical resemblance of identical twins. And Plomin and associates were probably correct when they commented that "self report data of older twins might yield different results."[1]

Matheny and colleague's 1976 twin sample consisted of 121 identical and 70 fraternal pairs (ages 3.5 to 13).[2] They did not provide twins' mean age, which is an important omission given that differences between 3-year-old and 13-year-old children are substantial. For personality measures, Matheny et al. calculated rank-order correlations between twins' physical resemblance and scores on the Children's Personality Questionnaire (CPQ). Of the 14 CPQ scales, two were significantly correlated in the direction of environmental expectations, one was significantly correlated in the genetic direction, and the remaining scales showed no significant correlation. The investigators concluded that "the perceived physical similarity of same-sex twins is not a significant determinant of behavioral outcome."[3] As Richardson and Norgate pointed out, however, "the unreliability of the study's difference measures casts doubt upon its conclusions."[4]

SUBSEQUENT EEA TEST PUBLICATIONS

I now briefly discuss six EEA test studies published after Kendler's 1995 review (I use italics to identify these studies as well).[5]

In 1997, *Michele LaBuda and colleagues* assessed the impact of the EEA on substance abuse disorders. They found that "identical twins reported significantly closer relations than fraternal twin pairs," and that closeness had an effect on "drug abuse and/or dependence," but not on alcohol dependence.[6] Because identical-fraternal differences remained significant even after supposedly controlling for closeness, they concluded that their results "support the validity of the equal environment assumption in twin studies of substance use disorders."[7]

In *Kendler and Gardner's 1998* investigation, identical twins were found to have socialized together more frequently as children, and to have had their simi-

---

1. Ibid., p. 51.
2. Matheny et al., 1976.
3. Ibid., p. 349.
4. Richardson & Norgate, in press.
5. The study of Eaves et al. (2003) discusses problems in testing for the EEA in individual studies, and is not itself a study testing the EEA.
6. LaBuda et al., 1997, p. 155.
7. Ibid., p. 163.

larities emphasized more by parents, teachers and others.[1] There was no significant difference in identicals' and fraternals' reported similarity of childhood
treatment. The authors concluded that these factors did not significantly predict
twin resemblance for a number of psychiatric disorders.

Klump and colleagues' 2000 EEA test study assessed the relationship between
physical resemblance and eating attitudes and behaviors. Like most twin
researchers I discuss in this chapter, they defined the EEA in the trait-relevant
sense:

> The Equal Environments Assumption (EEA) has been continually challenged
> by twin study critics. The EEA posits that both members of a monozygotic (MZ)
> twin pair are as likely to be treated the same by the environment as both members
> of a dizygotic (DZ) twin pair with respect to environmental influences that are of
> etiologic importance to the trait under study.[2]

Based on this definition, they concluded that because they found no significant associations between physical resemblance and eating attitudes, their
findings "provide support for the EEA in twin studies of eating attitudes and
behaviors."[3]

The German team of *Peter Borkenau and colleagues* published their EEA test
study in 2002. From their large twin sample (793 pairs), they found that identicals reported more similar experiences than fraternals and were also more
similar in personality. After assessing twins' reported treatment similarity, they
concluded, "Across twin pairs...treatment similarity was unrelated to personality resemblance, except in the combined group of MZ and DZ twins."[4]
However, if the combined group of twins showed that treatment similarity *was*
related to personality resemblance, it appears that their results failed to support
the EEA.

Nikole Cronk and her colleagues published a study in 2002 assessing the EEA
as it relates to "problem behavior" among female adolescent twins.[5] This may be
the only EEA-test study to have assessed the assumption in two different ways
(the second and fourth methods). The researchers concluded that twin resemblance on environmental similarity measures was not "strongly or consistently"
related to behavioral problems. In addition, they found that mothers' correct or
incorrect assessment of their twins' zygosity did not significantly alter the
results which, they concluded, "lend support for the validity of the EEA."[6] A
major problem with this study, however, is that the investigators relied on
mothers' recollections in diagnosing the four behavioral problems (separation

---

1. Kendler & Gardner, 1998.
2. Klump et al., p. 51.
3. Ibid., p. 51.
4. Borkenau et al., 2002, p. 261.
5. Cronk et al., 2002.
6. Ibid., p. 829.

anxiety disorder, ADHD, oppositional defiant disorder, and conduct disorder). As we have seen, parental recall measures are not reliable, and are subject to rating bias due to expectancy effects. According to the investigators, "Retrospective reports of lifetime symptoms are subject to selective memory problems and relatively low test-retest reliability relative to reports about current functioning."[1]

Finally, in what may be the only EEA-test study performed by researchers who questioned the validly of the twin method and who lacked professional or philosophical allegiances to behavior genetics and twin research, sociologists *Allan Horwitz and colleagues* published their results in 2003. They pointed out that the EEA is "fundamental to interpretations of the findings from twin studies," and observed that previous EEA test studies "do so only partially, usually examining only one or two broad measures of the environment."[2] Horwitz and colleagues analyzed data from 414 twin pairs (230 identical, 187 fraternal) and assessed the relationship between several environmental variables (which included twins' peer networks), and depression and "alcohol use and abuse." They concluded that "measures of the social environment sometimes reduce or eliminate apparent genetic affects," suggesting that "past twin studies could overstate the effect of genetic influences because some similarities in behavior among monozygotic compared to dizygotic twins stem from social influences."[3]

CONCLUSION

The body of research comprising the five ways that the equal environment assumption of the twin method has been tested does little to counter the views of critics, who have argued that genetic conclusions based on twin method data are confounded by environmental factors. The EEA test literature is more about twin researchers attempting to convince others of the validity of their methods than it is an objective assessment of the EEA.

Furthermore, the very notion that the EEA can be "tested" is faulty, since its validity can be determined only by looking at the larger picture of how identical and fraternal pairs exist in, and interact with, the social and familial environments in which they live. The authors of the EEA test studies looked at small segments of twins' experiences and attempted to generalize their findings to support the EEA and the twin method. In doing so, they lost sight of the fact that — like family studies — the twin method is unable to disentangle theorized genetic and environmental influences on psychopathology.

---

1. Ibid., p. 835.
2. Horwitz et al., 2003, p. 113.
3. Ibid., p. 111.

We have seen that the validity of drawing genetic inferences from iden-tical-fraternal concordance rate differences depends on the assumption that identical twins share their environment, experience similar treatments, and share a psychological bond to the same extent as same-sex fraternal twins. However, the flawed and narrowly focused EEA test literature provides little support for the EEA, regardless of how twin researchers have defined it. Moreover, any false theory or assumption can be "tested" and upheld as long as the "testers" (1) determine the hypotheses to be tested, (2) perform the tests, (3) draw the conclusions, and (4) remain blind to obvious real-world refutations of their conclusions.

Thus, despite its current widespread acceptance, the twin method remains an environmentally confounded research method whose results provide no sci-entifically acceptable evidence in support of genetic influences on psychiatric disorders or psychological trait variation.

Chapter 10. Bipolar Disorder and Genetics

According to the authors of a 2004 APA textbook chapter, bipolar affective disorder (BPD) "may have the strongest genetic component of any mental disorder,"[1] while other mainstream commentators have recognized "the vast areas of ignorance that still exist in our understanding of the causes of bipolar disorder...."[2] Generally speaking, the hereditary basis of BPD (also known as "manic-depressive disorder") is rarely if ever questioned in mainstream publications. As with autism, however, a closer look at the evidence reveals little support for genetic theories, which, as should by now be expected, are based on the results of family, twin, and adoption studies. And as we will see in Chapter 11, researchers have thus far failed, after more than three decades of research, to identify the genes they believe predispose people to bipolar disorder.

Because no neuropathological abnormalities have been discovered,[3] bipolar disorder is diagnosed, like most psychiatric disorders, on the basis of clinical observation. According to the DSM-IV-TR, "there appear to be no laboratory features that are diagnostic of Bipolar I Disorder or that distinguish Major Depressive Episodes found in Bipolar I Disorder from those in Major Depressive Disorder or Bipolar II Disorder."[4]

BPD is grouped in the "mood" or "affective" disorder diagnostic category, which includes major depressive disorders and bipolar disorders. The hallmark of bipolar disorder is the "manic episode," which according to the DSM-IV-TR includes symptoms such as "inflated self-esteem or grandiosity," "decreased need for sleep," being "more talkative than usual," "racing thoughts," "distractibility," "increase in goal-directed activity," and "excessive involvement in pleasurable

---

1. Weller et al., 2004, p. 454.
2. Fawcett, 2005, p. 2.
3. Cavanaugh, 2004.
4. APA, 2000, p. 384.

activities that have a high potential for painful consequences." Furthermore, the episode "is sufficiently severe to cause marked impairment in occupational functioning or in usual social activities or relationships with others, or to necessitate hospitalization to prevent harm to self or others, or there are psychotic features."[1] Most people diagnosed bipolar have a history of depressive episodes. In contrast, *unipolar* depression does not alternate with mania, nor do people diagnosed unipolar have a history of manic ("Bipolar I") or hypomanic ("Bipolar II") episodes, the latter being characterized by less severe symptoms than manic episodes. The lifetime prevalence of bipolar disorder is generally given as 1%.[2]

ORIGINS

The first modern description of bipolar disorder was by the French psychiatrist Jean-Pierre Falret, who in the mid-1850s developed the concept of "folie circulaire" (circular madness).[3] Emil Kraepelin is the founder of the "manic-depressive insanity" (MDI) concept, which in the early 20th century he described as consisting of "periodic and circular insanity," "simple mania," "melancholia," "amentia," and other changes in mood. For Kraepelin, these states were "manifestations of a single morbid process."[4] Thus, he viewed them as expressions of the same disorder. Regarding mania in particular, Kraepelin saw this behavior as one manifestation of manic-depressive insanity, not as a feature distinguishing BPD from unipolar depression, as it is viewed in contemporary diagnostic practice.[5]

In Kraepelin's view, dementia praecox (schizophrenia) and manic-depressive insanity were distinctly different disorders. As a commentator noted, "Bipolar affective disorder and schizophrenia constitute the twin pillars of the classically defined psychosis."[6] A major reason that Kraepelin decided to separate these conditions was his belief that they had differing genetic origins, based on his interpretation of his patients' family histories. Although lacking systematic family, twin, or adoption data, Kraepelin believed that he could demonstrate "hereditary taint" in the vast majority of the cases,[7] and that "compared to innate predisposition external influences only play a very subordinate part in the causation of manic-depressive insanity." Further evidence supporting the importance of heredity was, for Kraepelin, "the powerlessness of our efforts to

1. Ibid, p. 362.
2. Belmaker, 2004; Tohen & Angst, 2002.
3. Angst & Sellaro, 2000.
4. Kraepelin, 1976, p. 1.
5. Goodwin & Jamison, 1990, p. 4.
6. Cavanaugh, 2004, p. 203.
7. Kraepelin, 1976, p. 165.

cure," which "convince us that the attacks of manic-depressive insanity may be to an astonishing degree *independent of external influence*."[1]

In late 1950s several investigators, working independently on different continents, began to argue that, as opposed to Kraepelin, unipolar and bipolar disorders were distinct entities.[2] This position became more accepted over time, culminating in the official separation of the two diagnoses in DSM-III, published in 1980. This distinction has remained in the three subsequent revisions of the DSM (DSM-III-R, DSM-IV, and DSM-IV-TR).

THE SHAKY CASE FOR GENETICS

Family Studies

Mood disorders, like most types of behavior, tend to run in families. We have seen, however, that this finding is consistent with both genetic and environmental theories of causation. Basing their work on Kraepelin's unitary MDI concept, the authors of the early family studies (published before the 1960s), carried out mainly in Europe, did not distinguish between the unipolar and bipolar diagnoses. Moreover, these early studies failed to use structured diagnostic criteria, structured interviews, blind diagnoses, control groups, or standardized research techniques.[3]

Family studies performed since the 1960s making the unipolar/bipolar distinction have found a mean BPD rate of 7% among the first-degree biological relatives of people diagnosed with BPD, which is far higher than the population expectation or the control group rate.[4] Unfortunately, as with other disorders, some genetic researchers conclude that the results of BPD family studies suggest the operation of genetic factors. For example, Kalidindi and McGuffin wrote in 2003 that "Bipolar II disorder...is increased in the families of Bipolar I probands compared with the general population...indicating a genetic component to the milder Bipolar II disorder also."[5]

Turning to the popular media, the following exchange took place on the June 12, 2005, *Larry King Live* program on Cable News Network (CNN). A caller asked BPD expert Kay Redfield Jamison about genetics:

> CALLER: "I have a question. I am 52, and have probably suffered from depression from the age of late 20s on. I have a 29-year-old son, [who] as a young child was

---

1. Ibid., p. 181. Emphasis in original.
2. Akiskal, 2002.
3. Tsuang & Faraone, 1990.
4. Jones et al., 2003.
5. Kalidindi & McGuffin, 2003, p. 482.

diagnosed with ADHD, who has recently, within the past two years, been diagnosed with bipolar disorder. My question is, is this inherited or hereditary?"

JAMISON: "Well, bipolar illness is certainly hereditary. It's genetics, as also ADHD is genetic. We've known for hundreds and hundreds and hundreds of years that it is genetic, and there is a lot of actually very exciting research going on now."[1]

Maybe we've known for "hundreds and hundreds of years" that manic behavior runs in families, but twin and adoption studies have only been around for a few decades. Jamison unfortunately equated "running in the family" with "it's genetic," and passed this misinformed opinion on to millions of unwitting television viewers.

According to Tsuang and Faraone, in their 1990 book *Genetics of Mood Disorders*, "Overall, family studies of bipolar probands strongly support the hypothesis that their first-degree relatives are at greater risk of bipolar disorder than is the general population."[2] The family method constitutes, for Faraone and Tsuang, the first link of a "chain of psychiatric genetic research," which culminates in the identification of predisposing genes.[3] Although Faraone and Tsuang recognized that "'familial' and 'genetic' are not synonymous," they wrote that after "establishing that a disorder runs in families, the psychiatric geneticist asks: 'What are the relative contributions of genes and environment as causes of mental illness?"[4] To answer this question, they looked to twin and adoption studies. However, they skipped an important stage at this point, since the appropriate question to ask, after finding that a condition is familial, is *whether* genes play a role in the familial clustering or transmission. By arguing that the discovery of familial clustering of a disorder leads to an assessment of the "relative contribution" of genes and environment, they implied that family studies have already established a role for genetic influences.

Twin Studies

Like family research, twin studies performed prior to the 1960s utilized Kraepelin's broad manic-depressive insanity concept. Thus, while psychiatric geneticists view these studies as providing evidence for the genetic basis of affective disorders in general, their authors did not separate the bipolar and unipolar forms of affective disorder.[5]

A handful of twin studies have looked specifically at bipolar disorder. As expected, they found higher concordance among identical versus fraternal twins. Their results are seen in Table 10.1.

---

1. *Larry King Live*, "Panel Discusses Depression." Cable News Network program. Retrieved on 6/25/05 from: http://transcripts.cnn.com/TRANSCRIPTS/0506/12/lkl.01.html.
2. Tsuang & Faraone, 1990, p. 47.
3. Faraone, Tsuang, & Tsuang, 1999; Tsuang et al., 1994.
4. Faraone, Tsuang, & Tsuang, 1999, p. 11.
5. Tsuang & Faraone (1990) provided a detailed review of these studies.

Table 10.1

PAIRWISE BIPOLAR DISORDER (BPD) TWIN CONCORDANCE RATES

Authors	Year	Identical	%	Fraternal	%
Allen et al.	1974	1/5	20	0/15	0
Bertelsen et al.	1977	32/55	62	9/52	8
Torgersen	1986	3/4	75	0/6	0
Kendler et al. [a]	1993	5/10	39	1/19	5
Cardno et al. [b]	2002	8/22	36	2/27	7
Kieseppä et al.	2004	3/7	43	1/18	6
POOLED RATES		52/103	50%	13/137	9%

[a] Kendler, Pedersen, et al. (1993).
[b] In an expanded sample from this study, McGuffin et al. (2003) reported probandwise BPD rates of 40% identical, 5.5% fraternal.

The pooled BPD concordance rate is roughly 50% identical and 10% fraternal, a ratio of about 5:1. Researchers and reviewers invariably conclude in favor of genetic factors explaining these differing rates. While the original researchers at times mentioned the equal environment assumption, the authors of BPD review articles rarely mention it. The following quotations exemplify how reviewers interpret BPD twin data:

"Twin studies are consistent with family studies in suggesting that genetic factors play a substantial role in the mood disorders."[1]

"Family, twin, and adoption studies have provided strong evidence for a genetic etiology in BPD."[2]

"Twin studies of mood disorders reveal that genetic factors have a far greater etiologic role in bipolar disorder than in nonbipolar major depression."[3]

"The heritability of bipolar disorder as evidenced in twin studies was estimated as 0.59%.... Given this overwhelming evidence of a major genetic component, bipolar disorder became the first psychiatric disorder which was submitted to linkage analysis..."[4]

"Studies of twins suggest that the concordance for bipolar illness is between 40 percent and 80 percent in monozygotic twins and is lower (10 to 20 percent) in dizygotic twins, a difference that suggests a genetic component to the disorder."[5]

"Bipolar disorder...is a chronic psychiatric disorder with a worldwide lifetime prevalence of 0.5%-1.5% and a predominantly genetic etiology based on twin study data."[6]

---

1. Ibid., p. 91.
2. Potash & DePaulo, 2000, p. 8.
3. Merikangas et al., 2002, p. 459.
4. Maier, 2002, p. 39.
5. Belmaker, 2004, p. 478.

"Heredity appears to be the only etiological factor with a reasonably firm base, as evidenced by family, twin, and adoption studies. Heritability estimates exceed 50%."[1]

Early BPD genetic researchers placed their faith in the traditional definition of the EEA, which states, without qualification, that identical and fraternal environments are equal (see Chapter 1). In a 1976 review article, for example, Elliot Gershon and colleagues wrote that affective disorder twin studies, which they believed "strongly suggest" the importance of genetic influences, "assume that the intrapair differences of environment are the same for MZ and DZ twins."[2] And, according to psychiatric genetic researcher Julien Mendlewicz,

> The twin method allows comparison of concordance rates for a trait between sets of monozygotic (MZ) and dizygotic (DZ) twins. Both types of twins share a similar environment, but they are genetically different.[3]

Samuel Barondes's 1998 *Mood Genes: Hunting for Origins of Mania and Depression* contains an enthusiastic description of the search for predisposing "mood genes." Barondes saw twins as a "godsend for behavior geneticists," since "assuming that the shared environments of sets of identical twins and sets of fraternal twins are roughly equal (which appears to be the case), comparing their degree of similarity gives the indication of the relative contributions of nature and nurture."[4] Elsewhere, Barondes wrote that "the only major distinction between...identical and fraternal twins is their different degrees of genetic similarity."[5] The problem with Barondes's conception of the equal environment assumption, the validity of which he alludes to only parenthetically, is that it has been rejected not only by critics, but by most twin researchers as well (see Chapters 1 and 9). Yet, he describes a search for genes based largely on the acceptance of the widely-rejected traditional definition of the equal environment assumption. Similarly, another group of commentators, among whom was the Director of the US National Institute of Mental Health, wrote that when the identical twin concordance rate is higher than the fraternal rate, "it is strongly suggestive that the underlying cause of the illness has a genetic component because it is reasonable to assume that siblings living together are subject to the same environmental influences."[6]

Thus it is "reasonable," or "appears to be the case," to state without qualification that identical and fraternal twins experience the same environments. Yet,

---

6. Segurado et al., 2003, p. 50.
1. Baron, 2002.
2. Gershon et al., 1976, p. 233.
3. Mendlewicz, 1988, p. 198.
4. Barondes, 1998, p. 81.
5. Barondes, 1999, p. 129.
6. Cowan et al., 2002, p. 41.

there exists little evidence in support if this claim, and a mountain of evidence and common sense against it (including the EEA-test literature I discussed in Chapter 9).

Like other areas of psychiatry, some aspects of BPD genetic research have been misrepresented by influential secondary sources. Two examples are edited textbooks on genetic theories in psychiatry that appeared in 2003. *Psychiatric Genetics and Genomics*, edited by McGuffin, Owen, and Gottesman, included a chapter on the genetics of affective disorders by Jones et al., and Kalidindi and McGuffin had a similar chapter in Plomin and colleagues' *Behavioral Genetics in the Postgenomic Era*. According to Jones et al., Gottesman and Bertelsen performed a 1989 study of the offspring of discordant bipolar identical twins, just as they had with the offspring of discordant schizophrenia pairs:

> An interesting and illuminating approach to the study of bipolar twins is to fol-low the offspring of the unaffected members of discordant MZ pairs. Gottesman and Bertelsen (1989) found an elevated risk of bipolar illness in this group indistin-guishable from that in the offspring of individuals affected by bipolar disorder.[1]

Although Jones et al. cited Gottesman and Bertelsen's 1989 article on dis-cordant schizophrenia twin pairs[2] (discussed in Chapter 6 of this book, and in Chapter 6 of *The Gene Illusion*), Gottesman and Bertelsen, in fact, did not study discordant bipolar pairs.

Unfortunately, Kalidindi and McGuffin reproduced this mistake in their chapter:

> The incomplete concordance in MZ twins indicates that nongenetic factors play a role in the liability to bipolar disorder. Thus, discordant MZ pairs might arise because the affected proband has a nongenetic type of disorder. Gottesman and Ber-telsen (1989) tested this by studying the offspring of discordant bipolar twins. They found that in the offspring of the unaffected MZ twins, the risk of bipolar disorder was increased to the same degree as in the offspring of the affected proband.[3]

I do no know why Gottesman and Bertelsen were erroneously credited in 2003 with studying the offspring of discordant bipolar pairs. To my knowledge, no one prior to this had ever made such a claim, and according to Bertelsen himself, Gottesman "suggested that I should write a similar paper confirming unexpressed genotypes for manic-depressive disorders."[4] However, although he was able to obtain some data for discordant unipolar pairs, "the majority of the bipolar pairs were concordant, almost reaching 100% if broad concordance is considered. Offspring analysis therefore is not possible for the bipolar discordant pairs."[5]

---

1. Jones et al., 2003, p. 221.
2. Gottesman & Bertelsen, 1989a.
3. Kalidindi & McGuffin, 2003, p. 484.
4. Bertelsen, 2004, p. 129.
5. Ibid., p. 131.

I am not suggesting that anyone has intentionally provided false information. Still, we saw in Chapter 5 that inaccurate accounts are picked up by secondary sources and take on a life of their own. One can only hope that future textbook authors do not rely on sources such as those documented above, and avoid further references to Gottesman and Bertelsen's non-existent study of the "offspring of discordant bipolar twins."

* * *

As I have argued throughout this book and elsewhere, the results of psychiatric twin studies prove nothing about genetics. Thus, the greater resemblance of identical twins versus same-sex fraternal twins for bipolar disorder or manic-depression cannot establish the genetic basis of the condition.

Adoption Studies

Like schizophrenia and ADHD, but unlike autism, bipolar family and twin studies have been supplemented by adoption studies. Although several affective disorder adoption studies have been published since the 1970s, only three studied BPD adoptees: Mendlewicz and Rainer's 1977 Belgian study, the 1986 Danish-American study of Wender, Kety, Rosenthal and their Danish colleagues, and von Knorring and colleagues' 1983 Swedish study. Here, I focus on the first two studies, which reviewers often refer to as the "two major adoption studies of bipolar disorder."[1] As we will see, they provide no evidence in support of genetic influences on BPD. Briefly, von Knorring et al. found zero cases of affective disorder (bipolar or otherwise) among the eight biological relatives of their five BPD or "cycloid disorder" adoptees.[2]

Mendlewicz and Rainer, 1977

Here, the researchers used the Adoptees' Family method (see Figure 2.1; see also Chapter 3), studying the biological and adoptive relatives of a group of 29 index adoptees diagnosed with bipolar disorder. Mendlewicz and Rainer found a significantly higher rate of psychopathology and affective disorders among their index biological versus index adoptive parents, leading them to conclude that "the results demonstrate the importance of genetic factors in the aetiology of manic-depressive illness."[3]

---

1. Merikangas et al., 2002, p. 459.
2. Von Knorring et al., 1983, p. 947.
3. Mendlewicz & Rainer, 1977, p. 329. The investigators also used three control groups: (1) the parents of non-adopted people diagnosed BPD, (2) the adoptive and biological parents of normal adoptees, and (3) the parents of people who had contracted polio. However, they based their conclusions in favor of genetics primarily on a comparison within the index group.

According to Mendlewicz and Rainer, "Comparison of adoptive parents of persons with a psychiatric disorder with their biological parents provides a unique opportunity to separate the interacting aetiological roles of heredity and environment."[1] However, although the investigators and most subsequent reviewers I am aware of failed to address this, a higher rate of disorder in the biological parent group is easily explained on environmental grounds. The reason, as we saw in Chapter 2, is that adoptive parents constitute a population screened for mental health as a part of the adoption process. As I quoted David Rosenthal, "The screening with respect to adopting parents is well known, since adoption agencies have long taken the view that mentally ill people do not make the kinds of parents that serve the best interests of the child."[2] Rutter and colleagues observed that these results "could be no more than an artifactual consequence of the tendency to select mentally healthy individuals as suitable adopting parents."[3] And in Chapter 5, we saw Seymour Kety argue that comparisons between adoptees' biological and adoptive relatives are "improper" and "fallacious."[4]

One of the few researchers or reviewers in the BPD field discussing this obvious pitfall is Elliot Gershon, who, in his chapter in Goodwin and Jamison's authoritative 1990 *Manic-Depressive Illness*, wrote that "Comparisons of adoptive with biological relatives reveal little because of the careful screening that is traditional in adoption placements."[5] And in an earlier work, Gershon and colleagues made the following astute observation:

> Another comparison sometimes made is between biological and adoptive relatives (Bl vs. Al...). However, there are potential biases in this comparison that makes it unreasonable to infer a genetic component from Bl > Al. Parents who give up children for adoption may have a higher degree of psychopathology than the general population, whereas adoptive parents are a selected group and more likely to be psychologically healthy at the time of adoption. Thus if Bl > Al, this does not prove that there is a genetic contribution since base rates will differ from this and other selection variables.[6]

Therefore, for reasons having nothing to do with genetics, we would expect to find less psychopathology among adoptees' adoptive parents — who constitute a population screened for mental health — versus adoptees' (unscreened) biological parents, whose psychological distress may have been the reason they gave up their child in the first place. As Gershon argued, it is "unreasonable to infer a genetic component" from this comparison. Thus, Mendlewicz

---

1. Ibid., p. 327.
2. Rosenthal, 1971a, p. 194.
3. Rutter et al., 1990.
4. Kety, 1983b, p. 964.
5. Gershon, 1990, p. 377.
6. Clerget-Darpoux et al., 1986, p. 306.

and Rainer's discovery of more psychopathology or affective disorders among their adoptees' biological versus adoptive parents could reflect nothing more than adoption agencies having screened prospective adoptive parents for psychopathology, suggesting that the study's logic is invalid on its face. Indeed, the most reasonable conclusion we can draw from their results is that the Belgian adoption agencies were successful in screening prospective adoptive parents for mental disorders.[1]

Leaving aside this invalidating problem, Mendlewicz and Rainer's results *do not* support genetic theories of bipolar disorder. Their biological and adoptive parents groups consisted of 58 parents each, and we see in Figure 10.1 that there were 4 bipolar diagnoses among the biological parents (6.9%), and 1 bipolar diagnosis among the adoptive parents (1.7%).[2] This comparison is not statistically significant, meaning that, statistically speaking, there was no difference in bipolar diagnoses between these two groups.[3]

However, Mendlewicz and Rainer decided to count diagnoses they combined in an "affective spectrum" consisting of bipolar disorder, unipolar depression, schizoaffective disorder, and cyclothymic disorder. Because there were many more cases of unipolar depression among the biological parents, they were able find a statistically significant affective disorder clustering in this group. Unfortunately, many subsequent commentators followed Mendlewicz and Rainer in pointing to the significantly greater biological parent clustering of "psychopathology" or "affective disorders," while downplaying or ignoring the statistically non-significant clustering of bipolar disorder among these parents. Given that the bipolar-unipolar distinction was based on the perceived need to differentiate these disorders on the basis of several characteristics,[4] combining these diagnoses under the "affective disorder" heading in order to claim genetic influences on bipolar disorder — comparable to the "schizophrenia spectrum" concept we examined in Chapter 3 — is dubious indeed.

Wender, Kety, Rosenthal et al., 1986

The other "major adoption study of bipolar disorder" was performed by Danish-American adoption researchers Paul Wender, Seymour Kety, David Rosenthal, and colleagues. Like Mendlewicz and Rainer, Wender et al. started with adoptees, and then diagnosed their adoptive and biological relatives. There were 71 index "proband" and 71 control adoptees. However, only 10 (14%) of the 71 index adoptees were diagnosed with bipolar disorder. Other index adoptee diagnoses included "affect reaction" (N = 13), "neurotic depression" (N = 21), and

---

1. Sarbin & Mancuso (1980) made a similar point about the Danish schizophrenia adoption studies.
2. Mendlewicz & Rainer, 1977, p. 328, Table 3.
3. 4/58 vs. 1/58, p = .18, Fisher's Exact Test, one-tailed.
4. Akiskal, 2002; Goodwin & Jamison, 1990, Chapter 3.

Figure 10.1

MENDLEWICZ & RAINER'S 1977 BELGIAN BIPOLAR DISORDER ADOPTION STUDY

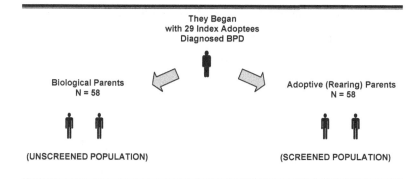

They Began
with 29 Index Adoptees
Diagnosed BPD

Biological Parents
N = 58

Adoptive (Rearing) Parents
N = 58

(UNSCREENED POPULATION)

(SCREENED POPULATION)

RESULTS: Parents Diagnosed with BPD

Biological
4/58 (6.9%)   vs.   *Adoptive*
1/58 (1.7%)

**Probability = .18, Fisher's Exact Test, One-Tailed.
Not Statistically Significant**

Source: Data from Mendlewicz & Rainer, 1977. BPD = Bipolar disorder.
"Screened" and "Unscreened" status refers to adoption agencies' practice of screening
prospective adoptive (rearing) parents for psychopathology.

"unipolar" (N = 27). The study's design, and results pertaining to BPD, are seen in Figure 10.2.

Whereas Mendlewicz and Rainer compared the diagnostic status of their index adoptees' biological and adoptive parents, Wender and colleagues (like Kety et al. in 1968 and 1975; see Chapters 3 & 5) compared the diagnostic status of their index versus control biological relatives. As seen in Figure 10.2, their results showed a higher (though statistically non-significant) rate of bipolar disorder among the *control* biological relatives. There were 2 (0.5%) bipolar diagnoses among 387 index biological relatives, versus 3 (0.9%) diagnoses among 344 control biological relatives.[1] Moreover, both rates are comparable to general population expectations. However, the investigators decided to combine all affective

---

1. Wender et al., 1986, p. 926, Table 3.

Figure 10.2

WENDER AND COLLEAGUES'
1986 DANISH-AMERICAN AFFECTIVE DISORDER
ADOPTION STUDY

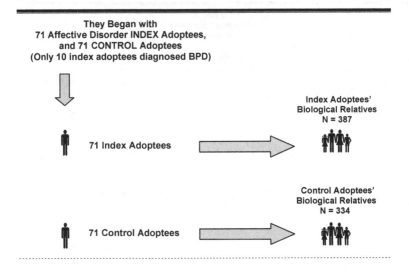

**They Began with
71 Affective Disorder INDEX Adoptees,
and 71 CONTROL Adoptees
(Only 10 index adoptees diagnosed BPD)**

71 Index Adoptees

Index Adoptees'
Biological Relatives
N = 387

71 Control Adoptees

Control Adoptees'
Biological Relatives
N = 334

RESULTS: Relatives Diagnosed with Bipolar Disorder

Index Biological   *Control Biological*
2/387 (0.5%)   **vs.**   **3/334 (0.9%)**

**This comparison is not statistically significant, and in both
groups, relatives are diagnosed with BPD at rates comparable to
the 1% general population rate**

Source: Data from Wender et al., 1986. BPD = Bipolar disorder.

disorders to create an "affective spectrum," and found many more index versus control biological relatives diagnosed with "uncertain major mood disorder" and "unipolar depression." Attempting to justify their decision to use a spectrum, they wrote, "These illnesses were considered together, since they have generally been regarded as the most dependent on internal (genetic or biological) factors."[1]

1. Ibid., p. 926.

This is yet another example of genetic researchers' use of circular reasoning, since their conclusions in favor of genetic influences on affective disorders depended on their prior assumption of both the genetic basis and genetic relationship of these disorders.

But even after deciding to combine affective disorders into a spectrum, Wender and colleagues still found no statistically significant results.[1] To find such results, they had to count "uncertain" cases. Moreover, in direct contrast to their schizophrenia studies (see Chapter 3 & 5), they decided to count half-siblings with one-half the weight of the full-sibs:

> The problem with half-siblings is how to weight them compared with the first-degree relatives (parents and siblings). One approach is to halve the number of half-siblings: the rationale is that a half-sibling is one half as likely as a full sibling to manifest a genetically transmitted disorder. Calculations based on that assumption were also made and will be discussed below.[2]

The investigators concluded that an "increased frequency of affective-spectrum disorder...has been found among the biological relatives of the adoptees with affective disorder. This finding is consistent with the hypothesis that genetic factors play a role in at least some of these disorders."[3] While this conclusion is highly questionable as it relates to affective disorders for the reasons I have already outlined (in addition to the potentially confounding influence of selective placement, which the investigators did not mention), their findings permit no valid conclusions in favor of genetic influences on bipolar disorder. Interestingly, although rarely mentioned in review articles and textbooks, Wender et al. more or less agreed with this assessment. While claiming to have found "a significant genetic contribution to unipolar depression and suicide,"[4] they recognized their "failure to find such a differential for bipolar illness..."[5] Only 10 of the 71 index adoptees carried this diagnosis, and the researchers did not disclose whether either of the two people diagnosed with BPD were biological relatives of these 10 BPD adoptees.

Interestingly, had Wender et al. compared the affective disorder status of their index biological versus index adoptive relatives, as Mendlewicz and Rainer had, they again would have found no statistically significant differences. According to their results, 7.5% of their index biological relatives (29/387) were diagnosed with a "broad affective-spectrum disorder," compared to 6.1% of index adoptive relatives (11/180).[6]

* * *

---

1. 7.5% index vs. 4.7% control, p = .074. See Wender et al., 1986, p. 927, Table 4.
2. Wender et al., 1986, p. 925.
3. Ibid., p. 929.
4. Wender et al., 1986, p. 923.
5. Ibid., p. 928.
6. Ibid., p. 927, Table 4. Using Fisher's Exact Test, one-tailed, the comparison is not statistically significant (29/387 vs. 11/180, p = .34).

As we have seen, no adoption study published to date has come close to providing scientifically acceptable evidence in support of genetic influences on bipolar disorder. Yet, textbooks and review articles by influential authors usually claim that adoption studies provide such evidence for BPD as a component of the "affective disorder" umbrella. Still others claim that adoption studies found genetic influences on BPD as a distinct diagnosis, which simply is not the case. Before documenting examples of this, I will quote a 2003 review article by genetically-oriented psychiatrists Jordan Smoller and Christine Finn, who provided a reasonably accurate assessment of the results of BPD adoption research. They concluded that in Mendlewicz and Rainer's study, "small numbers precluded meaningful analyses of bipolar disorder alone." And regarding Wender and colleagues' study, they wrote, "Again, the small number of cases of bipolar disorder did not permit meaningful statistical comparisons for this disorder per se."[1] In contrast, one might cite 26 reviewers who erroneously claimed that adoption studies found genetic influences on BPD:

Wender et al., 1986

> We did find a significant increase in both definite and uncertain unipolar and bipolar disorders taken together and in unipolar illness among the biological relatives of the index cases, compared with those of the controls. Our failure to find such a differential for bipolar illness, *in contrast with Mendlewicz and Rainer,* may be due to the small size of our sample compared with theirs [emphasis added].[2]

Davidson & Neale, 1998

> Several small-scale adoption studies have also supported the idea that both bipolar and unipolar disorder have a heritable component.[3] [They cited Mendlewicz & Rainer, and Wender et al. as examples.]

Faraone et al., 1999

> Drs. Mendlewicz and Rainer found more mood disorders among the biologic in comparison with the adoptive parents of bipolar adoptees or the parents of normal adoptees. Overall, the results of Drs. Mendlewicz and Rainer's adoption study indicate that genetic, not environmental, factors are implicated in the familial transmission of bipolar disorder.[4]

Paris, 1999

> In the best-designed study conducted to date, Mendlewicz and Rainer (1977) found that the morbid risk for affective illness was no greater in adopted-away children of parents with mood disorders than it was in adopted children whose biological parents did not have mood disorders.[5] [Although Paris made no claims about BPD specifically, his description of Mendlewicz & Rainer's "best designed" study is

---

1. Smoller & Finn, 2003, p. 54.
2. Wender et al., 1986, p. 928.
3. Davidson & Neale, 1998, p. 240.
4. Faraone, Tsuang, & Tsuang, 1999, p. 40.
5. Paris, 1999, p. 106.

incorrect. Mendlewicz & Rainer, von Knorring et al., and Wender et al. used the Adoptees' Family method. Therefore, they did not, as Paris claimed, study "the adopted-away children of parents with mood disorders." On the contrary, they studied the biological and adoptive relatives of adoptees with mood disorders.]

Report of the US Surgeon General, 1999

In studies of monozygotic twins reared separately ("adopted away"), the results also revealed an increased risk of depression and bipolar disorder compared with controls (Mendlewicz & Rainer 1977; Wender et al., 1986).[1] [These investigations did not study "monozygotic twins reared separately," or any other type of twins.]

DSM-IV-TR, 2000

Twin and adoption studies provide strong evidence of a genetic influence for Bipolar I Disorder.[2]

Gelernter & Goldman, 2000

Bipolar affective disorder was one of the first major psychiatric disorders approached through a linkage strategy. It is clearly familial, and has long been recognized as such; support for a genetic contribution comes from twin...adoption...and family...studies.[3]

Potash & DePaulo, 2000

The one adoption study which focused exclusively on BP [bipolar] probands, Mendlewicz and Rainer...showed that the biological relatives of BP adoptees were at greater risk for BPD than were the adoptive relatives, further supporting the importance of genetic factors in transmission of the disorder.[4]

Craddock & Jones, 2001

There have been only two adoption studies of bipolar disorder, neither of which was large — one had 30 probands (Mendlewicz & Rainer, 1977), the other 10 (Wender et al., 1986) — but they both showed that the risk of bipolar illness is greater in biological relatives than in adoptive relatives of the probands.[5]

Rehm et al., 2001

Mendlewicz and Rainer (1977) studied 29 bipolar adoptees. Thirty-one percent of biological parents of these patients had bipolar disorders, compared to 12% in the adoptive parents.[6]

---

1. Surgeon General, 1999, Chapter 4, p. 256.
2. APA, 2000, p. 386.
3. Gelernter & Goldman, 2000. Online edition of *Neuropsychopharmacology: The Fifth Generation of Progress*, retrieved on 6/11/05 from http://www.acnp.org/g4/GN401000091/ Default.htm.
4. Potash & DePaulo, 2000, p. 10.
5. Craddock & Jones, 2001, pp. s128-s129.
6. Rehm et al., 2001, p. 316.

Cowan, Kopnisky, & Hyman, 2002

That BPD runs in families has long been acknowledged. Twin and adoption studies have taken this a good deal further by indicating the high degree of heritability (on the order of 0.8) of the disorder...[1]

Maier, 2002

The two major adoption studies of bipolar disorder found an approximately 3-fold increased risk of bipolar disorder among biological compared to adoptive relatives of probands.[2] [This is not what Wender et al. found, and in Mendlewicz & Rainer's study the comparison was not statistically significant.]

Merikangas et al., 2002

Two major adoption studies of bipolar disorder [Mendlewicz & Rainer, 1977; Wender et al., 1986] yielded an approximately threefold increased risk of bipolar disorder among biological compared with adoptive relatives of probands.[3] [See my comments on Maier's review.]

Sklar, 2002

For bipolar disorder and schizophrenia, numerous family, twin, and adoption studies have identified a strong genetic component to these behavioral psychiatric disorders.[4]

Tohen & Angst, 2002

The evidence from family, adoption, and twin studies clearly supports the evidence that bipolar disorder is genetically transmitted.[5]

Faraone & Tsuang, 2003

Family, twin, and adoption studies clearly show bipolar disorder to be a highly heritable condition.[6]

Sadock & Sadock, 2003

Adoption studies have also produced data supporting the genetic basis for the inheritance of mood disorders. Two of three adoption studies have found a strong genetic component for the inheritance of major depressive disorder; the only adoption study for bipolar I disorder also indicated a genetic basis. These adoption studies have shown that the biological children of affected parents remain at increased risk of a mood disorder, even if they are reared in nonaffected adoptive families.[7] [These investigators did not study "the biological children of affected parents." See my comments on the Paris quotation.]

---

1. Cowan et al., 2002, p. 16.
2. Maier, 2002. Maier's formulation is very similar Merikangas and colleagues' 2002 account, which Maier cited in his chapter.
3. Merikangas et al., 2002, p. 459.
4. Sklar, 2002, p. 371.
5. Tohen & Angst, 2002, p. 440.
6. Faraone & Tsuang, 2003, p. 1.
7. Sadock & Sadock, 2003, p. 540.

Blackwood & Muir, 2004

It has long been suspected and is now firmly established that bipolar disorder is familial, and there is a 10-fold increase in the risk of illness in a first-degree relative of someone with the disorder compared to the population risk. That this is partly due to genetic rather than purely environmental factors is confirmed by adoption studies and the well-replicated observation that concordance rates are significantly higher in identical than fraternal twins.[1]

Macgregor et al., 2004

Bipolar disorder...and schizophrenia...are severe psychiatric illnesses.... There is strong evidence for a genetic aetiology in such disorders with high heritabilities reported in twin data and adoption studies.[2]

Shih et al., 2004

Only two adoption studies have been performed using a modern concept of bipolar disorder. Both studies found that biological parents of bipolar adoptees are more likely to have bipolar disorder than adoptive parents.[3]

Craddock, O'Donovan, & Owen, 2005

There is substantial evidence from family, twin, and adoption studies for the importance of genes in influencing susceptibility to schizophrenia and bipolar disorder...[4]

Kealey et al., 2005

Although family, twin and adoption studies indicate a strong genetic component, the etiology of BPD has yet to be determined.[5]

Maziade et al., 2005

Family, twin and adoption studies have shown that schizophrenia (SZ) and bipolar disorder (BP) are complex and highly heritable disorders.[6]

Raybould et al., 2005

Family, twin and adoption studies provide evidence that genetic factors are important in determining susceptibility to bipolar disorder.[7]

Shastry, 2005

BPD has a complex etiology.... Family, twin and adoption studies suggest that it is a heritable disorder.[8]

---

1. Blackwood & Muir, 2004, p. 224.
2. Macgregor et al., 2004.
3. Shih et al., 2004, p. 268.
4. Cradock et al., 2005, p. 193.
5. Kealey et al., 2005.
6. Maziade et al., 2005, p. 486.
7. Raybould et al., 2005, p. 696.
8. Shastry, 2005, p. 273.

Shink et al., 2005

> Twin, adoption and family studies have supported the importance of genetic factors in the predisposition to bipolar disorder but its aetiology remains unknown.[1]

Yet again, authoritative secondary sources have played an important role in creating myths about psychiatric genetic research, leading professionals and students alike to believe that there is "overwhelming evidence" that the major psychiatric disorders are influenced by genetic factors. In fact, there isn't.

CONCLUSION

Like ADHD, schizophrenia, autism, and other major psychiatric disorders, there exists little scientifically acceptable evidence in support of genetic influences on bipolar disorder, also known as manic-depressive disorder. As Gershon wrote in 1990, "We would conclude that the adoption data do not provide a broad base of supportive data on the hypothesis that [bipolar] disorders are transmitted before the age of adoption."[2] For him, twin studies provide the main support for hereditary influences, but as been shown, the results of these studies are easily interpreted on environmental grounds. Like others disorders, an emphasis on genetics diverts resources from research into other potentially fruitful areas. In this instance, studies suggest that a considerable percentage of adults diagnosed with BPD suffered severe abuse as children.[3]

As expected, the mistaken belief that genetic factors have been established has led to the search for BPD genes at the molecular level. In the following chapter, I discuss the decades-old failure to find these presumed genes, as well as the failure to find genes for other disorders in psychiatry.

---

1. Shink et al., 2005, p. 545.
2. Gershon, 1990, p. 378.
3. Garno et al., 2005; Leverich et al., 2002.

CHAPTER 11. GENOTYPE OR GENOHYPE? THE FRUITLESS SEARCH FOR GENES IN PSYCHIATRY[1]

"It is highly unlikely that spirochete-like big explanations remain to be discovered for major psychiatric disorders. We have hunted for big, simple neuropathological explanations for psychiatric disorders and have not found them. We have hunted for big, simple neurochemical explanations for psychiatric disorders and have not found them. We have hunted for big, simple genetic explanations for psychiatric disorders and have not found them."[2] — *American psychiatric geneticist Kenneth Kendler, 2005*

"The strong, clear, and direct causal relationship implied by the concept of 'a gene for ...' does not exist for psychiatric disorders. Although we may wish it to be true, we do not have and are not likely to ever discover 'genes for' psychiatric illness."[3] — *Kendler, 2005*

"Whereas genetically complex traits are being successfully pinned down to the molecular level in other fields of medicine, psychiatric genetics still awaits a major breakthrough."[4] — *German psychiatric geneticist Peter Propping, 2005*

"When are we going to be there [finding genes that cause psychiatric disorders]? Being an optimist, my response is 'soon.' But readers would be forgiven for being skeptical because they have heard this before...A small personal example of impatience and embarrassment about the slower-than-expected progress towards identifying QTLs [genes of varying effect sizes] is that my co-authors and I decided that we would not write the next edition of our behavioural genetics textbook... until we had some solid DNA results to present. The reason for this decision was that our 2001 edition had enthused about the field being on the cusp of a new post-

---

1. The phrase "Genotype or Genohype?" is taken from the title of a 2001 article by T. M. Marteau (Marteau, 2001).
2. Kendler, 2005c, p. 434-435.
3. Kendler, 2005a, p. 1250.
4. Propping, 2005, p. 2.

genomic era in which DNA risk indicators would add great value to behavioural research. We are still on that cusp." — *Behavior Geneticist Robert Plomin, 2005*[1]

In this final chapter I analyze and assess molecular genetic research in psychiatry.[2] My basic position on this topic was the subject of a chapter in *The Gene Illusion*, where I argued that the failure to find genes relates much more to researchers' misguided faith in the results of family, twin, and adoption studies than it does to the difficulty of finding genes. There is much more to say about this, and much has been published since the original edition of *The Gene Illusion* went to press in the spring of 2003. The results, however, remain the same despite worldwide efforts and untold millions of research dollars: No genes have been found that cause the major psychiatric disorders.[3] Unfortunately, it is widely believed that such genes *have* been discovered.

A major reason is that the media tends to report "gene discoveries" for abnormal behavior and psychiatric disorders (including those discussed in this chapter), but pays little attention to, or fails to report entirely, replication failures and retractions. According to Peter Conrad, who has studied media coverage of molecular genetic research:

> Articles reporting significant new genetic research are typically reported in prominent places in newspapers magazines. Yet, if subsequent research does not replicate the findings or disconfirms the first study's results, how does the news media cover the dissenting studies? The way the media reports on such subsequent studies contributes to an information flow problem in public discourse.[4]

Psychologist Carl Ratner wrote that "news articles are important because they shape the public's scientific knowledge," and that it is important "to ascertain whether the public is being educated or misled."[5] Clearly, Ratner argued, the public is being misled:

> Reports in the media and journals that psychological phenomena are genetically caused are dubious....They fail to point out obvious logical and empirical errors in the research which they report. They rarely present critiques by dissenting social scientists. News reporters headline any suggestion of genetic determination of psychology regardless of how preposterous and undocumented it is.[6]

The back cover of a recent book popularizing genetic research in psychiatry proclaimed, "New discoveries about the genetic underpinnings of human experience are now being reported at a furious rate."[7] But few such "discoveries"

---

1. Plomin, 2005, p. 1030.
2. Portions of this chapter appeared in Joseph, 2004a.
3. Except possibly Alzheimer's disease, which is more a brain disease than a "mental illness."
4. Conrad, 2002, p. 62.
5. Ratner, 2004, p. 29.
6. Ibid., p. 41.
7. Anonymous, 2005.

actually exist. A 2005 edition of the *Wall Street Journal* featured an article by veteran science writer Sharon Begley, who wrote as if genes for psychiatric disorders have been discovered when, in fact, the opposite is true:

> A gene on chromosome 9, linked to autism, seems to count only if it came from dad. One on chromosome 2 and one on 22 are associated with schizophrenia; only the copies from dad count. Having a family tree mostly free of these diseases is therefore no assurance of good health. If the disease runs on dad's side, his gene may be defective, and that is the one that matters.[1]

Reports of this type lead people to believe that initial positive reports that are later unsubstantiated are actual gene findings. Although journalists play their role in creating this false impression, in many cases they merely report the claims of the scientific investigators they interview. Internet searches for terms such as "schizophrenia gene," "bipolar gene," "ADHD gene," "autism gene," "depression gene" and so on encounter dozens of reported "gene discoveries" for the major psychiatric disorders. In virtually every case, however, these claims are followed by largely unpublicized failures to replicate.

* * *

I am a clinical psychologist who works with people seeking help for various problems, who may be in distress, and who may have been diagnosed with one or more psychiatric disorder. I claim no expertise in genetics, biology, or other "hard sciences." However, my clinical experience, and that of other clinicians, makes it clear that most psychiatrically diagnosed people have suffered abuse, neglect, or abandonment as children, and/or experience high levels of oppression, stress, or loss in their current lives. Critics will point out that such anecdotal evidence does not constitute scientific evidence in support of environmental causes of mental distress. This is true, but at times I find myself, when listening to people's painful and gut wrenching stories, thinking something along the lines of, "Obviously, genes can't possibly explain the pain this person in front of me is experiencing as she tells me her story. But her life history certainly can."

Fortunately, genetic expertise, while helpful, is not a prerequisite for understanding and analyzing molecular genetic research in psychiatry. What is more important is an understanding of the main issues involved, and a careful analysis of what the investigators and the authors of secondary sources conclude about the results of this research. This chapter will approach the topic in a relatively freewheeling style, keeping the use technical language and details on how molecular genetic research is performed to a minimum.

Undoubtedly, some readers will take exception to the way I discuss the research and researchers I highlight. And, although most molecular genetic

---

1. Begley, 2005.

researchers in psychiatry are motivated by their desire to alleviate human suffering, I draw a qualitative distinction between psychiatric conditions such as unipolar depression, bipolar disorder, schizophrenia, and ADHD, for which medical and biological explanations are dubious, and real diseases of the body, such as cancer and diabetes, for which medical explanations and interventions are essential. Although gene searches for true medical diseases (breast cancer, for example) are often performed at the expense of investigating crucial environmental factors, the tone of this chapter would be far different were I reporting on molecular genetic research in medicine.

In 1997, psychiatric genetic researcher David Comings wrote,

> Family, twin, and adoption studies have clearly indicated the important role of genes in a wide range of human behavioral disorders. However, in this era of molecular biology, where the genes for almost every important nonpsychiatric genetic disorder have been discovered, to simply say that genes play a role in behavior is inadequate. The longer we fail to identify *which* genes play a role, the more skeptics will come to believe that no genes play a role in determining how we behave [emphasis in original].[1]

Comings, like other psychiatric genetic researchers, believed that family, twin, and adoption studies "indicate the important role of genes." He then acknowledged that, whereas genes have been found for "almost every important nonpsychiatric genetic disorder," the complete opposite results have been obtained for behavioral disorders. This is a telling point, although it did not lead Comings to conclude that something may be fundamentally wrong with psychiatric genetic assumptions, methods, and approaches. He did warn his colleagues that their continuing failure to identify genes would give ammunition to critics "who believe that no genes play a role in determining how we behave."

However, it is not true, as Comings implies, that environmentalist critics argue that genes play no role in determining human behavior. That would be ridiculous, as genes obviously determine whether an organism becomes — and behaves like — a human, an alligator, a snapping turtle, an earthworm, a buffalo, a sea anemone, and so on. Obviously, humans are hardwired with some species-specific traits and instincts, or potentials for manifesting species-specific traits. The fact that I am expressing human language in the form of this book is proof enough of this. Popularizers of psychiatric genetic research, on the other hand, argue that genetic factors play an important role in determining differences *between humans*. The unfortunate caricature of the environmentalist position as claiming that genes play no role in determining human behavior became the straw person set up by Steven Pinker in his 2002 book *The Blank Slate*, where he argued in favor of an important role for genetics in explaining psychological and behavioral differences among humans.[2]

---

1. Comings, 1997, p. 236.
2. Pinker, 2002. For a critical review of Pinker's book, see Menand, 2002.

The discovery of a gene causing Huntington's disease in the early 1980s gave rise to expectations that investigators would soon discover genes causing the major psychiatric disorders as well. Psychiatric geneticists Melvin McInnis and James Potash described the era's enthusiasm as follows:

> The last two decades of the 20th century began with enthusiasm generated by the prospect of genetic mapping and positional cloning. The good fortune of finding linkage to Huntington's disease with the eighth genetic marker studied engendered unbridled enthusiasm in those of us graduating from medical school in the early 1980s as genetic mapping strategies were described and laboratory methods leapt forward with the discovery of the polymerase chain reaction (PCR). Family collections for many illnesses began, including psychiatric disorders, and the hunt was on. Some of us...thought the task would be rather straightforward: gather a collection of pedigrees, make categorical diagnoses, employ a genome-wide scan, achieve a LOD score of three, find the gene, and then study how it caused the disease. Quickly, promising linkages in schizophrenia and bipolar disorder were reported. Then just as quickly negative replication attempts were published. Though Mendelism, with its emphasis on single gene causation, dominated the field for a time, it has become apparent that the genetic risk to psychiatric disorders is likely to be a multifaceted problem, with little in common with Mendel's peas. Complexity is now a central theme in psychiatric genetics, and indeed genetics as a field.[1]

And German psychiatric geneticist Peter Propping recalled that, in the 1980s, "We were all optimistic that this approach [linkage] would lead to the rapid detection of genes predisposing to mental disorder. No other medical phenotype with a complex genetic background was tackled with as many genome scans as schizophrenia and bipolar disorder."[2]

The disappointment over failing to complete the "straightforward" task of finding genes continues as I write these lines, two decades later. Of course, this does not prove that genes for psychiatric disorders do not exist. What is striking, however, is that among those who continue to search for genes, the mere *possibility* that genes do not exist is rarely, if ever, considered.

THE STAGES OF MOLECULAR GENETIC RESEARCH IN PSYCHIATRY

The search for genes in psychiatry follows the same basic script, regardless of the disorder being studied. First, researchers and reviewers claim that previous epidemiological studies of families and twins (and in some cases adoptees) have established the genetic basis of the disorder. Rarely, however, do they offer a critical review of the methodological problems and questionable theoretical assumptions of these studies. In a sense this is understandable, given that molecular genetic researchers are in the business of looking for genes, not

---

1. McInnis & Potash, 2004, p. 243.
2. Propping, 2005, p. 5.

debunking previous research. Indeed, previous research on families, twins, and adoptees supplies their *raison d'être*. And for most disorders, genetic researchers believe (1) that genes exist and await discovery, and (2) that finding genes will aid in the treatment or prevention of the disorder.

The *second* stage involves speculation about what type of genetic transmission is operating. Single gene transmission is usually rejected after the initial failures to find single genes, and various theories about polygenic inheritance (the actions of several genes of various effect sizes) are put forward. This sometimes involves further speculation about how many genes may be involved, and on which chromosomes some of these might be located.

The *third* stage is what might be called the rhetoric stage. Rather than emphasize their failure to find genes for the major psychiatric disorders, molecular genetic researchers and reviewers, while believing that the task of finding genes for "complex disorders" is more difficult than they first imagined, speak in glowing terms of how discoveries in the 21rst century "post-genomic era" are coming soon, and of the direction their research will take after they find genes. As Boyle pointed out, rhetoric of this sort is used to support "the idea of progress waiting to happen."[1]

The *fourth* "throw in the towel stage" has yet to occur in psychiatric molecular genetic research. However, apart from the unlikely event that genes for autism, ADHD, schizophrenia, and bipolar disorder both exist and are eventually discovered, this will be the final stage of psychiatric molecular genetic research.

RESEARCH METHODS

The most popular methods in molecular genetic research are linkage studies, genome scans, and association studies. In a *linkage* study, researchers attempt to identify genetic markers associated with a presumed disease gene among consanguineous family members. Findings are often represented as a logarithm of odds (LOD) score, which expresses the probability that the linkage occurred by chance. Although there have been some recent modifications, in general an LOD score higher than 3 (1000:1 odds in favor of linkage) is needed to be able to claim statistically significant linkage. Linkage studies attempt to identify areas of the chromosome where relevant genes might be located, but they are unable to identify actual genes. This is the task of follow-up studies to those finding significant linkage. A *genome scan* (also known as "systematic mapping") analyzes the complete genome of an individual against a set of markers whose positions on the chromosomes are known. A genome scan looks

---

1. Boyle, 2002b, p. 204.

for common patterns of inheritance between these markers and the disease characteristics, and identifies linkage regions on the chromosomes. Unlike typical linkage analyses, which frequently are based on hypothesized "candidate genes" for the various disorders under study, genome scans make no assumptions about the possible location of genes.[1] *Association studies* compare the frequency of genetic markers among unrelated affected individuals and a control group, and are performed with population-based case-control, or family-based samples. A genetic marker is defined as a segment of DNA with an identifiable physical location on a chromosome, whose inheritance can be followed.

Psychiatric geneticists postulate two main types of genetic transmission for the disorders they study. The first is *mendelian inheritance,* in which a disease or trait is passed from parents to offspring by a single dominant, recessive, or sex-linked gene. Medical disorders such as Huntington's disease and PKU are caused by a person inheriting a single disease gene. No gene has been found to cause any of the major psychiatric disorders, and most researchers now agree that it is very unlikely that any is caused by a single gene. The second is the *polygenic* approach, meaning that many genes of varying effect sizes are believed to contribute to the appearance of a disorder. This means that investigators look for several genes, or individual genes thought to have a large effect size. Furthermore, most investigators recognize that environmental factors are necessary to bring about disorders in people presumed to be susceptible on the basis of polygenic inheritance.

CAUSE AND EFFECT

Researchers view many medical conditions (e.g., asthma, heart disease, diabetes mellitus), and psychiatric conditions (e.g., schizophrenia, depression, ADHD) as "multifactorial complex diseases," meaning that there is "a complex interacting admixture of multiple genes and multiple environmental risk factors."[2] Thus, molecular genetic researchers in psychiatry assume that psychiatric disorders (1) are valid entities that can be reliably diagnosed; (2) are caused by, among other factors, genetic variation (polymorphisms) or genetic mutations; and, (3) have corresponding biological defects in the brain. According to critically-minded geneticists Jonathan Beckwith and Joseph Alper, "An important assumption underlying genetic studies is that the trait as defined by observable or clinical manifestations corresponds to an actual entity that is influenced by genes."[3]

---

1. King et al., 2002a.
2. Rutter, 2001, p. 227.
3. Beckwith & Alper, 2002, p. 317.

Although psychiatric geneticists frequently use the term "complex disease," their position that complex diseases in psychiatry (1) are real diseases, (2) are caused by genetic defects, and (3) are caused by biological defects, is a theory, not a fact. More often, as Boyle observed, the term is used in psychiatric genetics "in order to maintain its credibility in the face of inconsistent and negative data.... the idea of schizophrenia as a complex disease is brought into service, with the implication that setbacks are to be expected."[1]

A psychiatric condition or syndrome is called a "complex disease" *because of* the failure to find genes, while subsequent failures are explained on the basis of the "complex" nature of the "disease." Circular reasoning of this type is seen in a 2003 review of autism research, where the authors claimed that the "current lack of success in finding genes for autism is similar to that of complex diseases."[2] More accurately, the "lack of success" in finding genes *defines* "complex diseases" in psychiatry.

Moreover, even if a gene is *associated* (correlated) with a condition, it does not necessarily mean that the gene *causes* the condition. Researchers often make this unwarranted leap, however, and the claim that a gene is found more often among affected people leads to the hasty conclusion that "a gene for" the condition has been identified.

African-Americans have higher blood pressure levels than white Americans, and researchers have found that blood pressure levels are indeed correlated with skin darkness.[3] However, although molecular genetic researchers would find that genes producing darker skin color correlate with blood pressure levels, it does not mean that these genes cause high blood pressure. Clearly, environmental factors affecting people with dark skin might completely explain the correlation. "If you follow me around Nordstrom's, and put me in jail at nine times the rate of whites, and refuse to give me a bank loan, I might get hypertensive," argued sociologist Troy Duster. "What's generating my increased blood pressure are the social forces at play, not my DNA."[4]

But even if a gene is necessary for the manifestation of a trait or condition, it still doesn't mean that it is a causative factor. As Ratner pointed out, "The fact that something is a necessary foundation for something does not mean that it causes it."[5] For example, we observe that all vehicles traveling over 60 miles per hour on the freeway have two or more tires. However, although tires are *necessary* for vehicles to move 60 miles per hour, tires do not *cause* vehicles to move forward; engines do. Ratner challenged claims that a defective gene causes language impairments. "Obviously, language requires a normal genetic substratum,"

---

1. Boyle, 2002b, p. 202.
2. Volkmar & Pauls, 2003, p. 1136.
3. Henig, 2004.
4. Quoted in Henig, 2004, p. 49.
5. Ratner, 2004, p. 30.

he wrote, "and a defective genome undermines the ability to use language — just as it undermines the ability to play Monopoly." He concluded, however, that "this does not mean that a gene causes or predisposes language, any more than it causes or predisposes me to play Monopoly."[1]

Still, behavior geneticists such as Robert Plomin argue that "correlations between DNA differences and behavioral differences can be interpreted causally," because "DNA differences cause the behavioral differences but not the other way around."[2] Plomin is mistaken, however, because even though behavior cannot change the DNA, a correlation could still be spurious. For example, there is a strong correlation between having a Y chromosome and being the chief executive officer (CEO) of a Fortune 500 corporation. This does not mean that having a Y chromosome causes or predisposes someone to become a CEO. Similar to Duster's example, the correlation is the result of social privileges granted to people with Y chromosomes, i.e., men, rather than the action of the chromosome itself. As psychiatric genetic researchers Harrison and Weinberger recognized, "allelic association *per se* is not evidence of gene identification."[3]

There are, according the Grier Page and colleagues, four possible explanations for finding a gene variation (polymorphism) associated with a disorder:

1. The variation "is actually causative for the disease or trait."
2. "The association is a false positive due to random chance."
3. The variation "is in disequilibrium with the true causative allele."
4. The variation is "associated because of some systematic bias in the biology, study, samples, or analysis."[4]

Most journalists and many scientists conclude, however, is that the variation is causative. Other explanations tend to be downplayed or ignored.

Page and colleagues believed that if "bias cannot be removed" from genetic research, "it is not appropriate to suggest causation."[5] And far from having established causation, there is little evidence from psychiatric molecular genetic research that even association has been established for psychiatric disorders.

In 2005, Peter Propping outlined some explanations for the rash of false-positive findings in psychiatric molecular genetic studies, which he divided into "scientific arguments," and "non-scientific arguments."[6] Among the former, he listed "Establishing the psychiatric diagnosis may be problematic," "Psychiatric symptoms may be epiphenomena, we need endophenotypes," "False positive findings may result from multiple testing," and "Problems may arise from ethnic differences." The four "non-scientific" problems he mentioned were, "Premature publication because of competition pressure," "Premature publication because of

---

1. Ibid., p. 30.
2. Plomin, 2004, p. 348.
3. Harrison & Weinberger, 2005, p. 5.
4. Page et al., 2003, p. 713.
5. Ibid., p. 716.
6. Propping, 2005, p. 6.

commercial interests," "Selective publication of positive findings," and the "Lower standard of investigators than in other fields." Among these, Propping saw "selective publication of positive findings to be the most threatening one for our field," and discussed the "danger that journals preferentially publish positive findings, because a silent coalition exists between author and editor: both are interested in publishing positive findings."[1] The problem of bias against the publication of negative findings is by no means limited to psychiatric genetics, and plagues many areas of science.[2] Given that "negative findings may be as valuable as the positive ones," Propping proposed that psychiatric genetics provide "a platform to publish all empirical findings — obviously with quality control — independent of their interpretation."[3]

According to Beckwith and Alper, statistical error is one source of bias that could explain the plethora of false-positive findings. As an example, they discussed early claims that a gene for "novelty seeking" had been discovered:

> How is it possible that two independent studies reaching the same conclusion might both be wrong? As is the case in the search for genes for mental illness, the answer lies in the statistics of gene research. Probably more than 20,000 genes are involved in the functioning of the human brain, any one of which could conceivably be associated with a trait such as novelty seeking. As a result, traditional statistical methods vastly overestimate the likelihood that an association between one of these genes and the trait is a real association rather than merely being the result of a statistical accident like finding five consecutive heads when tossing a coin.[4]

Harold Göring and colleagues argued that it is very difficult to identify the chromosomal positions of genes with small effects sizes "by a single set of data of currently realistic size...," and that an "LOD score, is itself a function of the parameter(s) characterizing the genotype-phenotype relationship. Statistical significance and the estimated parameter(s) therefore are not independent but highly correlated."[5] They argued that statistical results from genome scans and all other gene finding methods are "almost certainly biased upwards, probably to a large degree," and that "there appears to be no satisfactory way of correcting for the genomewide bias within a study."[6]

It turns out that even molecular genetic research depends on the acceptance of questionable theoretical assumptions, not only in the investigators' decision to perform this research in the first place, but also because they factor assumptions about genetics into mathematical models of familial transmission. According to psychiatric geneticist Peter McGuffin, "Unfortunately, conventional linkage requires several assumptions. These are that major gene effects

---

1. Ibid., p. 6.
2. Joseph & Baldwin, 2000.
3. Propping, 2005, p. 6.
4. Beckwith & Alper, 2002, p. 321.
5. Göring et al., 2001, p. 1357.
6. Ibid., p. 1367.

(rather than just multiple small gene effects) exist, that there is some way of assuring genetic homogeneity, and that the mode of transmission of the disorder is known."[1] And according to Faraone and colleagues, "The main drawback of the LOD score method is that we must specify the mode of genetic transmission."[2]

Clearly, if one assumes that some mode of genetic transmission is occurring, one assumes that genes must play a role. Although molecular genetic researchers test multiple genetic models in computer analyses of their findings,[3] all models assume that some type of genetic transmission is occurring. Thus, the large number of false positive linkage findings may be another example in psychiatric genetics of questionable assumptions leading researchers to the premature conclusion that genetic factors (or actual genes) exist, since these results may be the product of plugging false assumptions into their LOD score calculations. Geneticist Robert Elston and his colleagues recently discussed the "misuse of mathematical proofs by ignoring the underlying assumptions necessary for their validity" in sibling pair linkage analyses.[4] Moreover, testing multiple hypotheses increases the likelihood that significant LOD scores are mere chance results. According to Beckwith:

> What is crucial to note in [linkage] approaches is that each time a numerical value is chosen for parameters in calculating the LOD scores, a different hypothesis is being tested. Researchers are testing multiple hypotheses about how the condition is inherited. Because they are testing not one but many hypotheses, there is an increased likelihood that a high LOD score associated with one set of properties has occurred by chance. That is, the more hypotheses are tested, the probability that an association that is found is merely due to chance becomes stronger and stronger.[5]

* * *

The upcoming discussion focuses on the four diagnoses covered in the previous chapters: schizophrenia, attention-deficit hyperactivity disorder (ADHD), autism, and bipolar disorder (BPD). (Although I do not review unipolar depression molecular genetic research, the results in this area are similar to these conditions.[6]) Undoubtedly, numerous "gene findings" in psychiatry will be reported between the time this book goes to press and the time you will be reading it. However, my hope is that readers will be better able to view these claims with a healthy degree of skepticism, realizing that highly publicized yet unfounded claims of gene discoveries have been made since at least 1987–88,

---

1. McGuffin, 2004, p. 197.
2. Faraone, Tsuang, & Tsuang, 1999, p. 131.
3. Beckwith, 2006; Coon, 1999.
4. Elston et al., 2005, p. 152.
5. Beckwith, 2006.
6. Plomin & McGuffin, 2003. For a critique of brain disease theories of depression, see Lynch, 2004.

when (subsequently non-replicated) schizophrenia and bipolar disorder "gene findings" were announced by separate research teams.[1]

The latter part of 2002 witnessed the beginning of influential genetic researchers declaring victory in the schizophrenia gene hunt. According to psychiatric investigator C. Robert Cloninger, in an article published in the prestigious *Proceedings of the National Academy of Sciences*, susceptibility genes for schizophrenia were "discovered" in the summer of 2002 by three separate research teams: Straub et al., Chumakov et al., and Stefansson et al.[2] My evaluation of this claim was that Cloninger's position "was more the result of wishful thinking than objective scientific evaluation."[3] Nevertheless, Cloninger was echoed by schizophrenia researcher Daniel Weinberger, who, at the March–April, 2003 Ninth International Congress on Schizophrenia Research in Colorado Springs, claimed that six schizophrenia susceptibility genes had been identified.[4] The National Institute of Mental Health (NIMH) subsequently rewarded Weinberger with a new schizophrenia gene research program, at an annual cost of $6 million.[5]

Referring to the studies Cloninger mentioned, in September 2003 Kendler wrote of "several watershed events in the number of possible susceptibility genes...and replication of at least two of them."[6] Kendler hedged his bets in this passage, as the "watershed" events referred to the discovery of "possible" susceptibility genes. Although Kendler mentioned two replications, there have also been failures to replicate.[7] The Chumakov group worked for the French pharmaceutical company Genset S.A., which has a financial incentive to patent genes and to design drugs to act on particular genes. Moreover, neither Straub et al. nor Stefansson et al. claimed to have discovered susceptibility genes for schizophrenia. The latter group noted that the NRG1 gene "had not previously been considered in the context of schizophrenia," and proposed it as a "fascinating candidate gene for the disease."[8] And Straub et al., who found an association

---

1. The schizophrenia study was by Sherrington et al., 1988. The bipolar disorder study was by Egeland et al., 1987.
2. Cloninger, 2002, p. 13367; Chumakov et al., 2002; Straub et al., 2002; Stefansson et al., 2002.
3. Joseph, 2003, p. 283.
4. Arehart-Treichel, 2003.
5. National Institutes of Health, 2003.
6. Kendler, 2003a, p. 1549.
7. For example, see Lewis et al., 2003; Morris et al., 2003; Van Den Bogaert et al., 2003; Williams et al., 2003.
8. Stefansson et al., 2002, p. 888.

between schizophrenia and the "Dysbindin" gene on chromosome 6, cautioned that, due to the "well-documented difficulties" in finding genes for complex disorders, "The results presented here represent only an initial step towards an understanding of the possible etiologic role that dysbindin plays in schizophrenia."[1] Psychiatric genetic researchers Paul Harrison and Michael Owen, in a more guarded tone than Cloninger and Weinberger, wrote in February, 2003 that while recent findings "are potentially very important, they should be viewed with caution."[2] Given the history of non-replicated claims, they recommended that "stringent criteria [be] applied when evaluating reports of schizophrenia susceptibility genes."[3]

An April 2003 edition of *The American Journal of Psychiatry*, devoted entirely to "Psychiatry in the Genomics Era," mentioned no "schizophrenia susceptibility gene" discoveries. For example, veteran psychiatric researcher William Bunney and his colleagues recognized that "There is currently no fundamental understanding of the genes that increase the risk for psychiatric disorders."[4] In the same issue, genetic investigators Kathleen Ries Merikangas and Neil Risch described the "ping-pong game between linkage and association claims and disconfirmations" in psychiatric genetic research. They commented further, "there has been considerable debate regarding what constitutes acceptable evidence of a true replication."[5]

This April 2003 edition of the *Journal* was interesting in that, whereas most of its authors claimed enormous progress and predicted exciting future developments in psychiatric genetics, they downplayed the fact that genes for the major psychiatric disorders had not been discovered, despite more than two decades of research. Psychiatric genetics was placed at the cutting edge of psychiatry and was presented as being on the verge of great discoveries. All authors in this special edition of the *Journal* assumed that genes for psychiatric disorders both exist and will be discovered in the near future. "It will be incumbent upon clinicians," wrote Merikangas and Risch, "to become familiar with knowledge gleaned from genetic epidemiologic and genomics research."[6] The Nobel Laureate co-discoverer of DNA, James Watson, penned the issue's lead article, lending an air of authority to psychiatry as continuing in the tradition of watershed discoveries in genetics. However, Watson claimed no discovery of genes for psychiatric disorders, referring only to the "tools of molecular biology...now being applied to...a wide variety of complex disorders such as schizophrenia, mood disorders, and substance abuse."[7] In words echoing

---

1. Straub et al., 2002, p. 343.
2. Harrison & Owen, 2003, p. 418.
3. Ibid., p. 419.
4. Bunney et al., 2003, p. 657.
5. Merikangas & Risch, 2003b, p. 627.
6. Ibid., p. 632.
7. Watson, 2003, p. 614.

current predictions in psychiatry, he closed by stating, "I am confident that during the upcoming years the heritage of the double helix will help psychiatrists, neuroscientists, and behavioral scientists unlock many secrets of the mind and brain."[1]

Later in 2003, genetic investigators Lynn DeLisi and Timothy Crow cautioned that the genes cited by Cloninger and Weinberger "have not yet been determined to show specific modification in multiple members with schizophrenia within families."[2] In a letter published in the May 2003 edition of *The Lancet*, Crow observed that meta-analyses claiming replication for linkages to several chromosomal regions "agree with respect to only one chromosomal arm...of the nine they highlight. A reasonable conclusion is that the null hypothesis has not yet been disproved."[3] In an October 2003 article in *Science*, geneticist James Kennedy, psychiatrist Nancy Andreasen and their colleagues wrote that recent results from schizophrenia molecular genetic research "provide some grounds for cautious optimism."[4] Subsequently, the editors of *Science* declared the "identification" of "mental illness" genes to have been the second most important "scientific breakthrough" of 2003,[5] and Elkin, Kalidindi, and McGuffin wrote in 2004 that "Schizophrenia genes have been found at last."[6] Later in 2004, Kendler, on the basis of mixed yet "suggestive" replication results, wondered whether "a corner has been turned in our long struggle to understand the genetic basis of schizophrenia."[7]

In early 2005, Kendler concluded that, "the first generation of linkage and association studies in schizophrenia has succeeded in identifying replicated susceptibility genes."[8] I disagree, and it is likely that these claims will share the same fate as other such unsubstantiated claims we have seen over the past decades. On a website promoting the 2005 World Congress on Psychiatric Genetics XIII, where the announcement of gene discoveries would seem to be the order of the day, Faraone stated modestly, "These are intriguing times for psychiatric genetic research. For some disorders, meta-analyses and large collaborative projects have begun to define linkage regions and candidate genes worthy of further study."[9] And in Kirov and colleagues' June, 2005 review article, they pointed merely to molecular genetic studies having found "a number of potential regions of linkage and 2 associated chromosomal abnormalities" for

---

1. Ibid., p. 614.
2. DeLisi & Crow, 2003, p. 599.
3. Crow, 2003, p. 1829.
4. Kennedy et al., 2003, p. 822.
5. Anonymous, 2003, p. 2039.
6. Elkin et al., 2004, p. 107.
7. Kendler, 2004, p. 1535.
8. Fanous & Kendler, 2005, pp. 10-11.
9. Faraone, S. V. Welcome to 2005 World Congress on Psychiatric Genetics XIII. Retrieved on 6/11/05 from http://www.wcpg.org/welcome.htm.

schizophrenia.[1] And as we saw at the outset of this chapter, by mid-2005 Kendler would write, "we do not have and are not likely to ever discover 'genes for' psychiatric illness."

The Search for Genes is Based on Unsound Evidence

According to Ridley, although linkage studies "have largely failed for psychoses," the "role of genes in schizophrenia is proved by the twin studies and adoption studies, not by finding or failing to find particular genes."[2] This is merely a journalistic echo of schizophrenia molecular genetic researchers' fundamental error, which is their mistaken belief that the genetic basis of the condition has been definitively established. Given the long history of claims followed by retractions or failures to replicate, the *American Journal of Psychiatry* published a letter by critic S. J. Pittelli in 2003, who asked, "At what point is it safe to say that no...linkages are likely to be found for schizophrenia or other mental disorders?" He continued,

> Perhaps schizophrenia and other mental disorders, even though shown to be heritable in twin studies, are not actually genetic. This, of course, is paradoxical and would require a paradigm shift in our understanding of mental illness and genetics. But the fact of the matter is that the evidence is pointing in that direction, despite what our current scientific bias leads us to believe. The mind is a complex thing, and our attempts to reduce it to genetic loci or epigenetic expression is surely an oversimplification, if not a complete misrepresentation.[3]

In response, DeLisi and Crow wrote of having "some sympathy with [Pittelli's] opinion if it refers to the gene sequence rather than its expression," even if "twin and adoption study evidence cannot be discarded."[4] But no one is asking them, as yet, to simply "discard" twin and adoption studies. However, it is incumbent upon them to perform a critical reexamination of the original documents produced by the authors of these studies — in addition to the works of critics — and to carefully assess these studies' methods and the validity of their theoretical assumptions. At this point they may well conclude that these studies should indeed be discarded, and that their search for "schizophrenia genes" was based on the erroneous assumption that family, twin, and adoption studies demonstrate the genetic basis of schizophrenia.

In a 2003 textbook chapter, schizophrenia researchers Alastair Cardno and Robin Murray wrote,

> In the mid-1980s, it was widely predicted that family, twin and adoption studies of schizophrenia would be rendered obsolete by the application of molecular genetic techniques to this condition. However, the anticipated advances have yet to

1. Kirov et al., 2005, p. 1440.
2. Ridley, 2003, p. 107.
3. Pittelli, 2003, p. 597.
4. DeLisi & Crow, 2003, p. 598.

occur. Consequently, there has been a revival of interest in the classical genetic epidemiological approach to schizophrenia.[1]

Rather than simply reproduce previously faulty research, often based on false theoretical assumptions, this "revival of interest" must focus on earlier *criticisms* of genetic claims derived from these previous "genetic epidemiological" approaches. I urge psychiatric geneticists and molecular genetic researchers to read, in addition to my previous book, works such as Don Jackson's chapter in *The Etiology of Schizophrenia*, Theodore Lidz and colleagues' critiques of schizophrenia adoption research, Lewontin and colleagues' *Not in Their Genes*, the second edition of Boyle's *Schizophrenia: A Scientific Delusion?*, and Ross and Pam's *Pseudoscience in Biological Psychiatry: Blaming the Body*. Only then will they understand that their predictions of the mid-1980s may have been based on mistaken interpretations of previous kinship studies. Only then will it be possible for Cardno and Murray to reconsider their position that "there can be no doubt that schizophrenia is under considerable genetic influence."[2]

It is unclear whether most molecular genetic researchers, in addition to ignoring the critics, have made a careful examination of the original kinship studies upon which they base their work (see Chapters 5 and 10). In Merikangas and Risch's article, for example, they wrote in reference to a 1953 review article by Franz Kallmann, of "a sophisticated series of twin and family studies in the United States...and Europe...that corroborated the genetic roots of schizophrenia or manic depressive psychosis that had been demonstrated in the early part of the 20th century."[3] In their reference list, however, they cited no schizophrenia or manic depression twin studies. These studies were far from "sophisticated," and their often crude biases and methodological problems prompted schizophrenia twin researchers in the 1960s to make many improvements in their work (see Chapter 6). Moreover, no one discovered the "genetic roots" of schizophrenia and manic depression "in the early part of the 20th century," although many at the time *believed* that they had. There were only a handful of family studies that were carried out nonblinded by fanatical racial hygienists such as Kallmann and Ernst Rüdin.[4] *Today*, unlike Kallmann in 1953, most genetic researchers would concede that even the most perfectly performed family study cannot disentangle possible genetic and environmental influences on psychopathology.

Another example is Cowan, Kopnisky, and Hyman's article in the 2002 *Annual Review of Neuroscience*. These authors wrote that family, twin and adoption studies "have established beyond doubt that schizophrenia, manic-depressive disorder, autism, and several other illnesses have a large genetic component." For

---

1. Cardno & Murray, 2003, p. 195.
2. Ibid., p. 211.
3. Merikangas & Risch, 2003b, p. 625.
4. Rüdin, 1916.

them, "Kallman's [*sic*] twin studies were the first to establish that the concordance rate for schizophrenia among monozygotic (MZ) twins was on the order
of 50% but only 5%-15% among dizygotic twins and other siblings (Kallman
[*sic*], 1938)."[1] In fact, Kallmann's 1946 twin study (not 1938, as Cowan et al.
reported, apparently confusing Kallmann's twin and family studies) was preceded by the studies of Luxenburger and Rosanoff (see Table 6.2), who reported
identical twin concordance rates of over 50%. And Kallmann himself reported an
age-corrected identical concordance rate of 86%, not 50%.

Cowan and colleagues then moved on to adoption research:

> Adoption studies (in which children, one of whose biological parents suffered
> from schizophrenia or a related disorder, were adopted shortly after birth and
> brought up by different adoptive parents) strongly suggest that these risks are due
> to genetic rather than environmental factors.[2]

As evidence, Cowan et al. cited "Kety et al., 1976, Kendler & Gruenberg,
1984." However, we have seen that their description matches Rosenthal's
Adoptees method, but not Kety's Adoptees' Family method (see Figure 2.1).
These reviewers, who held posts at the US National Institute of Mental Health,
ask us to accept their claim that the genetic basis of schizophrenia and other
psychiatric disorders has been "established beyond doubt," yet their accounts
raise questions of whether they are familiar with many of the original research
publications that form the basis of their claims.

Optimistic reports are driven partly by researchers' need to maintain
funding. Rather than emphasize over 20 years of *failures*, they emphasize hoped
for *future* discoveries. In framing their results in this way, they hope to maintain
funding sources that might dry up if they emphasized the possibility that they
won't find anything in the future. In the words of Hastings Center researcher
Erik Parens, "Some of this 'genes for' language is run-of-the-mill hype. The language is intended to attract attention, and ultimately dollars."[3] It may also
attract attention and dollars for the publishers of scientific journals. According
to Propping, "A critical observer might comment that the scientific journals have
been the big winners, with their increased impact factors caused by repeated
citations in the non-replication studies. Through this mechanism psychiatric
genetics has certainly raised the impact of the high-ranking journals."[4]

Researchers have resorted to meta-analyses (combining the results of
several previous studies) of investigations finding negative and positive results
in an attempt to produce significant results in support of associations between
disorders and chromosomal regions. An example in schizophrenia research is
Glatt, Faraone, and Tsuang's 2003 meta-analysis of the catechol O-methyltrans-

---

1. Cowan et al., 2002, p. 8.
2. Ibid., p. 8.
3. Parens, 2004, p. s4.
4. Propping, 2005, p. 6.

ferase (COMT) linkage data literature.[1] The investigators included 18 previous studies in their analysis, yet only 4 of these had found significant linkage for the COMT polymorphism. They concluded that the Val allele of the COMT polymorphism "may be considered a risk factor for schizophrenia..."[2] In response, Pittelli wrote, "I find this trend of using meta-analysis to resurrect largely negative genetic linkage studies disturbing. It appears to be nothing more than a manipulation of data to obtain a desired result."[3] It does indeed appear to be such a manipulation, yet readers relatively unsophisticated in genetic research and terminology may well conclude that yet another "schizophrenia gene" has been identified. As 2005 rolled around, additional research groups were reporting a lack of association between schizophrenia and the Val allele.[4]

The Fruitless Search for Schizophrenia Biological Markers (Endophenotypes)

Biological markers in psychiatry (also known as "endophenotypes," "subclinical traits," "intermediate phenotypes," and "vulnerability markers"[5]) have been defined as "any neurobiological measure related to the underlying molecular genetics of the illness, including biochemical, endocrinological, neurophysiological, neuroanatomical, or neuropsychological markers."[6] Some characteristics researchers require for a marker are (1) that it can be reliably measured, (2) that it is manifest among all people with a susceptibility locus (independent of state), (3) that it is specific to the disorder under study, and (4) that it is inherited.[7] Gottesman and Shields introduced this concept into psychiatry in 1972, hoping that one day researchers would discover biological or behavioral schizophrenia endophenotypes "which would not only discriminate schizophrenics from other psychotics, but will also be found in all the identical cotwins of schizophrenics whether concordant or discordant."[8] For example, cancer is diagnosed through a biopsy, hypertension with a blood pressure test, and diabetes by measuring blood glucose levels.

In psychiatry, however, diagnoses are made by the observation and assessment of people's behaviors and mood, or by reports of these behaviors and moods. There are no biological markers or laboratory tests for psychiatric disorders because, with rare possible exceptions (e.g., autism), it is unlikely that they have a biological basis. Rather, human psychological distress (other than normal reactions to loss) is largely the result of environmental factors such as

---

1. Glatt et al., 2003.
2. Ibid., p. 474.
3. Pittelli, 2004, p. 1134.
4. Fan et al., 2005; Ho et al., 2005.
5. Gottesman & Gould, 2003.
6. Egan et al., 2003, p. 277.
7. Lenox et al., 2002; Skuse, 2001.
8. Gottesman & Shields, 1972, p. 336.

abusive and inadequate family rearing environments, social and political oppression (such as racism and sexism), and the difficulty in coping with the demands and messages of advanced industrial societies.[1] Since these and other environmental factors are indeed the main cause of psychopathology, then, unlike real diseases, there will be no biological markers identifying people with mental disorders (other than those caused by environmental stressors). Nevertheless, advocates of biological and genetic theories believe that it is only a matter of time until biological markers are discovered. In the words of psychiatrist Joel Paris (who came to psychiatric genetics relatively late in his career), "psychiatrists would eventually like to be able to make diagnoses much in the same way as internists, confirming their clinical impressions by conducting laboratory tests."[2] Of course, they would. But it will never happen because, unlike internists, psychiatrists treat people with emotional — not physical — conditions.

Given the ongoing failure to find genes for psychiatric disorders, several investigators now believe that discovering biological markers is necessary for molecular genetic research to go forward, acknowledging that "defining the phenotype solely in terms of an operational diagnosis of schizophrenia may be inadequate."[3] In their contribution to the April 2003 *American Journal of Psychiatry* genomics edition, Gottesman and Gould wrote that because "multiple genetic linkage and association studies using current classification systems [such as the DSM]...have all fallen short of success, the [endophenotype] term and its usefulness have reemerged.... Endophenotypes are being seen as a viable and perhaps necessary mechanism for overcoming the barriers to progress."[4] More forcefully, behavior geneticist Richard Rende wrote that the failure to identify biological markers for psychiatric disorders "would make the search for actual genetic susceptibility loci nearly impossible..."[5] Researchers have discussed several "candidate" biological markers for schizophrenia, such as eye tracking dysfunction, clinical and cognitive phenotypes, neurological phenotypes, electrophysiological markers, and neuroimaging phenotypes.[6]

The current emphasis on identifying biological markers is an indication that schizophrenia molecular genetic research is reaching the desperation stage. It is supremely ironic that DSM-type "operational diagnoses" are now seen as inadequate for diagnosing schizophrenia and other disorders when, for decades,

---

1. See Bentall, 2003, Read et al., 2004, for recent reviews of environmental causes of psychosis.
2. Paris, 1999, p. 22.
3. Cardno & Murray, 2003, p. 212.
4. Gottesman & Gould, 2003, p. 637. In 2001 Kendler listed biological marker research as one of the "five areas of development for twin studies [that] are likely to be particularly fruitful in the coming years" (Kendler, 2001, p. 1012).
5. Rende, 2004, p. 121.
6. Egan et al., 2003.

psychiatry has emphasized the supposed reliability and validity of its diagnoses. In other words, if psychiatrists can reliably identify people with "schizophrenia," as they frequently claim, why do they need biological markers to enable them to identify schizophrenia? According to Merikangas and Risch, "Psychiatric disorder phenotypes, based solely on clinical manifestations without pathognomonic markers, still lack conclusive evidence for the validity of classification and the reliability of measurement."[1] And Kennedy and colleagues wrote that, "the psychoses have nonspecific pleomorphic phenotypes, making diagnosis and nosology difficult."[2] But if schizophrenia and other psychiatric diagnoses are of questionable validity and reliability, then previous family, twin, and adoption study results might be worth little more than the paper they were printed on. The current perceived necessity of identifying biological markers underscores the need to re-examine family, twin, and adoption studies because, if "schizophrenia" can't be reliably identified in the "genomics era," we can safely assume that investigators such as Rüdin, Luxenburger, Kallmann, Gottesman, Kety, Rosenthal, Slater, Tienari and others couldn't reliably identify it, either.

Another problem in biological marker research is that, if a DSM-type diagnosis is inadequate for the purpose of finding genes, it is also inadequate for the purpose of finding biological markers. In other words, the same "schizophrenia phenotype" utilized in the search for genes is used for marker searches as well. According to schizophrenia researchers M. F. Egan and colleagues, "Most studies of intermediate phenotypes begin by looking for a difference between first-degree relatives and controls."[3] First-degree relatives of whom? The answer is, of course, of people diagnosed with schizophrenia according to the DSM, that is, the same faulty diagnostic scheme that necessitated the search for biological markers in the first place. As an analogy, suppose we wish to identify gold particles in rocks with "Instrument A," which we suspect might be unreliable. Naturally, we would seek a second gold assaying "Instrument B" in order to verify the original finding. But if we use Instrument A to validate Instrument B, we are simply repeating the same error. This, in a nutshell, is the logic behind the search for biological markers in psychiatry.

Another factor in schizophrenia biological marker research is that the effects of neuroleptic (also known as "antipsychotic") drugs potentially confound the findings. These drugs are known, in some cases, to produce neurological malfunction and irreversible brain damage (for example, tardive dyskinesia).[4] Even in cases where this does not occur, people using these drugs behave differently from when they are not taking them, meaning that attempts to look for trait markers such as eye tracking and memory function must take

---

1. Merikangas & Risch, 2003b, pp. 627-628.
2. Kennedy et al., 2003, p. 825.
3. Egan et al., 2003, p. 280.
4. Breggin & Cohen, 1999; Cohen, 1997.

the effects of neuroleptics into account. This is a difficult task, because most people diagnosed with schizophrenia in the industrialized world are taking neuroleptic drugs. In a 2004 schizophrenia biological marker study of 24 people diagnosed with schizophrenia, for example, 19 were taking "atypical antipsychotic" medication, 4 were taking "typical antipsychotics," and only 1 was not medicated.[1]

Egan and colleagues recognized that biological marker research is "plagued with problems of phenocopies," and that "neurobiological abnormalities may occur because of medication, drug abuse or other problems associated with chronic mental illness."[2] Nevertheless, psychiatric geneticist Ming Tsuang and colleagues published a 2005 preliminary study in which they claimed to be able to differentiate people diagnosed with schizophrenia from "normal control subjects" on the basis of blood cell-derived gene expression profiles.[3] However, critics have pointed out for many years that the use of "normal controls" in schizophrenia research leaves studies much more vulnerable to the presence of confounding variables.[4]

Ironically, the continuing failure to identify endophenotypes calls into question claims that schizophrenia is a brain disorder. If it is really a brain disorder, then studies claiming to have discovered brain dysfunction should supply numerous schizophrenia endophenotypes. For example, if a researcher discovers that people diagnosed with schizophrenia have enlarged brain ventricles, then enlarged brain ventricles would become an excellent endophenotype for schizophrenia. However, although nothing like this has occurred, dubious brain dysfunction and genetic theories continue to cross-validate each other. Meanwhile, in the face of no evidence of their existence, some researchers are content to simply proclaim that schizophrenia endophenotypes exist: "Specific endophenotypes probably occur at fairly high frequency in the broad schizophrenia patient population."[5]

The Future

The question may not be whether or when schizophrenia genes will be identified, but when, and accompanied by which conclusions, the search will be called off. It is likely that at some point those financing this research will realize that they may be wasting their money. But as long as funding remains, researchers will continue to look for schizophrenia genes and, most likely, will continue to not find them. At some point the more thoughtful among them will reexamine the entire "schizophrenia as genetic disorder" fiasco and may begin to

1. Ettinger et al, 2004, p. 178.
2. Egan et al., 2003, p. 292.
3. Tsuang et al., 2005.
4. Boyle, 2002a.
5. Heinrichs, 2005, p. 239.

question some of its basic tenets. For others, the challenge will be to explain negative finding in novel ways, while continuing to argue that although schizophrenia is more "complex" than originally believed, discoveries are just around the corner.[1] Genetic researchers Hans Moises and colleagues wrote in 2004, "The polygenic model of schizophrenia has now survived more than 30 years of testing by family, twin, adoption, and linkage studies."[2] What is unclear is how the polygenic model is supported by 30 years of *failures* to find linkage.

The "inconsistencies and uncertainties regarding the linkage findings" have led, according to Harrison and Weinberger in 2005, to the formation of "two schools of thought" in the research community.[3] The "skeptical" school "doubts all results and views the strategy as nonproductive," while the "optimistic" school "sees the results as consistent with predictions of weak effect of multiple genes and with genetic heterogeneity, and which looks forward to ever larger-scale studies." One error of the "optimistic school" is that its "predictions" are *based on* the failure to discover genes. This is analogous to the US government's 2001–2004 claim that the Iraqi government possessed "weapons of mass destruction" ("WMDs"). After invading Iraq and failing to find WMDs, the US government claimed that these weapons had been destroyed, or that they were cleverly hidden. The continuing failure to find WMDs was then presented as being "consistent" with the "prediction" that these weapons would not be found for the above-stated reasons. Yet, these predictions were based on *ad hoc* hypotheses developed after the embarrassing failure to find WMDs. Apparently, US government leaders preferred this course to simply admitting that they had been wrong all along.[4]

In the original edition of *The Gene Illusion*, I quoted Nancy Andreasen's statement that schizophrenia is caused by an "'invisible lesion' that cannot be seen with the naked eye or under a microscope."[5] I then asked rhetorically whether researchers might someday claim that schizophrenia is caused by "invisible genes." Little did I realize how soon this would come to pass. In 2003, Timothy Crow discussed the possibility that "the relevant variation is epigenetic — i.e., involves modifications such as methylation of the sequence rather than alterations of the DNA sequence itself. For this reason, the modification is invisible in terms of the linkage strategy."[6]

---

1. In one case, a Nobel Laureate geneticist claimed to have created "schizophrenia" in genetically modified mice. See Smith, D. *Schizophrenia gene discovery breakthrough.* Retrieved on 12/12/03 from http://www.smh.com.au/articles/2003/07/08/1057430208709.html
2. Moises et al., 2004, p. 148.
3. Harrison & Weinberger, 2005.
4. A difference is that, whereas the U.S. government may have known all along that Iraq had no WMDs, genetic researchers believe that the genes they are looking for actually exist.
5. Andreasen, 2001, p. 209.

In 1992, Michael Owen wondered aloud if schizophrenia "will become a graveyard for molecular geneticists?" Because molecular genetic research in schizophrenia was "still in its infancy," he reasoned, "talk of graveyards is premature."[1] Owen and other investigators were beginning their work in a "multi-centre collaborative programme" of schizophrenia molecular genetic research, supported by European foundations and the American NIMH. "Both programmes," wrote Owen, "will take up to five years to complete but should be successful in detecting and locating a major gene (or major genes) *if such exist* [emphasis added]."[2] These words were written in 1992, and schizophrenia genes have yet to be found. Real progress in understanding, treating, and preventing psychosis will be made only when the adjectives describing schizophrenia genes are changed from "invisible" or "elusive," to "probably nonexistent."

In 1999, I published a critical review of schizophrenia twin and adoption research, which ended with the following prediction: "Based on the weight of the evidence, it is predicted here that a gene for schizophrenia will not be found, because it does not exist."[3] As of this writing, and despite some recent claims of gene discoveries,[4] I see no reason to modify this prediction.

THE FRUITLESS SEARCH FOR ADHD GENES

ADHD is the poster child for psychiatry's transformation of misbehavior into medical diagnosis, epitomized by a 2002 "Consensus Statement on ADHD" signed by Russell Barkley and more than 80 other ADHD researchers.[5] In their statement, Barkley et al. claimed that there is "no disagreement" among "scientists who have devoted years, if not entire careers" to the study of ADHD, that it is a "real medical condition." Those opposing this position were portrayed as "social critics and fringe doctors whose political agenda would have you and the public believe there is no real [medical] disorder here." Undoubtedly, they had in mind critics such as *Talking Back to Ritalin* author Peter Breggin. The Statement went on to claim that ADHD has been linked to "several specific brain regions," and that children diagnosed with ADHD "demonstrate relatively smaller areas of brain matter...." Naturally, twin studies were said to have provided evidence that ADHD is "primarily inherited," and that the genetic contribution to deficits in

---

6. Crow, 2003, p. 1929.
1. Owen, 1992, p. 292.
2. Ibid., p. 291.
3. Joseph, 1999b, p. 137.
4. For example, see Egan et al., 2004; Elkin et al., 2004. See also the August 11th, 2004 NIMH press release entitled *Schizophrenia Gene Variant Linked to Risk Traits*, retrieved on 9/10/2004 from http://www.nimh.nih.gov/press/prschiz-gene.cfm.
5. Barkley et al., 2002. For a response to the Statement, see Timimi et al., 2004.

attention and inhibition are "nearly approaching the genetic contribution to human height." Finally, Barkley et al. claimed that one (unnamed) gene "has recently been reliably demonstrated to be associated with this disorder," and that "the search for more is underway by more than 12 different scientific teams worldwide at this time." No citations were provided for any claim about brain defects or genetics, although readers were informed that "a full list of references can be obtained from Professor Russell Barkley."

Like schizophrenia, previous kinship research has laid the basis for molecular genetic investigations in ADHD (see Chapter 2). According to Faraone and colleagues, "Family, twin, and adoption studies show attention deficit hyperactivity disorder (ADHD) to have a substantial genetic component...."[1] And others have written that these studies' results provide a "compelling argument for now searching for susceptibly genes at the molecular level."[2] Also like schizophrenia, the failure to discover genes has led to speculation that many genes are involved.[3]

The search for genes is based on mainstream psychiatry's beliefs about ADHD, which include (1) that ADHD is a valid diagnostic category that can be reliably diagnosed, (2) that ADHD is a familial disorder, (3) that ADHD involves a malfunction of the brain, (4) that the greater resemblance of identical versus fraternal twins on ADHD-related measures is the result of the former's greater genetic similarity, (5) that ADHD adoption studies suggest the importance of genetic factors, (6) that researchers possess the technology to find genes, and (7) that gene discoveries would aid in the treatment or prevention of ADHD. These points roughly parallel Faraone and colleagues' "chain of psychiatric genetic research" approach (discussed briefly in Chapter 10), which requires positive results from family studies, twin studies, adoption studies, segregation analysis, and linkage and association studies as a prerequisite for searching for actual genes.[4] It is a truism, however, that a chain is only as strong as its weakest link. There is little evidence supporting points 1, 3, 4, and 5; and point 7 is debatable. Thus, the heretofore fruitless search for genes should be understood as a demonstration of the misplaced faith that researchers have in conclusions drawn by ADHD twin and adoption researchers, rather than as a result of the difficulty of finding genes.

Researchers currently focus on genes involved with the brain's dopamine receptors, which they view as "candidate genes" on the basis of an a priori hypothesis derived from neurochemical and neuropharmacological research.[5] The major areas of interest have been the DRD4 dopamine receptor gene and the

---

1. Faraone, Biederman, et al., 1999, p. 768.
2. Thapar et al., 1999, p. 108.
3. Zametkin et al., 2001.
4. Faraone, Tsuang, & Tsuang, 1999, p. 11.
5. Asherson & Curran, 2001; Barr, 2001.

DAT1 dopamine transporter gene. In their 2000 response to my article on the genetics of ADHD, Faraone and Biederman claimed that "molecular genetic studies have implicated these two genes...in the etiology of ADHD."[1] However, subsequent studies have failed to replicate an association between ADHD and the DRD4 or DAT1 genes,[2] with one reviewer concluding in 2004 that the associations have been "refuted."[3]

A 2002 study by University of California, Los Angeles (UCLA) geneticist Susan Smalley and her colleagues linking ADHD to a locus on a region of chromosome 16p13 was, as by now should be expected, accompanied by media reports announcing a new "ADHD gene discovery."[4] The title of a UCLA press release read, "UCLA geneticists find location of major gene in ADHD; targeted region also linked to autism."[5] In this press release, Smalley claimed that her team's findings "have narrowed our search for a risk gene underlying ADHD to some 100 to 150 genes," while cautioning that "we must wait for independent replication of our results to confirm these findings."[6] However, a 2003 ADHD genome scan published by the same research group, as well as a 2003 genome scan by another group, found no statistically significant results for the 16p13 region (or for the DRD4 or DAT1 genes).[7] In a 2004 follow-up, the investigators made new tentative linkage-finding claims.[8] By mid-2005, Faraone and colleagues concluded that "the handful of genome-wide scans that have been conducted thus far show divergent findings and are, therefore, not conclusive."[9]

Many years ago, psychiatry critic Thomas Szasz observed that "Much of what passes for scientific advance in psychiatry is, in fact, rhetorical innovation."[10] Nowhere is this seen more clearly than in the psychiatric genetics literature in general, and in the ADHD genetic literature in particular. Rather than face the sobering reality that they have found no ADHD genes — and may never find presumed ADHD genes — psychiatric geneticists and their supporters instead write optimistically about the great strides they have made, and how ADHD genes will soon be identified. They write as if they were searching for the

---

1. Faraone & Biederman, 2000, p. 573.
2. Bakker et al., 2005; Barkley, 2003; Mill et al., 2005; Ogdie et al., 2003; Shastry, 2004; Thapar, 2003.
3. Shastry, 2004. A study published in early 2004 (Langley et al., 2004) found an association between the DRD4 7-repeat allele and neuropsychological test performance among ADHD children.
4. Smalley et al., 2002.
5. UCLA press release. (2002). *UCLA geneticists find location of major gene in ADHD; targeted region also linked to autism.* Retrieved on 10/1/03 from http://www.vaccinationnews.com/DailyNews/October2002/UCLAGeneticists23.htm
6. Ibid.
7. Bakker et al., 2003; Ogdie et al., 2003.
8. Ogdie et al., 2004.
9. Faraone et al., 2005.
10. Szasz, 1964, p. 525.

cure of a deadly disease or the virus causing an epidemic. But ADHD is simply a grouping of socially disapproved behaviors falsely passed off as a disease, and finding genes would do little if anything to "cure" these behaviors. Still, the ADHD genetic literature is replete with claims that gene discoveries are imminent, and that finding genes would be an important discovery.

Few prominent psychiatric genetic researchers have been willing to put their cards on the table. One who did was Lynn DeLisi, who in 2000 wrote frankly that psychiatric genetics "appears to be at a crossroads or crisis" as investigators continue to look for the "elusive gene or genes" for schizophrenia and other disorders.[1] In contrast, recent statements by ADHD genetic researchers fail to emphasize that they have found no ADHD genes. Instead, they emphasize what they *hope* to find, often writing in ways that falsely suggest that genes have been discovered. Some examples include: The "initial studies of candidate genes suggest that mutations of genes within the dopamine system that richly innervates frontal-striatal circuits may increase the susceptibility of ADHD"[2]; "There have been rapid advances in psychiatric genetics research and much recent interest in the genetics of ADHD"[3]; "Some of the specific genes involved [in ADHD] are just beginning to be identified"[4]; "The identification of susceptibility genes is only the initial step. Describing the molecular mechanisms involved...will be a far greater challenge"[5]; "Molecular genetic studies have implicated the dopamine D4 and the dopamine transporter as candidate genes"[6]; and "Molecular genetic studies have discovered genes that explain some of the disorder's genetic transmission."[7]

Generally speaking, these investigators substitute *language* for real gene findings. Thus, when they scan the genome and find no ADHD genes, they can say that genes are "implicated," or that researchers are making "enormous advances," or that genes are "just beginning to be identified," or that studies "suggest" the finding of genes, and so on. Ultimately, however, optimistic statements cannot eliminate the necessity of finding actual genes. "Psychiatry as a discipline," wrote psychiatric geneticists Zerbin-Rüdin and Kendler in words that apply perfectly to the current discussion, "has too often been characterized by many speculations based on few facts."[8]

A formula for explaining away negative results was put forward in 2000 by psychiatric geneticists Comings and Blum. Although admitting that "despite enormous effort using lod and linkage techniques...the number of disease specific

---

1. DeLisi, 2000, p. 190.
2. Tannock, 1998, p. 89.
3. Thapar et al., 1999.
4. Comings et al., 2000, p. 178.
5. Asherson & Curran, 2001, p. 126.
6. Wilens et al., 2002, p. 119.
7. Faraone, 2005, p. 6.
8. Zerbin-Rüdin & Kendler, 1996, p. 332.

genes identified" for diagnoses such as schizophrenia, manic-depressive disorder, autism, and panic disorder "is virtually zero," they put forward replication failures as a normal aspect of research on polygenic disorders.[1] For Comings and Blum, in polygenic disorders "it is very likely that if one group of investigators reports a significant association with a specific gene, that association may not be replicated in another study." Thus, "the variation between studies is the expected outcome in studies of polygenic disorders."[2] In other words, whether researchers find gene associations for conditions such as ADHD, or do *not* find gene associations, they can still claim that their results are expected and that the trait is genetically transmitted. This "heads I win, tails you lose" approach to molecular genetic research means that genetic predictions are changed to fit disappointing data, as opposed to the standard practice of collecting data in order to test a prediction. "The argument that ADHD is 'mediated by many genes acting in concert,'" observed Pittelli, "is rather circular in that it is based primarily on the complete failure of molecular genetic studies to find such genes and replicate those findings."[3]

Biological Markers

As with schizophrenia, ADHD molecular genetics researchers seek to identify biological markers (endophenotypes). A group of researchers investigating biological markers for ADHD argued that "traditional nosological categories described in the DSM-IV...and ICD-10...are suboptimal when it comes to describing who is affected and carrying susceptibility genes and who is not," and that to "unravel the genetic constellation of ADHD, emphasis should be on the description of endophenotypes."[4] Several cognitive traits have been proposed as possible markers to be studied. But as was discussed earlier, if a DSM-type diagnosis is inadequate for gene searches, it is also inadequate for marker studies. Once again, molecular genetic research has entered a theoretical blind alley.

We recall from Chapter 2 that Peter Breggin has observed that ADHD is "simply a list of behaviors that require extra attention from teachers." Despite my respect for Breggin's work, when I first read this I felt Breggin couldn't be right. I then looked up the DSM-IV ADHD diagnostic criteria and discovered that he was completely right. In fact, most of the criteria, such as "fidgeting," "forgetting," and "having difficulty awaiting turn" are found among most "normal" children.[5] The difference between "normal" and "ADHD" children, according to the DSM-IV and DSM-IV-TR, is the *frequency* of these behaviors, denoted by the word "often" (for example, "is often forgetful in daily activities").

1. Comings & Blum, 2000, p. 327.
2. Ibid., p. 327.
3. Pittelli, 2002, p. 496.
4. Slaats-Willemse et al., 2003, pp. 1242-1243.
5. APA, 2000, p. 92.

Given these criteria, what type of ADHD "endophenotypes" could we expect to find? If both "normal" and "ADHD" children exhibit ADHD symptoms, albeit in differing degrees, how can a gene or biological marker know the difference between "normal" and "often" in a given culture?

Are Gene Findings Necessary in Order to Study Environmental Factors?

In 2003, Faraone commented, as quoted in an article by Kathryn Brown in *Science*, "My hope is that once we've discovered those [ADHD] genes, we'll be able to do a prospective study of kids at high versus low genetic risk. That's when you'll see environmental factors at work."[1] What is difficult to understand is why genes must be discovered as a prerequisite for pinpointing environmental factors. In response to this statement, critic Jonathan Leo commented, "But certainly one can still see environmental factors at work in children without knowing their genotype." According to Brown, Faraone added that "eventually...environmental changes could play an important role in treating some ADHD patients."[2] Leo responded, "Eventually? What are we waiting for? Why not implement the changes right now? Changing the environment is exactly what many people opposed to Ritalin have been saying for years."[3]

In their 2000 psychiatric genetic "Millennium Article," Owen and colleagues wrote about genetic and environmental factors in psychiatric disorders as follows:

> Indeed, even when all susceptibility genes for a given disorder have been identified, it will still not be possible to predict the development of disease with certainty until the relevant environmental risk factors have also been identified and the nature of the various interactions understood.[4]

But if we are able to identify environmental risk factors, there is a good chance that *everyone* can be protected from a psychiatric disorder in the same way that people — regardless of their genetic makeup — are protected from pellagra by eating vitamin-enriched diets. Owen and colleagues recognized that the (future) discovery of susceptibility genes would still require the identification of environmental triggers in order to "predict the development" of a disorder. Yet, ironically, the discovery of these triggers could render the genetic findings largely irrelevant. Thus genetic investigators, while hoping for the discovery of environmental factors, understand that the supremacy of genetic explanations depends on environmental factors remaining mysterious. As sociologist Adam Hedgecoe observed, "environmental factors are usually classed [by geneticists] in terms of being 'non-specific' and hard to identify; unless one or two specific

1. Quoted in Brown, 2003, p. 160.
2. Ibid., p. 160.
3. Leo, 2003, p. 412.
4. Owen et al., 2000, p. 29.

environmental influences can be identified, then the role of environmental factors, while accepted, cannot be given priority."[1]

In summary, fantastic claims about ADHD gene discoveries have been made, are being made as I write this, and will continue to be made. The common denominator is that nothing has come, or will likely come, of any of these claims. Like schizophrenia, in 2000 I published a prediction that ADHD genes will not be discovered, because they do not exist.[2] I see no reason to modify this prediction either.

THE FRUITLESS SEARCH FOR AUTISM GENES

One can make a distinction between looking for autism genes versus looking for ADHD genes. The reason is that autism, which touches the core of what it means to be human, may be a true physical disease of the brain and usually lasts a lifetime. ADHD, on the other hand, is a pseudo disease reserved for misbehaving children who typically grow out of the behavior as adults (although we are currently witnessing a drug company-inspired campaign to convince perhaps millions of potential customers that they may suffer from "Adult ADD"). Autism is lifelong and heartbreaking; ADHD is temporary and annoying (although many adults may have various motives, including personal or vocational failures or difficulties, to believe that they "are" or have "always been" ADHD). Yet for both, something clearly is going on. For autism it may be (nongenetic) biological, whereas for ADHD-type behaviors it may be sociological.

Chapter 7 demonstrated that, contrary to popular and professional opinion, there is little scientifically acceptable evidence supporting autism as a genetic disorder. Perhaps because of this, the search for autism genes has followed a course similar to that of other disorders. First, there is uncritical reliance on family and twin study results, then well-publicized early claims of gene findings, followed by failures to replicate accompanied by rationalizations and optimistic statements about exciting discoveries to come.

Various autism candidate genes, linkage studies, genome scans, chromosome studies and so on have failed to produce genes for autism. Like the other disorders, the more failures that pile up, the more "complex" autism becomes. Indeed, the authors of a recent review called autism a "paradigmatic complex genetic disorder."[3] Others are content to list subsequently non-replicated linkage findings as if they provided evidence in support genetics. In a 2004 textbook chapter, for example, L. Y. Tsai identified the majority of human chromosomes as harboring autism genes: "Strong evidence for linkage to autism has

---

1. Hedgecoe, 2001, p. 892.
2. Joseph, 2000e.
3. Veenstra-VanderWeele et al., 2004.

been identified for chromosomes 1p, 2q, 3, 4p, 4q, 5p, 6q, 7q, 8, 10q, 11, 12, 15q11-q13, 16p, 17q, 18q, 19p, 19q, 22q, and Xp."[1] However, as a group of genetically-oriented researchers acknowledged in 2004, "Though numerous linkages and associations have been identified, they tend to diminish upon closer examination or attempted replication."[2]

Investigators have been searching for autism genes since the mid-1990s. At that time many were "wildly optimistic" about finding genes, with one investigator recalling, "I thought this would be an easy find. I really did."[3] A 1997 University if Chicago Hospital press release proclaimed, "Researchers Discover First Autism Susceptibility Gene."[4] According to genetic researcher Ed Cook, "This is just one of at least three to five genes whose interactions result in autism.... But nailing the first one confirms the value of the genetic approach and may provide clues about where to look for others." (This claim did not hold up to replication attempts.)

In 2000, the Associated Press reported, "Crucial Autism Gene Discovered."[5] According to this report, "Scientists have long theorized that about 15 different genes play a role in who is born with the severe brain disorder autism — and now they've finally found one of those genes." The newly "discovered" gene was HOXA1, the mutated version of which was said to be found in 40% of autistic subjects. "Children," continued the report, "need to inherit just one copy of the mutated gene from one parent to have autism." According to a 2000 statement by Duane Alexander, Director of the National Institute of Child Health and Human Development, and chair of National Institutes of Health Autism Coordinating Committee, HOXA1 gene research "strongly suggest[s] that a gene controlling early brain formation may underlie the development of autism in a large number of cases."[6] And in the September 2003 edition of *Scientific American*, NIMH Director Steven Hyman wrote that "a variation of HOXA1, a gene related to early brain development, seems to boost susceptibility to autism."[7] In the same year, however, investigators finding negative results were writing that the "potential significance" of the HOXA1 gene "to the etiology of autism must be reconsidered."[8] And in 2004, Veenstra-VanderWeele and Cook

1. Tsai, 2004, p. 277.
2. Wassink et al., 2004, p. 272.
3. Ritter, M. (2001). *Scientists optimistic about hunt for autism genes.* Retrieved on 10/1/03 from http://www.canoe.ca/Health0101/29_autism-ap.html
4. University of Chicago Hospitals. (1997). *Researchers discover first autism-susceptibility gene.* Retrieved on 10/1/03 from http://www.uchospitals.edu/news/1997/19970501-autism-gene.html
5. Associated Press. (2000). *Crucial autism gene discovered.* Retrieved on 10/1/03 from http://www.wired.com/news/technology/0,1282,40379,00.html
6. National Institutes of Health news release (11/27/2000). Retrieved on 12/12/03 from http://www.specialabilities.org/autism%20gene%20article.htm
7. Hyman, 2003, p. 100.
8. Collins et al., 2003, p. 346. A HOXA1 replication failure was published by Li and colleagues in 2002 (Li et al., 2002).

wrote that "attempts to replicate this [HOXA1] finding in larger samples have shown no supporting evidence."[1]

Autism molecular genetic researchers tend to rationalize their failures and look to the future. Following the trend in psychiatric genetics, after failing to find the expected gene or few genes of large effect size for autism, there is speculation that more genes are involved than originally expected. While such speculation is plausible, the opposing hypothesis — that there are no genes for autism — is never considered. According to Margaret Pericak-Vance, a leading autism molecular genetic investigator, "Gene discovery in autistic disorder...has been delayed by the fact that autistic disorder is a complex genetic disorder."[2] Similarly, Folstein and Rosen-Sheidly wrote that most cases of autism are "idiopathic and apparently due to complex inheritance patterns. This has made the identification of susceptibility genes difficult."[3] Finally, Veenstra-VanderWeele and Cook wrote in 2004 that, although "no gene variant has been identified yet as contributing to autism susceptibility...," this "is not unexpected in a complex genetic disease."[4]

However, it is not a "fact" that autism is a "complex genetic disease" — it is merely a *hypothesis*. Nevertheless, researchers can always fall back on the questionable "fact" they have created in order to explain their failures. The original investigators' opinions are then reported by secondary sources, such as the following conclusion found in a 2003 textbook: "It appears that there is no single gene that can account for the autism syndrome. Rather, there appear to be multiple genes involved..."[5] But isn't it also possible to state that there appear to be *no* genes involved? There is no doubt that these and other investigators are dedicated scientists attempting to make discoveries that could lead to successful treatments for autism, a most laudable goal. At the same time, they are unwilling to acknowledge that their theories and speculations are just that...theories and speculations. Most importantly, they are unwilling to entertain the possibility that previous family and twin studies can be interpreted in ways other than in the genetic direction.

Nine autism genome scans published through mid-2005 have produced no autism genes, although autism molecular genetic investigators continue to look at areas "suggested" by these studies.[6] According to Mark Blaxill, who examined the first seven studies in detail, "no major genome scan has produced significant *and* reproducible results...no candidate gene from a genome scan has shown a reproducible and statistically significant association with autism...[and] no candidate gene that has inspired multiple studies has shown a robust and repro-

---

1. Veenstra-VanderWeele & Cook, 2004.
2. Pericak-Vance, 2003, p. 268.
3. Folstein & Rosen-Sheidly, 2001, p. 963.
4. Veenstra-VanderWeele & Cook, 2004.
5. Klinger et al., 2003, p. 435.
6. Barnby et al., 2005; Cantor et al., 2005.

ducible connection to autism."[1] A major effort was launched in the summer of 2004 by the National Alliance for Autism Resea (NAAR) organization.[2] The NAAR Autism Genome Project, which drew samples from 1,500 affected families, pooled the resources of 50 academic and research institutions worldwide. As of this writing, the results of this undertaking have not been published.

One area of interest identified in the genome scans is a region on chromosome 7q, where research is currently focused. This region, however, has been identified only in some of the scans, and typically not in statistically significant elevations. In their 2002 review of the autism genetic literature, genetic investigators Catherine Lord and Anthony Bailey reported that while "several chromosomal regions [including 7q] appear particularly promising," because none of the genome scans "reported findings that meet the accepted threshold for proven linkage...it remains possible that the results simply represent the chance co-occurrence of a number of false-positives."[3] This illustrates how psychiatric genome scans, by chance alone, are likely to find something. As Folstein and Rosen-Sheidley wrote in 2001, "several genome scans have now been published and have yielded a fairly large number of suggestive linkage signals, only a few of which overlap from one study to the next."[4] However, we could interpret the lack of "overlap" between studies as showing that "suggestive linkage signals" are mere chance findings or are due to methodological errors. Furthermore, even if a finding is consistently replicated, it still doesn't prove that a gene *causes* autism, but only that the gene and autism are *correlated* (associated).

The search for autism genes, while seemingly a worthwhile scientific endeavor, can actually be counterproductive and diversionary. In a 2003 article, Chloe Silverman and Martha Herbert spoke of the alarming increase of autism cases and the accompanying "near pathological denial of both the fact of increasing rates and the role of extra-genetic factors implicated in the upsurge."[5] And as we saw in Chapter 7, environmental factors are implicated in this upsurge since "there is no such thing as a 'genetic epidemic.'"

Now that autism is seen as a genetic disorder, as much as $60 million has been earmarked in the search for the genetic roots of autism, while "other research approaches languish."[6] Silverman and Herbert spoke of the "twin denials" in current research as "the denial of increasing incidence and the denial of non-genetic biological and environmental factors." These twin denials, they wrote, are fueled by "the ideological role of genes and genetic research in con-

---

1. Blaxill, Defeat Autism Now (DAN) presentation.
2. National Alliance for Autism Resea. Retrieved on 2/12/05 from http://www.naar.org/news/render_pr.asp?intNewsItemID=176.
3. Lord & Bailey, 2002, p. 646.
4. Folstein & Rosen-Sheidley, 2001, p. 947.
5. The following quotes attributed to Silverman & Herbert are found in Silverman & Herbert, 2003.
6. Ibid.

temporary America." Not only does the pharmaceutical industry potentially benefit economically from such research, they wrote, but the denial of environ-mental causes helps protect corporations from liability, exemplified by vaccina-tions containing mercury-laden Thimerosal.[1] Finally, they observed that "genetic reductionism is a roadblock to developing the forceful science and social policy called for by the epidemic.... it weakens our response to a disaster that has already begun."

Unfortunately, genetic researchers de-emphasize contemporary theories about environmental causes of autism, and avoid words such as "epidemic." To do so would reduce the perceived importance of finding genes. Like the polio genetic researchers before them, they continue to believe that the cause of autism "lies in the host," and that environmental factors are, at best, of secondary importance.

Still, it is theoretically possible that some children carry genetic variations that reduce their ability to excrete harmful levels of mercury or other chemicals from their bodies, which might lead to autism or other disorders. As stated in Chapter 7, millions of American children were exposed to high levels of mercury in their vaccinations for many years.[2] While it is possible that children with genetic variations develop autism when overexposed to mercury, the decision to emphasize genes or exposure depends, as usual, on how one approaches the problem. Like PKU, pinpointing autism's environmental trigger and developing interventions to prevent it (such as limiting the intake of mercury during sen-sitive developmental periods, or reducing pre-natal exposure) could transform a "quintessentially highly heritable condition" into a preventable one. (Although I list mercury here as a possible environmental agent, it may well turn out that mercury has nothing to do with autism. Clearly, though, the same point applies to whichever factors are found to cause the disorder.)

The time has come to take a second look at the evidence put forward in support of genetic influences on autism, and to prioritize the pinpointing of environmental etiological factors regardless of whether genes are involved or not. Hanson and Gottesman may have been right when they concluded in 1976 that "biological, probably congenital (but not genetic) etiological agents" cause autism.[3] Emphasizing genetics and "heritability" delays the discovery of these possible biological causative agents.

---

1. Kirby, 2005.
2. Ibid.
3. Hanson & Gottesman, 1976, p. 226.

THE FRUITLESS SEARCH FOR BIPOLAR GENES

According to Samuel Barondes, current pharmacological interventions for mood (affective) disorders treat symptoms, but leave "the fundamental problem...unaddressed":

> Finding mood genes will change all this. Their discovery will help us understand the underlying differences in brain chemistry of people who develop severe mood swings, and these differences will become new targets for biological and behavioral treatments. Eventually, their discovery will even teach us how to prevent the development of this terrible source of suffering that haunts many millions of people around the world and that poses an awful threat to their descendants.[1]

The goal of alleviating human suffering is admirable. Still, good intentions notwithstanding, it is possible that directing research dollars and human resources in the genetic direction can impede the development of other types of interventions more likely to alleviate this suffering.

Because it is seen as one of the "most genetic" diagnosis in psychiatry, bipolar disorder (BPD) holds the distinction of being the first psychiatric condition subjected to molecular genetic analysis. After more than 30 years of research, however, investigators have arrived at the same impasse as seen in other areas of psychiatric genetics.[2] I will highlight the history of BPD molecular generic research briefly, while mostly steering clear of analyzing recent claims of linkage since they repeat a well-established pattern. Genetic researcher Miron Baron, for example, wrote a 1997 review article describing a "recent surge in linkage reports," which "may well be the harbinger of a new era in the quest for genes in bipolar disorder."[3] Although Baron cautioned that "it remains to be seen which, if any, of the claimed linkages will lead to gene discovery, the various studies display greater sophistication than their ill-fated predecessors.... The prospects for identifying and sequencing genes that predispose to bipolar disorder are brighter than ever."[4] Current claims about the importance of recent linkage investigations should be viewed in this historical context; now more than ever, we should adopt the Missouri state motto: "Show me."

A "Manic-Depressive History"

Risch and Botstein wrote in 1996, ironically, of the "manic-depressive" history of BPD molecular genetic research, which included "repeated claims for a variety of different loci followed by counter-claims and even retractions."[5] One

---

1. Barondes, 1998, pp. 2-3.
2. DePaulo, 2004.
3. Baron, 1997, p. 205.
4. Ibid., p. 208.
5. Risch & Botstein, 1996, p. 351.

of the first claims, put forward in the 1970s, was that BPD is linked to the X chromosome. Although this research was not replicated, there remains some interest in this area.[1]

The Huntington's disease (HD) gene discovery in the early 1980s inspired other groups to use similar methods in the BPD gene search,[2] with a highly publicized linkage claim published in a 1987 edition of *Nature*. Janice Egeland and associates' study focused on the Old Order Amish (OOA) of Pennsylvania, who were seen as an ideal group from which to obtain affected family pedigrees. The OOA, descendents of 30 founders who had emigrated from Europe in the early 19th century, constituted a closed, genetically isolated community possessing detailed genealogical records.[3] In addition, their religious and social prohibition against alcohol and drug use was believed to increase the reliability of BPD diagnoses. In their study of "Pedigree 110," Egeland et al. reported "compelling evidence for tight linkage between two DNA sequences located on chromosome 11 and a locus conferring a strong predisposition to bipolar affective disorders."[4] They concluded that their finding had "broad implications for research in human genetics and psychiatry," and should "provide an impetus to analogous research on other common clinical conditions."[5] However, despite the accompanying media frenzy, subsequent evidence cast doubt on the validity of the investigators' claims. Two years later, the same group of investigators published a retraction article describing their reanalysis of the data, which included two newly diagnosed members of Pedigree 110. This reduced the LOD score to a level below 3.0.[6] They concluded that the evidence for linkage they had previously reported "is not as strong as was initially suggested."[7]

Although this episode set the field back to some degree, most researchers never wavered in their belief that bipolar disorder carried an important genetic component. The new consensus held that it would be more difficult to find genes for "non-mendelian complex disorders" than originally believed. The events of the era were captured by Samuel Barondes in his 1998 book, *Mood Genes*. Barondes, himself a molecular genetic researcher, argued that although there had been setbacks, it was just a matter of time until "mood genes" are identified.

Returning to Risch and Botstein's 1996 article, they noted that the "euphoria of early linkage findings" such as Egeland's was "replaced by the dysphoria of non-replication," which had already become a "regular pattern, creating a roller coaster-type existence for many psychiatric genetics practitioners

---

1. Blackwood & Muir, 2004.
2. Barondes, 1998.
3. Egeland et al., 1987.
4. Ibid., p. 784.
5. Ibid., p. 786.
6. Kelsoe et al., 1989.
7. Ibid., p. 238.

as well as their interested observers."[1] The "distress engendered by the numerous reversals and non-replications," they continued, "has led many to rethink the paradigms being employed."[2] Unfortunately, instead of rethinking the paradigms of kinship studies of twins and adoptees, they focused on improving the science of molecular genetic research, arguing for a "complex basis for the diseases being studied."[3]

The New Millennium

The unexpected failure to replicate linkage claims or to find genes by the year 2000 was the subject of several commentaries by those active in the field. Risch, for example, believed that "Human genetics is now at a critical juncture," as the methods used to identify Mendelian syndromes "are failing to find the numerous genes causing more common, familial, non-mendelian diseases."[4] The questionable logic behind this claim, of course, is that the failure to find *any* genes must mean that there exist *numerous* genes.

This period also witnessed frustrated investigators beginning to pin their hopes on the then ongoing Human Genome Project (HGP) to help unravel the mystery of BPD genetics. In a 1997 article, Gottesman and his colleagues wrote that "The task of identifying genes [for behavioral disorders] will inevitably follow from the results of the Human Genome Project."[5] And a pair of researchers wrote in 2000, "The physical mapping of the human genome is proceeding at a rapid pace," voicing their belief that having this data in hand "will accelerate the process of finding [BPD] genes, because linked regions will no longer require local mapping efforts."[6] According to Hyman, the HGP is "critical...to the solution of complex disorders."[7] In another article focusing on the genetics of BPD, he wrote that mapping the complete sequence of the human genome "will ultimately permit successful identification of risk genes for bipolar disorder and other genetically complex disorders..."[8] As we know, however, the draft of the human genome was published in early 2001, and no psychiatric gene findings have resulted to date. Still, psychiatric genetic researchers such as John Kelsoe continue to believe that the HGP "promises revolutions in all aspects of medicine including psychiatry,"[9] while others see it as "a kind of new Rosetta Stone."[10]

1. Risch & Botstein, 1996, p. 351.
2. Ibid., p. 351.
3. Ibid., p. 352.
4. Risch, 2000, p. 847.
5. Goldsmith, et al., 1997, p. 380.
6. Potash & DePaulo, 2000, p. 20.
7. Hyman, 1999, p. 520.
8. Hyman, 2000, p. 436.
9. Kelsoe, 2004, p. 294.
10. McInnis, & Potash, 2004, p. 243.

A fascinating result from the Human Genome Project was the finding that humans possess far fewer genes than the 100,000 or so that textbooks published prior to 2001 usually reported. At the time of the sequencing of the genome in 2001, it was estimated that humans have about 35,000 genes, and by 2004 leading genetic researchers had reduced the number to 20,000–25,000.[1] Thus, humans "have about the same number of genes as a small flowering plant or a tiny worm."[2]

At the dawn of the new millennium, Hyman argued that it "is absolutely critical that we discover the genes that confer risk for bipolar disorder and other mood disorders." Furthermore, he wrote, "findings from twin studies and also from molecular genetic linkage analyses" demonstrate "that bipolar disorder is genetically complex. It appears that multiple genetic loci, each contributing relatively small increments of risk, interact with non-genetic factors...to produce illness and to modify its course." He concluded that the "failure, to date, to replicate any genetic linkages to bipolar disorder with adequate certainty is no different from the situation with most other common, genetically complex disorders that afflict humanity."[3]

If one could select a passage exemplifying problems with the biological/ genetic view of mental disorders, it would be this one. First, even if a genetic component to mood disorders existed, gene discovery would not necessarily be "critical" at all, since the proper environmental interventions could still be used to treat or prevent BPD. Second, Hyman believed that twin studies and molecular genetic studies demonstrate that BPD is "genetically complex." But twin studies depend on the dubious equal environment assumption, and linkage studies, as Hyman admits, have not been replicated. Thus, it is puzzling how a series of linkage *failures* supports the genetic "complexity" of BPD. Third, Hyman wrote that genes reside at "multiple loci," while at the same time admitting that research has failed to discover *any* loci. Finally, he argued that the failure to find genes is "no different" from "most other common, genetically complex disorders." However, although molecular genetic researchers in psychiatry usually fail to grasp this obvious point, we would also expect the same result for disorders with *no* genetic component.

Now and Beyond

As I write these lines, the story remains pretty much the same as it did 20 years ago: despite researchers' near unanimous agreement that BPD is heavily influenced by genetics, and that gene discoveries would be important, no BPD genes have been discovered, and no chromosomal loci have been identified and

---

1. Ritter, 2004.
2. Ibid., p. A12.
3. Hyman, 2000, p. 436.

consistently replicated.[1] To date, 25+ BPD genome scans have produced no consistent evidence for linkage.[2] Ricardo Segurado and colleagues published one of the most ambitious reports of genome scan data in 2003.[3] This study, whose co-authors read like a "Who's Who" of psychiatric genetic investigators, performed a meta-analysis of 18 previous genome scans. The result of this meta-analysis, according to the investigators, was that "No [chromosomal] region achieved genomewide statistical significance by several simulation-based criteria,"[4] which they viewed as "negative results."[5] In a 2004 review, another group of genetic researchers understatedly concluded that "overall the [BPD] linkage evidence has been less than striking."[6]

Like schizophrenia and ADHD, research on BPD biological markers is underway. Some proposed endophenotypes include abnormal regulation of circadian rhythms, response to sleep deprivation, electroencephalogram recordings of the scalp, behavioral responses to psychotropic drugs, and an increase in white matter hyperintensities.[7] According to one group of researchers, there "is hope that endophenotypes can potentially narrow down the genetic determinants, increasing the power we have to discover the genes that are both protective and/or responsible for complex behavioral disorders."[8] And in their 2002 NIMH workgroup report on the genetics of mood disorders, Merikangas and her colleagues recognized that "Mood disorder phenotypes, even bipolar I disorder" lack "conclusive evidence" of reliability and validity if diagnosed solely on the basis of clinical manifestations.[9]

Thus, bipolar disorder molecular genetic research has arrived at the same impasse we have seen in other psychiatric disorders, and for mostly the same reasons. And like the other disorders, most investigators have failed to take the logical step of questioning the existence of genetic influences. It seems the more failures that are recorded, the more "genetically complex" BPD becomes. And when even this speculation-turned-fact does not suffice, the rhetoric is cranked up a notch, as seen in a 2002 article where researchers described BPD as an "*extremely* complex disease" [emphasis added].[10] (In a similar vein, a pair of autism genetic investigators described autism as a "strikingly complex disorder,"[11] while another group called autism a "particularly complex disorder."[12])

1. Belmaker, 2004; Jones et al., 2003; Pittelli, 2005; Shastry, 2005.
2. Macgregor et al., 2004; Maziade et al., 2005; Segurado et al., 2003; Venken et al., 2005.
3. Segurado et al., 2003.
4. Ibid., p. 50.
5. Ibid., p. 58.
6. Detera-Wadleigh & McMahon, 2004, p. 310.
7. Lenox et al., 2002.
8. Ibid., p. 401.
9. Merikangas et al., 2002, p. 464.
10. Lenox et al., 2002, p. 410.
11. Veenstra-VanderWeele & Cook, 2004.
12. Bartlett et al., 2005, p. 221.

Perhaps, after further disappointments, researchers may begin to understand that a reassessment of BPD family, twin and adoption research would be a far better option than inserting a string of adjectives in front of an already questionable designation.

DISCUSSION

The crisis in psychiatric genetic molecular genetic research is growing due to the inability, after many years of searching, to find the genes presumed to cause psychiatric disorders. This failure may be due not to the difficulty of finding genes, but rather, to the mistaken belief that previous kinship studies have demonstrated the genetic basis of psychiatric disorders. Kendler wrote in 2003 that "gene finding methods," once viewed with skepticism, are "without doubt, the dominant scientific paradigm."[1] But what paradigm holds up when its predictions remain unfulfilled after more than 20 years of intense worldwide investigation?

Kendler tried to tackle this problem in a 2005 article entitled "Psychiatric Genetics: A Methodologic Critique."[2] He identified "four major research paradigms," consisting of (1) "Basic genetic epidemiology" and (2) "Advanced genetic epidemiology," which are based on family, twin and adoption studies, and the (3) "Gene finding" and (4) "Molecular genetics" paradigms, which respectively determine the genomic location of susceptibility genes, and the pathways from DNA variants to disorder. Kendler argued that, although "a substantial portion" of psychiatric gene finding claims "do not survive the test of replication,"[3] family, twin and adoption studies have found "genetic risk factors...for nearly all psychiatric and drug abuse disorders examined to date..."[4] Moreover,

> Unless there are strong and consistent methodologic biases operating across study designs, this body of work indicates that genetic risk factors are of substantial etiologic importance for all major psychiatric and drug disorders.[5]

However, as I have argued throughout this book and elsewhere, family, twin and adoption studies do indeed suffer from "strong and consistent methodologic biases operating across study designs."

Kendler then noted that the "low" replication level for linkage findings "contrasts strikingly with the high level of consistency seen in the results of genetic epidemiologic studies — for example, the results of family and twin

---

1. Kendler, 2003b, p. 763.
2. Kendler, 2005c. My response to Kendler's article, seen in the following paragraphs, can also be found in Joseph, 2005c.
3. Kendler, 2005c, p. 7.
4. Ibid., p. 6.
5. Ibid., p. 6.

studies of schizophrenia."[1] In fact, there is no striking contrast between these results *if they are viewed as evidence supporting a purely environmental etiology for psychiatric disorders.* With respect to a particular disorder, environmental theories of causation predict (a) familial clustering, (b) higher concordance of identical versus fraternal twins (see Chapter 1), and (c) the failure to find genes. And this is precisely what we find.

Rather than consider a purely environmental explanation as a competing paradigm, Kendler argued that linkage and association studies cannot be used to test "whether a twin or adoption study was correct in its conclusion that disorder x is heritable..." I agree, but negative results could at least compel researchers to take a second look at these methods. Unfortunately, they rarely do. Although Kendler viewed the four strategies he outlined as "competing paradigms," all four are components of the *same* biological/genetic paradigm, in contrast to what we might call the "environment/treatment/stress" paradigm.

Finally, Kendler called for integrating the four "paradigms" he identified, which would "require an appreciation of the complementary sources of information obtained by genetic epidemiologic and gene identification approaches."[2] (It is difficult to understand how a "striking contrast" can become "complementary sources of information" in the space of one article.) Kendler called this "explanatory pluralism," but what this means in practice is falling back on family, twin, and adoption results to explain the unexpected failure to find genes. It would be far better, in my view, to re-examine the assumptions, methods, and biases of these studies in the context of considering the possibility — just the possibility — that genes for the major psychiatric disorders do not exist.

In 2000, Risch attributed the "unfulfilled promise" of molecular genetic research in psychiatry to the "century-old debate between Mendelists and biometricists":

> The gene mutations studied by Mendel, and those more recently discovered by positional cloning, are those with large effect and strong genotype-phenotype correlations. They are effectively the "low-hanging fruit" that are easy to harvest. Now, however, we are left with the great majority of the fruit at the top of the tree with no obvious way to reach it. In genetics terms, these are the numerous genes of smaller effect that are likely to underlie most common, familial traits and diseases in humans — that is, the genes more closely related to the biometrical view of the world.[3]

---

1. Ibid., p. 7.
2. Ibid., p. 9.
3. Risch, 2000, p. 850.

The reason that Risch and others believe that there is "fruit at the top of the tree" can be summed up in two words: *twin research*. Due to their unshaken faith in this research and its underlying assumptions, they refuse to consider the possibility that there is *no* fruit at the top of the tree. It is ironic that, whereas Gregor Mendel did not need twins to make his landmark discoveries, psychiatric geneticists' misguided faith in twin research has led them into a blind alley. Fortunately, the continuing failures in psychiatric molecular genetic research have value in that they provide further evidence against the twin method's validity as an instrument for the detection of genetic influences on psychiatric disorders.

By the latter part of 2003, Merikangas and Risch seemed to be calling for a strategic retreat in the pages of *Science*. "Given the continuing difficulty of identifying genes for complex disorders in a robust, replicable manner, and the extensive resources devoted to this effort," they proposed that "complex diseases with the strongest evidence for genetic etiology, limited ability to modify exposure or risk factors, and highest public health impact should have the highest priority for genetic research."[1] While continuing to advocate "gene hunting" in "non-malleable" disorders such as schizophrenia and autism (ADHD was not mentioned), they called for de-emphasizing genetic research in "malleable" disorders such as AIDS and Type 2 diabetes.

This may signal the beginning of the end of molecular genetic research in psychiatry since, as I have already pointed out, continuing to not find genes cannot go on forever. Predictably, Merikangas and Risch failed to consider the possibility that they have found no genes because there *are* no genes, instead calling for emphasizing public health campaigns over genetics in order to reduce the prevalence of disorders "highly amenable to environmental modification." These investigators recognized that, in some cases, campaigns targeting known environmental factors are a preferred alternative to gene searching. One might suggest that genetic researchers shift their attention to environmental factors and attempt to transform currently "non-malleable" disorders into malleable ones, preventable or curable by public health campaigns. By the latter part of 2004, Risch recognized that "reports of linkage" for "complex traits" such as psychiatric disorders "are rarely replicated."[2]

Beckwith and Alper belong to a tiny handful of researchers to publicly entertain the possibility that the reason "so little progress has been made in associating genes with complex mental illnesses genes" is researchers' acceptance of false assumptions:

> One explanation for this failure that is rarely considered is that these illnesses are not in fact heritable. The evidence for heritability comes from behavior genetics familial studies involving, for example, comparisons between identical and fraternal twins. Perhaps the critics who maintained that these familial studies were suspect

---

1. Merikangas & Risch, 2003a, p. 599.
2. Mountain & Risch, 2004, p. S51.

have been right along...These critics have argued that the crucial assumptions of these studies are flawed, including the assumption that both identical and fraternal twins share environments to the same degree.[1]

CONCLUSIONS

Psychiatric genetic investigators have come up with every explanation under the sun for why they have found no genes, except perhaps the most obvious one, which goes something like this: There have been no gene discoveries for the major psychiatric disorders because there is little evidence that they have a genetic basis, and the diagnoses themselves are of questionable reliability and validity. Thus, the failure to find genes provides further evidence that biased and environmentally confounded family, twin, and adoption studies provide no scientifically acceptable data in support of genetics, and that this entire body of research must be critically reevaluated. It also calls into question neo-Kraepelinian claims that differing manifestations of psychological distress can be grouped into discrete entities, as opposed to the idea that all human distress falls on a continuum.[2] These are, perhaps, the most important conclusions we can draw from molecular genetic research in psychiatry to date.

* * *

The psychiatric genetics field is now suffering the consequences of nearly 100 years of enormously flawed and biased research, carried out for the most part by people strongly devoted to the genetic position well before they carried out their studies. The "items on the bill" include the frequent denial of, or failure to mention, the fact that the origins of psychiatric genetics lie in eugenics and racial hygiene; the reliance on highly questionable theoretical assumptions; changing the definition of particular mental disorders to ensure results in support of genetics; non-blinded diagnoses and zygosity determination; unwarranted assumptions about the reliability and validity of psychiatric diagnoses; arbitrary and biased methods of counting relatives; viewing statistically non-significant

1. Beckwith & Alper, 2002, p. 320. An alternative study might be conducted in order to test psychiatric molecular genetic research methods. We could select a trait that everyone agrees has no genetic basis. For example, there is an old baseball rivalry between the New York Yankees and the Boston Red Sox. Perhaps a group of Manhattan-based researchers could perform a linkage study in New England to determine whether they can identify chromosomal regions harboring suspected pathological Red Sox rooter genes. Like psychiatric disorders, computer models would test for chromosomal regions on the basis if several postulated modes of genetic transmission. A critic of psychiatric linkage research might predict that, if enough studies were performed, the Manhattan research team would detect chromosomal regions with high LOD scores for a Red Sox allegiance gene.
2. Bentall, 2003.

results as evidence in favor of genetics; failing to take potential environmental confounds seriously; ignoring, distorting, and dismissing important observations by critics; overlooking critical methodological flaws; ignoring, attempting to discredit, or twisting the results of studies whose results do not fit genetic predictions; conclusions drawn more from researchers' beliefs than from the data itself; the interpretation of family data as evidence in support of genetics; textbooks' creation of myths about "landmark" psychiatric genetic studies and the "overwhelming" evidence in support of genetic influences on mental disorders; the conversion of hypotheses into "facts"; a reliance on secondary sources' interpretation of previous research; the premature conclusion that previous kinship research proves that genes for mental disorders must exist; basing linkage results on models assuming a genetic transmission of the condition under study; the use of rhetoric as a means covering up the unexpected and disappointing failure to find genes; and, finally, the transformation of years, if not decades, of fruitless gene finding efforts into evidence of the "complex genetic nature" of psychiatric disorders.

As we now see playing out before us, the sum total of these items amounts to a bill that reads: *Genes for the major mental disorders are unlikely to exist.*

GLOSSARY

Ad Hoc Hypothesis. A hypothesis or explanation developed after the fact, to explain results that do not fit into an existing theoretical framework.

ADHD (ADD). "Attention-deficit hyperactivity disorder." A psychiatric diagnosis given most often to children exhibiting poor concentration, distractibility, hyperactivity, and impulsiveness.

Adoptees Method. A research method that begins with parents (usually mothers) diagnosed with the disorder in question. Researchers then determine the rate of this diagnosis among their adopted-away biological offspring. This rate is then compared with a control group consisting of the adopted-away biological offspring of parents not diagnosed with the disorder. If the study is methodologically sound, and all assumptions are valid, a statistically significant rate of diagnoses among index versus control adoptees suggests a role for genetic factors in causing the disorder (see Figure 2.1).

Adoptees' Family Method. A research method that begins with children given up for adoption who are later diagnosed with the disorder in question. A control group of non-diagnosed adoptees is also established. The investigators then attempt to identify and diagnose the biological and adoptive relatives in each group, and make statistical comparisons between groups (see Figure 2.1).

Adoption Study. A study of individuals adopted away as children, that attempts to disentangle the potential influences of genes and environment. In theory, these influences can be disentangled because adoptees are raised in the environment of one family, but received their genes from another family. In reality, factors such as selective placement, late placement, and attachment disturbance frequently make this theoretical separation difficult.

Adoptive Parents Method. A research method that compares the psychiatric status of three (and sometimes four) types of families. Unlike the Adoptees' Family method, this method does not investigate the biological relatives of

265

adopted-away children (see Figure 2.1). The Adoptive Parents method has been used mainly in the study of ADHD.

Allele. Variant forms of the same gene. Different alleles produce variations in inherited characteristics such as eye color or blood type.

Association Study. Compares the frequency of genetic markers among unrelated affected individuals and a control group, and is performed with population-based case-control, or family-based samples.

Assumption. A statement or belief that has not been verified as true, but which the project or investigation treats as if it were true, and conclusions are reached as if it were true.

Autism. A developmental disorder appearing by age three. It is variable in expression, but is recognized and diagnosed on the basis of (1) marked communication abnormalities, (2) impairment of the ability to communicate with others, and (3) an excess of stereotyped, ritualistic, and repetitive behaviors.

Behavior Genetics. A discipline, rooted in the field of psychology, that uses family, twin, and adoption studies to assess possible genetic influences on "continuously distributed" psychological traits such as personality and I.Q, and also on psychiatric disorders. In other areas of behavior genetics, researchers work primarily with non-human animals.

Biological Marker. (see *Endophenotype*).

Bipolar Disorder (Manic Depression). A psychiatric disorder characterized by moods that alternate between mania (see *Mania*) and depression (feeling sad and hopeless).

Borderline (Latent, B3) Schizophrenia. A crucial yet loosely defined component of the Danish-American schizophrenia adoption researchers' "schizophrenia spectrum of disorders." These investigators used symptoms such as "strange or atypical" thinking, "feelings of depersonalization," "anhedonia — never experiences intense pleasure," "lacking in depth," "mixture hetero- and homosexuality," and "multiple neurotic manifestations…severe widespread anxiety" to diagnose someone with borderline schizophrenia. Although Danish-American investigators such as Seymour Kety and David Rosenthal recognized that people receiving this diagnosis had few if any uniquely psychotic symptoms, and that in many cases they could not distinguish such people from others having disorders they believed were unrelated to schizophrenia, in statistical procedures they counted all such people the same as those they diagnosed with chronic (B1) schizophrenia. After performing such procedures they concluded that both diagnoses were genetically-based, and that they were genetically related to each other.

Chromosome. A threadlike structure in cells that carries genes (DNA).

Chronic Schizophrenia. (See *Schizophrenia*).

Complex Disorder. A disorder or syndrome believed to be caused by an interacting combination of multiple genes and multiple environmental risk factors.

The term is frequently used to explain the failure to find genes for psychiatric disorders.

Concordance. When both members of a twin pair are diagnosed with the same disorder.

Confound. An unforeseen or uncontrolled-for factor that threatens the validity of conclusions researchers draw from their studies. Although twin and adoption researchers usually interpret their findings as supporting genetic factors, uncontrolled-for environmental influences might lead others to interpret their findings solely in terms of environmental influences.

Control Group. In an experiment, the control group consists of subjects (people or objects) that exist in the same conditions as the experimental group, but do not undergo the treatment, or experience the factor being studied. In psychiatric genetic research, the control group frequently consists of people who are not diagnosed with the disorder being studied.

Diathesis-Stress Theory. The theory that a given disorder is caused by an inherited biological (genetic) predisposition in combination with environmental conditions or events.

Discordance. One member of a twin pair is diagnosed with a disorder, while the other twin is not.

DNA. Deoxyribonucleic acid. The material that in most cells is localized on chromosomes, and that carries genetic information.

DSM. "Diagnostic and Statistical Manual of Mental Disorders." A book produced by the American Psychiatric Association and revised periodically. Attempts to define and standardize diagnostic categories. The categories used in the DSM are accepted by most official organizations including hospitals, insurance companies, and other institutions.

Dual Mating Study. A study assessing the prevalence of a disorder among the biological offspring of two parents diagnosed with the same disorder.

Endophenotype (Biological Marker). A genetically-based neurobiological observable trait related to the assumed or proven molecular genetic basis of a disorder. Molecular genetic researchers seek to identify postulated endophenotypes in order to more precisely diagnose the disorders that are the subject of their research.

Equal Environment Assumption (EEA). The most important, and most controversial, assumption of the twin method. It holds that reared-together identical and same-sex fraternal twins experience the same environments. All conclusions in favor of genetics derived from twin method data depend on the validity of this assumption. The *traditional* equal environment assumption states, without qualification, that identical and same-sex fraternal twin environments are equal. After belatedly recognizing that identical twins do indeed experience more similar environments than fraternals, twin researchers added the qualification that these environments need only be equal regarding *trait relevant* features of the environment.

Environment. All non-genetic factors that could contribute to, or cause, diseases, psychiatric disorders, or psychological trait variation. Environmental factors include parenting styles, treatment, abuse, oppression, chemicals, viruses, accidents, and so on.

Eugenics. A doctrine which holds that humans can be "improved" by selective breeding to eradicate "undesirable" traits in society. Eugenicists argue that many social problems and psychiatric disorders are caused by inherited genetic traits in people, which can be bred out of the population for the benefit of future generations. Many German eugenicists of the first half of the 20th century preferred the term *racial hygiene* to eugenics.

Euthanasia. In the context of the history of eugenics, refers to a euphemism, first popularized by the Nazis, for the deliberate killing of institutionalized physically and mentally handicapped people, as well as people diagnosed with severe psychiatric disorders.

Experimental Group. (See *Index Group*).

Family Study. A study assessing the resemblance or diagnostic status of biological relatives who live together, and may include other biological relatives in different branches of the family. Usually compared to a control group or the general population. A family study can establish that a disorder runs in families (i.e., that it is *familial*), but cannot disentangle the possible roles of genetic and environmental influences in causing familial transmission or clustering.

Gene. Components of chromosomes composed of segments of DNA that are the basic functional units of heredity.

Genetic Counseling. Provides information, advice, and testing to prospective parents assumed to be at risk of having a child with a birth defect or genetic disorder.

Genetic Marker. A segment of DNA with an identifiable physical location on a chromosome and whose inheritance can be followed. Genetic markers are not necessarily (or usually) identified as the actual gene (s) influencing a trait, but only a segment of DNA (or visible section of a chromosome) that can be followed from one generation to the next.

Genetic Predisposition. (See *Diathesis-Stress Theory*).

Genetics. The study of the patterns of inheritance of specific traits.

Genome. The total genetic material of an organism or species.

Genome Scan. An analysis of the complete genome of an individual against a set of markers whose positions on the chromosomes are well known. A genome scan looks for common patterns of inheritance between these markers and a disorder's characteristics. This identifies linkage regions on the chromosomes.

Genomics. The analysis of the entire genome of a given organism or comparatively among different organisms. Genomics is a highly computer-based science comparing DNA over large stretches of DNA for classification or evolutionary purposes.

Genotype. An organism's individual genetic composition at a specified chromosomal location.

Fraternal (Dizygotic, DZ) Twins. Twins who develop from two separately fertilized eggs. Like ordinary siblings, these twins share a 50% average genetic resemblance.

Heritability. According to psychiatric genetic and behavior genetic theory, heritability measures the variance of a disorder or trait within a specified population of known genetic relatedness, in a specific environment, that can be attributed to genetic factors. Heritability estimates, which range from 0.0 to 1.0 (0% to 100%), are calculated from the results of twin studies, and to a lesser degree adoption studies. Some critics argue that heritability statistics are potentially misleading, while others call for abandoning the heritability concept entirely in psychiatry and psychology.

Human Genome Project. The international research project designed to map each human gene and to completely sequence human DNA. A draft of the human genome was published in 2001.

Huntington's Disease. A degenerative brain disorder, caused by a single dominant gene, that usually appears in mid-life. Its symptoms, which include involuntary movement of the face and limbs, mood swings, and forgetfulness, get worse as the disease progresses. It is generally fatal by age 60.

Index (Experimental) Group. A group of people diagnosed with the psychiatric disorder in question, and their relatives. Usually compared to a control group consisting of people not diagnosed with the disorder, and their relatives.

Identical (Monozygotic, MZ) Twins. Twins produced by the division of a single fertilized egg. Both have identical genotypes.

Inferential Statistics. Procedure whereby researchers make inferences or judgments about a population on the basis of data collected from a small representative sample drawn from that population.

Kinship Research. The study of relatives sharing various genetic relationships, for the purpose of assessing possible familial or genetic transmission of psychiatric disorders. Examples of kinship research include family, twin, and adoption studies.

Latent Schizophrenia. (See Borderline Schizophrenia).

Linkage. Close physical proximity of genetic loci on a chromosome. Linkage analysis is a statistical method for detecting linkage between a disease and markers of known location by following their inheritance in families.

LOD. A statistical estimate of whether two loci are likely to lie near each other on a chromosome, and therefore are likely to be inherited together. A LOD score of three or more is generally taken to indicate that the two loci are close.

Mania. An affective state marked by symptoms of elevated and expansive or irritable mood. May include decreased need for sleep, increased goal directed activities, impulsive behavior, and racing thoughts.

Mendelian Inheritance. An inherited disease or trait that is passed from parents to offspring by a single or few dominant, recessive, or sex-linked genes. For example, Huntington's disease and PKU are caused by a person inheriting a single disease gene, while blood groups are determined by the interaction of several different genes.

Meta-Analysis. A statistical analysis of data combined from two or more studies.

Molecular Genetics. The study of the structure and function of genes at the molecular level.

Munich School. (See Psychiatric Genetics).

Pedigree. A record of one's ancestors, offspring, and siblings through several generations. Although historically used to determine the pattern of genetic inheritance within a family, it is now understood that traits or diseases can cluster in family pedigrees for purely environmental reasons. Usually presented in graphic form through the use of standard symbols.

Pellagra. A disease caused by a deficiency of niacin (vitamin B6) or tryptophan. Characterized by gastrointestinal disturbances, skin disease (erythema), and nervous or mental disorders. Caused by malnutrition or other nutritional impairments.

Pharmacogenomics. A field that studies how genetic inheritance affects the way that the body responds to drugs.

Phenocopy. An environmentally-induced phenotype that mimics one caused by, or is believed to be caused by, genetic factors.

Phenotype. An observable trait or characteristic of an organism. For example eye color, weight, or the presence or absence of a disease.

PKU (Phenylketonuria). An inherited metabolic disorder in which the body cannot metabolize an amino acid called phenylalanine. Can result in mental retardation and other neurological problems. If detected early enough, the condition can be prevented by means of a special diet.

Poliomyelitis (Polio). A contagious viral disease of the central nervous system that can lead to paralysis.

Polygenic Disorder. A disorder believed to result from the combined action of many genes. Hereditary patterns are usually more complex than those of single-gene disorders.

Polymorphism. Multiple forms of a gene that are found in a given population, and often persist for many generations. In humans, skin color is a polymorphic trait.

Proband. In psychiatric genetic research terminology, a proband the first identified person from whom other family members are identified.

Pseudoscience. A set of ideas or claims based on theories purporting to be scientific, but are not scientific. According to psychologist Scott Lilienfeld and his colleagues, the ten warning signs of pseudoscience are "an overuse of ad hoc hypotheses designed to immunize claims from falsification," "absence of self-correction," "evasion of peer review," "emphasis on confirmation rather than refutation," "reversed burden of proof," "absence of connectivity," "over reliance on testimonial and anecdotal evidence," "use of obscurantist language," "absence of boundary conditions," and "the mantra of holism."

Psychiatric Genetics. A field founded by Ernst Rüdin and his German colleagues in the early part of the 20th century. German psychiatric geneticists used family and twin studies in an attempt to establish the genetic basis of psychiatric disorders. Their primary goal was to promote the eugenic program (called "racial hygiene" in Germany) of curbing the reproduction of people they viewed as carrying the "hereditary taint of mental illness," by sterilization or other means. After the Nazi seizure of power in 1933, the leaders of Rüdin's "Munich School" of psychiatric genetics supported and helped popularize Hitler's program of forcibly sterilizing "eugenically undesirable" people. Contemporary psychiatric geneticists investigate the causes of mental disorders in order to better treat and prevent them. Unlike the previous era, they usually avoid discussions of eugenics in relation to their findings. The implications of their theories, however, are obvious, and they often promote the use of genetic counseling.

Psychiatry. A branch of medicine dealing with the diagnosis and treatment of "mental disorders."

Psychopathology. Mental distress or the manifestation of abnormal behavior. Frequently grouped into diagnostic categories.

Racial Hygiene. (See Eugenics).

Schizophrenia. A psychiatric disorder characterized by symptoms such as hallucinations and delusions. It may also involve social withdrawal and affective flattening. Like many psychiatric disorders, the reliability and validity of this diagnosis is questionable. In the Danish-American adoption studies, this diagnosis was called "chronic (B1) schizophrenia."

Schizophrenia Spectrum Disorders. A group of psychiatric disorders or behavioral constellations believed by some to be genetically related to schizophrenia, but which are characterized by milder symptoms than schizophrenia.

Selective Placement. A potentially confounding aspect of adoption studies whereby children are systematically placed into adoptive homes sharing some characteristics (e.g., socioeconomic status, psychiatric diagnostic status) of their biological families. Researchers must assume that factors relating to the adoption process, including the policies of adoption agencies, did not lead to the placement of index adoptees into environments contributing to a higher rate of the disorder in question. If this "no selective placement assumption" is false, a higher rate of the disorder among index adoptees could be entirely the result of environmental factors.

Statistical Significance. The threshold at which researchers consider it unlikely that a result occurred by chance. By convention, a difference between two groups is usually considered statistically significant if it would be expected by chance less than 5% of the time.

Twin Method (Classical Twin Method). Compares the concordance rate or correlation of identical (monozygotic) versus same-sex fraternal (dizygotic) twins. Assuming that all twin method assumptions are true, a significantly greater resemblance among identical versus same-sex fraternal twins is attributed to the former's greater genetic resemblance (see Figure 1.1). Twin researchers then generalize this conclusion to apply to all people.

Zygosity. Genetic status of a pair of twins. That is, whether they are identical (monozygotic) or fraternal (dizygotic).

Zygosity Determination. Method used to determine whether a twin pair is identical or fraternal.

References

Abbott, A. (2000). German science begins to cure its historical amnesia. *Nature, 403,* 474-475.

Abrams, R., & Taylor, M. A. (1983). The genetics of schizophrenia: A reassessment using modern criteria. *American Journal of Psychiatry, 140,* 171-175.

Addair, J., & Snyder, L. H. (1942). Evidence for an autosomal recessive gene for susceptibility to paralytic poliomyelitis. *Journal of Heredity, 33,* 307-309.

Adelman, G. (Ed.). (1987). *Encyclopedia of Neuroscience* (Vol. 2). Boston: Birkhäuser.

Ainslie, R. C. (1985). *The psychology of twinship.* Lincoln: University of Nebraska Press.

Akiskal, H. S. (2002). Classification, diagnosis and boundaries of bipolar disorders: A review. In M. Maj, H. Akiskal, J. López-Ibor, & N. Sartorius (Eds.), Bipolar disorders (pp. 1-52). Chichester, UK: John Wiley and Sons.

Alberts-Corush, J., Firestone, P., & Goodman, J. T. (1986). Attention and impulsivity characteristics of the biological and adoptive parents of hyperactive and normal control children. *American Journal of Orthopsychiatry, 56,* 413-423.

Alford, J. R., Funk, C. L., & Hibbing, J. R. (2005). Are political orientations genetically transmitted? *American Political Science Review, 99,* 153-167.

Allen, M. G., Cohen, S., Pollin, W., & Greenspan, S. I. (1974). Affective illness in veteran twins: A diagnostic review. *American Journal of Psychiatry, 131,* 1234-1239.

Alper, J. (2002). Genetic complexity in human disease and behavior. In J. Alper, C. Ard, A. Asch, J. Beckwith, P. Conrad, & L. Geller (Eds.), *The double-edged helix: Social implications of genetics in a diverse society* (pp. 17-38). Baltimore: The Johns Hopkins University Press.

American Psychiatric Association. (1968). *Diagnostic and statistical manual of mental disorders* (2nd ed.). Washington, DC: Author.

American Psychiatric Association. (1980). *Diagnostic and statistical manual of mental disorders* (3rd ed.). Washington, DC: Author.

American Psychiatric Association. (1994). *Diagnostic and statistical manual of mental disorders* (4th ed.). Washington, DC: Author.

American Psychiatric Association. (2000). *Diagnostic and statistical manual of mental disorders* (4th ed., text revision). Washington, DC: Author.

American Psychological Association. (1938). *Psychological abstracts* (Vol. 12). W. Hunter & R. Willoughby (Eds.). Lancaster, PA: American Psychological Association.

American Psychological Association. (1939). *Psychological abstracts* (Vol. 13). W. Hunter & R. Willoughby (Eds.). Lancaster, PA: American Psychological Association.

Andreasen, N. C. (2001). *Brave new brain.* Oxford: Oxford University Press.

Angst, J., & Sellaro, R. (2000). Historical perspectives and natural history of bipolar disorder. *Biological Psychiatry, 48,* 445-457.

Anonymous. (1942). Euthanasia. *American Journal of Psychiatry, 99,* 141-143.

Anonymous. (2001, November). Irving I. Gottesman. Award for distinguished scientific contributions. *American Psychologist,* 864-867.

Anonymous. (2003). Scientific breakthrough of the year: The runners-up. *Science, 302,* 2039-2045.

Anonymous (2005). Back cover from Jang, K. L., *The behavioral genetics of psychopathology: A clinical guide.* Mahwah, NJ: Lawrence Erlbaum.

Arehart-Treichel, J. (2003, May 16th). With schizophrenia genes identified, what's next? *Psychiatric News, 38* (10). Retrieved 11/10/03, from http://pn.psychiatryonline.org/cgi/content/full/38/10/51.

Arnold, L. E., & Jensen, P. S. (1995). Attention-deficit disorders. In H. Kaplan & B. Sadock (Eds.), *Comprehensive textbook of psychiatry* (6th ed., Vol. 1, pp. 2295-2310). Baltimore: Williams & Wilkins.

Asherson, P. J., & Curran, S. (2001). Approaches to gene mapping in complex disorders and their application in child psychiatry and psychology. *British Journal of Psychiatry, 179,* 122-128.

Bailey, A., Le Couteur, A., Gottesman, I., Bolton, P., Simonoff, E., Yuzda, E., & Rutter, M. (1995). Autism as a strongly genetic disorder: Evidence from a British twin study. *Psychological Medicine, 25,* 63-77.

Bakker, S. C., van der Meulen, E. M., Buitelaar, J. K., Sandkuijl, L. A., Pauls, D. L., Monsuur, A. J., van 't Slot, R., Minderaa, R. B., Gunning, W. B., Pearson, P. L., & Sinke, R. J. (2003). A whole-genome scan in 164 Dutch sib pairs with attention-deficit/hyperactivity disorder: Suggestive evidence for linkage on chromosomes 7p and 15q. *American Journal of Human Genetics, 72,* 1251-1260.

Bakker, S. C., van der Meulen, E. M., Oteman, N., Schelleman, H., Pearson, P. L., Buitelaar, J. K, & Sinke, R. J. (2005). DAT1, DRD4, and DRD5 polymorphisms are not associated with ADHD in Dutch families. *American Journal of Medical Genetics Part B (Neuropsychiatric Genetics), 132B,* 50-52.

Barkley, R. A. (1998a). *Attention-deficit hyperactivity disorder: A handbook for diagnosis and treatment* (2nd ed.). New York: The Guilford Press.

Barkley, R. A. (1998b, September). Attention-deficit hyperactivity disorder. *Scientific American,* 66-71.

Barkley, R. A. (2003). Attention-deficit/hyperactivity disorder. In E. Mash & R. Barkley (Eds.), *Child psychopathology* (2nd ed., pp. 75-143). New York: The Guilford Press.

Barkley, R. A., Cook, E. H., Dulcan, M., et al. (2002). Consensus statement on ADHD. *European Child and Adolescent Psychiatry, 11,* 96-98.

Barnby, G., Abbott, A., Sykes, N., Morris, A., Weeks, D. E., Mott, R., Lamb, J., Bailey, A. J., & Monaco, A. P. (2005). Candidate-gene screening and association analysis at the autism-susceptibility locus on chromosome 16p: Evidence of association at *GRIN2A* and *ABAT*. *American Journal of Human Genetics, 76*, 950-966.

Baron, M. (1997). Genetic linkage and bipolar affective disorder: Progress and pitfalls. *Molecular Psychiatry, 2*, 200-210.

Baron, M. (1998). Psychiatric genetics and prejudice: Can the science be separated from the scientist? *Molecular Psychiatry, 3*, 96-100.

Baron, M. (2002). Manic-depression genes and the new millennium: Poised for discovery. *Molecular Psychiatry, 7*, 342-358.

Baron, M., Gruen, R., Rainer, J. D., Kane, J., Asnis, L., & Lord, S. (1985). A family study of schizophrenic and normal control probands: Implications for the spectrum concept of schizophrenia. *American Journal of Psychiatry, 142*, 447-455.

Barondes, S. H. (1998). *Mood genes: Hunting for origins of mania and depression.* New York: Oxford University Press.

Barondes, S. H. (1999). *Molecules and mental illness.* New York: Scientific American Library.

Barr, C. L. (2001). Genetics of childhood disorders: XXII. ADHD, Part 6: The dopamine D4 receptor gene. *Journal of the American Academy of Child and Adolescent Psychiatry, 40*, 118-121.

Bartlett, C. W., Gharani, N., Millonig, J. H., & Brzustowicz, L. M. (2005). Three autism candidate genes: A synthesis of human genetic analysis with other disciplines. *International Journal of Developmental Neuroscience, 23*, 221-234.

Beckwith, J. (2006). Whither human behavioral genetics? In E. Parens, A. Chapman, & N. Press (Eds.), *Wrestling with behavioral genetics: Implications for understanding selves and society* (pp. 74-99). Baltimore: Johns Hopkins University Press.

Beckwith, J., & Alper, J. S. (2002). Genetics of human personality: Social and ethical implications. In J. Benjamin, R. Ebstein, & R. Belmaker (Eds.), *Molecular genetics and the human personality* (pp. 315-331). Washington, DC: American Psychiatric Press.

Begley, S. (2005, June 24). Imprinted genes offer key to some diseases — and to possible cures. *Wall Street Journal*, p. B1.

Bellack, L. (Ed.). (1979). *Disorders of the schizophrenic syndrome.* New York: Basic Books.

Belmaker, R. H. (2004). Bipolar disorder. *New England Journal of Medicine, 351*, 476-486.

Benjamin, L. S. (1976). A reconsideration of the Kety and associates study of genetic factors in the transmission of schizophrenia. *American Journal of Psychiatry, 133*, 1129-1133.

Bentall, R. P. (2003). *Madness explained: Psychosis and human nature.* London: Allen Lane.

Bernard, S., Enayati, A., Redwood, L., Roger, H., & Binstock, T. (2001). Autism: A novel form of mercury poisoning. *Medical Hypotheses, 56*, 462-471.

Bertelsen, A. (2004). Contributions of Danish registers to understanding psychopathology: Thirty years of collaboration with Irving I. Gottesman. In L. DiLalla (Ed.), *Behavior genetic principles* (pp. 123-133). Washington, DC: American Psychological Association.

Bertelsen, A., Harvald, B., & Hauge, M. (1977). A Danish twin study of manic-depressive disorders. *British Journal of Psychiatry, 130*, 330-351.

Bertrand, J., Mars, A., Boyle, C., Bove, F., Yeargin-Allsopp, M., & Decoufle, P. (2001). Prevalence of autism in a United States population: The Brick Township, New Jersey, investigation. *Pediatrics, 108*, 1155-1161.

Betancur, C., Leboyer, M., & Gillberg, C. (2002). Increased rate or twins among affected sibling pairs with autism [Letter to the editor]. *American Journal of Human Genetics, 70*, 1381-1383.

Biederman, J. (2005). Attention-deficit/hyperactivity disorder: A selective overview. *Biological Psychiatry, 57*, 1215-1220.

Biederman, J., Faraone, S. V., Keenan, K., Knee, D., & Tsuang, M. T. (1990). Family-genetic and psychosocial risk factors in DSM-III attention deficit disorder. *Journal of the American Academy of Child and Adolescent Psychiatry, 29*, 526-533.

Biederman, J., Faraone, S. V., Mick, E., Spencer, T., Wilens, T., Kiely, K., Guite, J., Ablon, J. S., Reed, E., & Warburton, R. (1995). High risk for attention deficit hyperactivity disorder among children of parents with childhood onset of the disorder: A pilot study. *American Journal of Psychiatry, 152*, 431-435.

Biederman, J., Munir, K., Knee, D., Habelow, M., Autor, S., Hoge, S. K., & Waternaux, C. (1986). A family study of patients with attention deficit disorder and normal controls. *Journal of Psychiatric Research, 20*, 263-274.

Black, E. (2003). *War against the weak: Eugenics and America's campaign to create a master race.* New York: Four Walls Eight Windows.

Blackwood, D., & Muir, W. (2004). Biological theories of bipolar disorder. In M. Power (Ed.), *Mood disorders: A handbook of science and practice* (pp. 220-233). Chichester, UK: John Wiley and Sons.

Blaxill, M. F. (2004). What's going on? The question of time trends in autism. *Public Health Reports, 119*, 536-551.

Blaxill, M. F., Redwood, L., & Bernard, S. (2004). Thimerosal and autism? A plausible hypothesis that should not be dismissed. *Medical Hypotheses, 62*, 788-794.

Bleuler, E. (1950). *Dementia praecox or the group of schizophrenias.* New York: International Universities Press. (Originally published in 1911)

Bolton, P., Macdonald, H., Pickles, A., Rios, P., Goode, S., Crowson, M., Bailey, A., & Rutter, M. (1994). A case-control family history study of autism. *Journal of Child Psychology and Psychiatry, 35*, 877-900.

Borkenau, P., Riemann, R., Angleitner, A., & Spinath, F. M. (2002). Similarity of childhood experiences and personality resemblance in monozygotic and dizygotic twins: A test of the equal environments assumption. *Personality and Individual Differences, 33*, 261-269.

Bouchard, T. J., Jr. (1993). Genetic and environmental influences on adult personality: Evaluating the evidence. In J. Hettema & I. Deary (Eds.), *Basic issues in personality* (pp. 15-44). Dordrecht, The Netherlands: Kluwer Academic Publishers.

Bouchard, T. J., Jr. (1997). IQ similarity in twins reared apart: Findings and responses to critics. In R. Sternberg & E. Grigorenko (Eds.), *Intelligence, heredity, and environment* (pp. 126-160). New York: Cambridge University Press.

Bouchard, T. J., Jr., Lykken, D. T., McGue, M., Segal, N. L., & Tellegen, A. (1990). Sources of human psychological differences: The Minnesota Study of Twins Reared Apart. *Science, 250*, 223-228.

Boyle, M. (1990). *Schizophrenia: A scientific delusion?* New York: Routledge.

Boyle, M. (2002a). It's all done with smoke and mirrors. Or, how to create the illusion of a schizophrenic brain disease. *Clinical Psychology, 12*, 9-16

Boyle, M. (2002b). *Schizophrenia: A scientific delusion?* (2nd ed.). Hove, UK: Routledge.

Boyle, M. (2004). "Schizophrenia" and genetics: Does critical thinking stop here? *Journal of Critical Psychology, Counselling and Psychotherapy, 4,* 78-85.

Bradburn, N. M., Rips, L. J., & Shevell, S. K. (1987). Answering autobiographical questions: The impact of memory and inference on surveys. *Science, 236,* 157-161.

Breggin, P. R. (1991). *Toxic psychiatry.* New York: St. Martin's Press.

Breggin, P. R. (1998). *Talking back to Ritalin.* Monroe, ME: Common Courage Press.

Breggin, P. R. (2001a). Empowering social work in the era of biological psychiatry. *Ethical Human Sciences and Services, 3,* 197-206.

Breggin, P. (2001b). What people need to know about the drug treatment of children. In C. Newnes, G. Holmes, & C. Dunn (Eds.), *This is Madness Too: Critical perspectives on mental health services* (pp. 47-58). PCCS Books: Ross-on-Wye, UK.

Breggin, P. R., & Cohen, D. (1999). *Your drug may be your problem.* Reading, MA: Perseus.

Broberg, G., & Tydén, M. (1996). Eugenics in Sweden: Efficient care. In G. Broberg & N. Roll-Hansen (Eds.), *Eugenics and the welfare state: Sterilization policy in Denmark, Sweden, Norway, and Finland* (pp. 77-149). East Lansing, MI: Michigan State University Press.

Brown, K. (2003). New attention to ADHD genes. *Science, 301,* 160-161.

Bulmer, M. G. (1970). *The biology of twinning in man.* Oxford: Clarendon Press.

Bunney, W. E., Bunney, B. G., Vawter, M. P., Tomita, H., Li, J., Evans, S. J., Choudary, P. V., Myers, R. M., Jones, E. G., Watson, S. J., & Akil, H. (2003). Microarray technology: A review of strategies to discover candidate vulnerability genes in psychiatric disorders. *American Journal of Psychiatry, 160,* 657-666.

Burleigh, M. (1994). *Death and deliverance.* Cambridge, UK: Cambridge University Press.

Byerley, W., & Coon, H. (1995). Strategies to identify genes in schizophrenia. In J. Oldham & M. Riba (Eds.), *Review of psychiatry* (Vol. 14, pp. 361-381). Washington, DC: American Psychiatric Press.

Cantor, R. M., Kono, N., Duvall, J. A., Alvarez-Retuerto, A., Stone, J. L., Alarcón, M., Nelson, S. F., & Geschwind, D. H. (2005). Replication of autism linkage: Fine-mapping peak at 17q21. *American Journal of Human Genetics, 76,* 1050-1056.

Cantwell, D. P. (1972). Psychiatric illness in the families of hyperactive children. *Archives of General Psychiatry, 27,* 414-417.

Cantwell, D. P. (1975). Genetic studies of hyperactive children: Psychiatric illness in biologic and adopting parents. In R. Fieve, D. Rosenthal, & H. Brill (Eds.), *Genetic research in psychiatry* (pp. 273-280). Baltimore: The Johns Hopkins Press.

Cantwell, D. P. (1989). Comments by mentor. In L. Bloomingdale & J. Swanson (Eds.), *Attention deficit disorder, current concepts and emerging trends in attentional and behavioral disorders of childhood* (Vol. IV, pp. 79-84). New York: Pergamon Press.

Cardno, A. G., Frühling, V. R., Sham, P. C., Murray, R. M., & McGuffin, P. (2002). A twin study of genetic relationships between psychotic symptoms. *American Journal of Psychiatry, 159,* 539-545.

Cardno, A., & McGuffin, P. (1999). Psychiatric genetics. In H. Freeman (Ed.), *A century of psychiatry* (pp. 343-347). London: Moseby.

Cardno, A., & Murray, R. M. (2003). The "classical" genetic epidemiology of schizophrenia. In R. Murray, P. Jones, E. Susser, J. van Os, & M. Cannon (Eds.), *The epidemiology of schizophrenia* (pp. 195-219). Cambridge, UK: Cambridge University Press.

Carey, G. (2003). *Human genetics for the social sciences*. Thousand Oaks, CA: Sage.

Carey, G., & DiLalla, D. L. (1994). Personality and psychopathology: Genetic perspectives. *Journal of Abnormal Psychology, 103*, 32-43.

Carpenter, K. J. (Ed.). (1981). *Pellagra: Benchmark papers in biochemistry/2*. Stroudsberg, PA: Hutchinson Ross Publishing Company.

Carson, R. C., & Sanislow, C. A. III. (1993). The schizophrenias. In P. Sutker & H. Adams (Eds.), *Comprehensive handbook of psychopathology* (2nd ed., pp. 295-333). New York: Plenum Press.

Cassou, B., Schiff, M., & Stewart, J. (1980). Génétique et schizophrénie: Réévaluation d'un consensus [Genetics and schizophrenia: Reevaluation of a consensus]. *Psychiatrie de l'Enfant, 23*, 87-201.

Cavanagh, J. (2004). Epidemiology and classification of bipolar disorder. In M. Power (Ed.), *Mood disorders: A handbook of science and practice* (pp. 203-219). Chichester, UK: John Wiley and Sons.

Chase, A. (1975, February). The great pellagra cover-up. *Psychology Today, 8 (9)*, pp. 82-86.

Chase, A. (1980). *The legacy of Malthus: The social costs of the new scientific racism*. Urbana, IL/ Chicago: University of Illinois Press. (Originally published in 1977)

Chumakov, I., Blumenfeld, M., et al. (2002). Genetic and physiological data implicating the new human gene G72 and the gene for D-amino acid oxidase in schizophrenia. *Proceedings of the National Academy of Sciences, 99*, 13675 - 13680.

Clerget-Darpoux, F., Goldin, L. R., & Gershon, E. (1986). Clinical methods in psychiatric genetics. *Acta Psychiatrica Scandinavica, 305*-311.

Clifford, C. A., Hopper, J. L., Fulker, D. W., & Murray, R. M. (1984). A genetic and environmental analysis of a twin family study of alcohol use, anxiety, and depression. *Genetic Epidemiology, 1*, 63-79.

Cloninger, C. R. (2002). The discovery of susceptibility genes for mental disorders. *Proceedings of the National Academy of Sciences, 99*, 13365-13367.

Cohen, D. (1997). A critique of the use of neuroleptic drugs in psychiatry. In S. Fisher & R. Greenberg (Eds.), *From placebo to panacea: Putting psychotropic drugs to the test* (pp. 173-228). New York: John Wiley & Sons.

Cohen, D., Dibble, E., Grawe, J., & Pollin, W. (1975). Reliably separating identical from fraternal twins. *Archives of General Psychiatry, 32*, 1371-1375.

Coleman, J. C., Butcher, J. N., & Carson, R. C. (1984). *Abnormal psychology and modern life*. Glenview, IL: Scott, Foresman and Company.

Collins, J. S., Schroer, R. J., Bird, J., & Michaelis, R. C. (2003). The HOXA1 A218G polymorphism and autism: Lack of association in white and black patients from the South Carolina Autism Project. *Journal of Autism and Developmental Disorders, 33*, 343-348.

Comer, R. J. (1998). *Abnormal psychology*. New York: W. H. Freeman.

Comings, D. E. (1997). Polygenic inheritance in psychiatric disorders. In K. Blum & E. Noble (Eds.), *Handbook of psychiatric genetics* (pp. 235-260). Boca Raton, FL: CRC Press.

Comings, D. E., & Blum, K. (2000). Reward deficiency syndrome: Genetic aspects of behavioral disorders. *Progress in Brain Research, 126*, 325-337.

Comings, D. E., Gade-Andavolu, R., Gonzalez, N., Wu, S., Muhleman, D., Blake, H., Dietz, G., Saucier, G., & MacMurray, J. P. (2000). Comparison of the role of dopamine, serotonin, and noradrenaline genes in ADHD, ODD and conduct disorder: Multivariate regression analysis of 20 genes. *Clinical Genetics, 57,* 178-196.

Conrad, P. (2002). Genetics and behavior in the news: Dilemmas of a rising paradigm. In J. Alper, C. Ard, A. Asch, J. Beckwith, P. Conrad, & L. Geller (Eds.), *The double-edged helix: Social implications of genetics in a diverse society* (pp. 58-79). Baltimore: The Johns Hopkins University Press.

Cook, E. H., Jr. (1998). Genetics of autism. *Mental Retardation and Developmental Disabilities Research Reviews, 4,* 113-120.

Cook, E. H., Jr. (1999). Genetics of attention-deficit hyperactivity disorder. *Mental Retardation and Developmental Disabilities Research Reviews, 5,* 191-198.

Coon, H. (1999). Genetic analysis of complex diseases. In M. Labuda & E. Grigorenko (Eds.), *On the way to individuality: Methodological issues in behavioral genetics* (pp. 185-208). Commack, NY: Nova Science.

Cowan, W. M., Kopnisky, K. L., & Hyman, S. E. (2002). The Human Genome Project and its impact on psychiatry. *Annual Review of Neuroscience, 25,* 1-50.

Craddock, N., & Jones, I. (2001). Molecular genetics of bipolar disorder. *British Journal of Psychiatry, 178,* s128-s133.

Craddock, N., O'Donovan, M. C., & Owen, M. J. (2005). The genetics of schizophrenia and bipolar disorder: Dissecting psychosis. *Journal of Medical Genetics, 42,* 193-204.

Croen, L. A., Grether, J. K., Hoogstrate, J., & Selvin, S. (2002). The changing prevalence of autism in California. *Journal of Autism and Developmental Disorders, 32,* 207-215.

Cronk, N. J., Slutske, W. S., Madden, P. A. F., Bucholz, K. K., Reich, W., & Heath, A. C. (2002). Emotional and behavioral problems among female twins: An evaluation of the equal environment assumption. *Journal of the American Academy of Child and Adolescent Psychiatry, 41,* 829-837.

Crow, T. (2003). Genes for schizophrenia [Letter to the editor]. *Lancet, 361,* 1829.

Cunningham-Burley, S., & Kerr, A. (1999). Defining the "social": Towards an understanding of scientific and medical discourses on the social aspects of the new human genetics. *Sociology of Health and Illness, 21,* 647-668.

Davenport, C. B. (1911). *Heredity in relation to eugenics.* New York: Henry Holt and Company.

Davenport, C. B. (1916). The hereditary factor in pellagra. *Eugenics Record Office Bulletin No. 16.* Cold Spring Harbor, New York.

Davidson, G. C., & Neale, J. M. (1990). *Abnormal Psychology.* New York: John Wiley and Sons.

Davidson, G. C., & Neale, J. M. (1998). *Abnormal Psychology* (7th ed.). New York: John Wiley and Sons.

Davis, J. O., Phelps, J. A., & Bracha, H. S. (1995). Prenatal development of monozygotic twins and concordance for schizophrenia. *Schizophrenia Bulletin, 21,* 357-366.

D'haenen, H., den Boer, J. A., & Willner (Eds.). (2002). *Biological psychiatry* (Vol. 1). New York: John Wiley & Sons.

DeGrandpre, R. (1999). *Ritalin nation.* New York: W.W. Norton.

DeLisi, L. E. (2000). Critical overview of current approaches to genetic mechanisms in schizophrenia research. *Brain Research Reviews, 31,* 187-192.

DeLisi, L. E., & Crow, T. J. (2003). Drs. DeLisi and Crow reply [Letter to the editor]. *American Journal of Psychiatry, 160,* 598-599.

DePaulo, J. R. (2004). Genetics of bipolar disorder: Where do we stand? *Archives of General Psychiatry, 161*, 595-597.

Detera-Wadleigh, S. D., & McMahon, F. J. (2004). Genetic association studies in mood disorders: Issues and promise. *International Review of Psychiatry, 16*, 301-310

Deutsch, C. K. (1989). Adoption and attention deficit disorder. In L. Bloomingdale & J. Swanson (Eds.), *Attention deficit disorder, current concepts and emerging trends in attentional and behavioral disorders of childhood* (Vol. IV, pp. 67-79). New York: Pergamon Press.

Deutsch, C. K., Swanson, J. M., Bruell, J. H., Cantwell, D. P., Weinberg, F., & Baren, M. (1982). Overrepresentation of adoptees in children with the attention deficit disorder. *Behavior Genetics, 12*, 231-238.

Diller, L. H. (1998). *Running on Ritalin.* New York: Bantam Books.

Diller, L. H., Tanner, J. L., & Weil, J. (1995). Etiology of ADHD: Nature or Nurture? [Letter to the editor]. *American Journal of Psychiatry, 153*, 451-452.

Dinwiddie, S. H., Hoop, J., & Gershon, E. S. (2004). Ethical issues in the use of genetic information. *International Review of Psychiatry, 16*, 320-328.

Dusek, V. (1987). Bewitching science. *Science for the People, 19*, 19-22.

Eaves, L., Foley, D., & Silberg, J. (2003). Has the "equal environments" assumption been tested in twin studies? *Twin Research, 6*, 486-489.

Eaves, L., Silberg., J., Hewitt, J. K., Meyer, J., Rutter, M., Simonoff, E., Neale, M., & Pickles, A. (1993). Genes, personality, and psychopathology: A latent class analysis of liability to symptoms of attention-deficit hyperactivity disorder in twins. In R. Plomin & G. McClearn (Eds.), *Nature, nurture, and psychology* (pp. 285-303). Washington, DC: American Psychological Association Press.

Edelbrock, C., Rende, R., Plomin, R., & Thompson, L. (1995). A twin study of competence and problem behavior in childhood and early adolescence. *Journal of Child Psychology and Psychiatry, 36*, 775-785.

Egan, M. F., Leboyer, M., & Weinberger, D. R. (2003). Intermediate phenotypes in genetic studies of schizophrenia. In S. Hirsch & D. Weinberger (Eds.), *Schizophrenia* (2nd ed., pp. 277-297). Malden, MA: Blackwell.

Egan, M. F., Straub, R. E., et al. (2004). Variation in *GRM3* affects cognition, prefrontal glutamate, and risk for schizophrenia. *Proceedings of the National Academy of Science, 101*, 12604-12609.

Egeland, J. A., Gerhard, D. S., Pauls, D. L., Sussex, J. N., & Kidd, K. K. (1987). Bipolar affective disorders linked to DNA markers on chromosome 11. *Nature, 325*, 783-787.

Elkin, A., Kalidindi, S., & McGuffin, P. (2004). Have schizophrenia genes been found? *Current Opinion in Psychiatry, 17*, 107-113.

Elston, R. C., Song, D., & Iyengar, S. K. (2005). Mathematical assumptions versus biological reality: Myths in affected sib pair linkage analysis. *American Journal of Human Genetics, 76*, 152-156.

Essen-Möller, E. (1941). Psychiatrische Untersuchungen an einer Serie von Zwillingen [Psychiatric investigations on a series of twins]. *Acta Psychiatrica et Neurologica* (Suppl. 23). Copenhagen: Munksgaard.

Essen-Möller, E. (1970). Twenty-one psychiatric cases and their MZ cotwins. *Acta Geneticae Medicae et Gemellologiae, 19*, 315-317.

Etheridge, E. W. (1972). *The butterfly caste: A social history of pellagra in the south.* Westport, CT: Greenwood Publishing Company.

Ettinger, U., Kumari, V., Crawford, T., Corr, P. J., Das, M., Zachariah, E., Hughes, C., Sumich, A. L., Rabe-Hesketh, S., & Sharma, T. (2004). Smooth pursuit and antisaccade eye movements in siblings discordant for schizophrenia. *Journal of Psychiatric Research, 38,* 177-184.

Eysenck, H. J., vs. Kamin, L. J. (1981). *The intelligence controversy.* New York: John Wiley & Sons.

Fabsitz, R. R., Garrison, R. J., Feinleib, M., & Hjortland, M. (1978). A twin analysis of dietary intake: Evidence for a need to control for possible environmental differences in MZ and DZ twins. *Behavior Genetics, 8,* 15-25.

Falciglia, G. A., & Norton, P. A. (1994). Evidence for a genetic influence on preference for some foods. *Journal of the American Dietetic Association, 94,* 154-158.

Fan, J., Zhang, C., Gu, N., Li, X., Sun, W., Wang, H., Feng, G., St. Clair, D., & He, L. (2005). Catechol-O-methyltransferase gene Val/Met functional polymorphism and the risk for schizophrenia: A large-scale association study plus meta-analysis. *Biological Psychiatry, 57,* 139-144.

Fanous, A. H., & Kendler, K. S. (2005). Genetic heterogeneity, modifier genes, and quantitative phenotypes for psychiatric illness: Searching for a framework. *Molecular Psychiatry, 10,* 6-13.

Faraone, S. V. (1996). Discussion of "Genetic influence on parent-reported attention-related problems in a Norwegian general population twin sample." *Journal of the American Academy of Child and Adolescent Psychiatry, 35,* 596-598.

Faraone, S. V. (2004). Genetics of adult attention-deficit/hyperactivity disorder. *Psychiatric Clinics of North America, 27,* 303-321.

Faraone, S. V. (2005). The scientific foundation for understanding attention-deficit/hyperactivity disorder as a valid psychiatric disorder. *European Child and Adolescent Psychiatry, 14,* 1-10.

Faraone, S. V., & Biederman, J. (2000). Nature, nurture, and attention deficit hyperactivity disorder. *Developmental Review, 20,* 568-581.

Faraone, S. V., Biederman, J., Keenan, K., & Tsuang, M. T. (1991). A family-genetic study of girls with DSM-III attention deficit disorder. *American Journal of Psychiatry, 148,* 112-117.

Faraone, S. V., Biederman, J., Weiffenbach, B., Keith, T., Chu, M. P., Weaver, A., Spencer, T. J., Wilens, T. E., Frazier, J., Cleves, M., & Sakai, J. (1999). Dopamine D4 gene 7-repeat allele and attention deficit hyperactivity disorder. *American Journal of Psychiatry, 156,* 768-770.

Faraone, S. V., Perlis, R. H., Doyle, A. E., Smoller, J. W., Goralnick, J. J., Holmgren, M. A., & Sklar, P. (2005). Molecular genetics of attention-deficit/hyperactivity disorder. *Biological Psychiatry, 57,* 1313-1323.

Faraone, S. V., & Tsuang, M. T. (2003). Heterogeneity and the genetics of bipolar disorder. *American Journal of Medical Genetics (Series C, Semin. Med. Genet.), 123C,* 1-9.

Faraone, S. V., Tsuang, M. T., & Tsuang, D. W. (1999). *Genetics of mental disorders.* New York: The Guilford Press.

Farber, S. L. (1981). *Identical twins reared apart: A reanalysis.* New York: Basic Books.

Farmer, A., & McGuffin, P. (1999). Ethics and psychiatric genetics. In S. Bloch, P. Chodoff, & S. Green (Eds.), *Psychiatric ethics* (pp. 479-493). Oxford: Oxford University Press.

Fawcett, J. (2005). What do we know for sure about bipolar disorder? *American Journal of Psychiatry, 162,* 1-2.

Fischer, M. (1973). *Genetic and environmental factors in schizophrenia.* Copenhagen: Munksgaard.

Folstein, S. E., & Piven, J. (1991). Etiology of autism: Genetic influences. *Pediatrics, 87* (Suppl), 767-773.

Folstein, S. E., & Rosen-Sheidley, B. (2001). Genetics of autism: Complex aetiology for a heterogeneous disorder. *Nature Reviews Genetics, 2,* 943-955.

Folstein, S. E., & Rutter, M. (1977a). Genetic influences on infantile autism. *Nature, 265,* 726-728.

Folstein, S. E., & Rutter, M. (1977b). Infantile autism: A genetic study of 21 twin pairs. *Journal of Child Psychology and Psychiatry, 18,* 297-321.

Frangos, E., Athanassenas, G., Tsitourides, S., Katsanou, N., & Alexandrakou, P. (1985). Prevalence of DSM-III schizophrenia among the first-degree relatives of schizophrenic probands. *Acta Psychiatrica Scandinavica, 72,* 382-386.

Garno, J. L., Goldberg, J. F., Ramirez, P. M., & Ritzler, B. A. (2005). Impact of childhood abuse on the clinical course of bipolar disorder. *British Journal of Psychiatry, 186,* 121-125.

Gelder, M., Gath, D., Mayou, R., & Cowen, P. (1996). *Oxford textbook of psychiatry* (3rd ed.). Oxford: Oxford University Press.

Gelder, M., Lopez-Ibor, J., & Andreasen, N. (Eds.). (2000). *New Oxford textbook of psychiatry.* New York: Oxford University Press.

Gerlai, J., & Gerlai, R. (2003). Autism: A large unmet medical need and a complex research problem. *Physiology and Behavior, 79,* 461-470.

Gershon, E. S. (1990). Genetics. In F. Goodwin & K. Jamison, *Manic-depressive illness* (pp. 373-401). New York: Oxford University Press.

Gershon, E. S. (1997). Ernst Rüdin, a Nazi psychiatrist and geneticist [Letter to the editor]. *American Journal of Human Genetics (Neuropsychiatric Genetics), 74,* 457-458.

Gershon, E. S., Bunney, W. E., Leckman, J. F., Van Eerdewegh, M., & DeBauche, B. A. (1976). The inheritance of affective disorders: A review of data and of hypotheses. *Behavior Genetics, 6,* 227-261.

Gilger, J. W., Pennington, B. F., & DeFries, J. C. (1992). A twin study of the etiology of comorbidity: Attention-deficit hyperactivity disorder and dyslexia. *Journal of the American Academy of Child and Adolescent Psychiatry, 31,* 343-348.

Gillberg, C., Steffenburg, S., & Schaumann, H. (1991). Is autism more common now than ten years ago? *British Journal of Psychiatry, 158,* 403-409.

Gillis, J. J., Gilger, J. W., Pennington, B. F., & DeFries, J. C. (1992). Attention deficit disorder in reading-disabled twins: Evidence for a genetic etiology. *Journal of Abnormal Child Psychology, 20,* 303-315.

Gjone, H., Stevenson, J., & Sundet, J. M. (1996). Genetic influence on parent-reported attention-related problems in a Norwegian general population twin study. *Journal of the American Academy of Child and Adolescent Psychiatry, 35,* 588-596.

Glasson, E. J., Bower, C., Petterson, B., de Klerk, N., Chaney, G., & Hallmayer, J. F. (2004). Perinatal factors and the development of schizophrenia. *Archives of General Psychiatry, 61,* 618-627.

Glatt, S. J., Faraone, S. V., & Tsuang, M. T. (2003). Association between a functional catechol O-methyltransferase gene polymorphism and schizophrenia: Meta-analysis of case-control and family-based studies. *American Journal of Psychiatry, 160,* 469-476.

Goldberger, J., & Wheeler, G. A. (1915). Experimental pellagra in the human subject brought about by a restricted diet. *Public Health Reports, 30*, 3336-3339.

Goldberger, J., Wheeler, G. A., & Sydenstricker, E. (1918). A study of the diet on nonpellagrous and of pellagrous households. *Journal of the American Medical Association, 71*, 944-949.

Goldman, L. R., Shannon, M. W., et al. (2001). Technical report: Mercury in the environment: Implications for pediatricians. *Pediatrics, 108*, 197-205.

Goldsmith, H. H., Gottesman, I. I., & Lemry, K. S. (1997). Epigenetic approaches to developmental psychopathology. *Development and Psychopathology, 9*, 365-387.

Goodman, R., & Stevenson, J. (1989a). A twin study of hyperactivity — I. An examination of hyperactivity scores and categories derived from the Rutter Teacher and Parent Questionnaires. *Journal of Child Psychology and Psychiatry, 30*, 671-689.

Goodman, R., & Stevenson, J. (1989b). A twin study of hyperactivity — II. The etiological role of genes, family relationships and perinatal adversity. *Journal of Child Psychology and Psychiatry, 30*, 691-709.

Goodwin, F. K., & Jamison, K. R. (1990). *Manic-depressive illness.* New York: Oxford University Press.

Göring, H. H., Terwilliger, J. D., & Blangero, J. (2001). Large upward bias in estimation of locus-specific effects from genomewide scans. *American Journal of Human Genetics, 69*, 1357-1369.

Gottesfeld, H. (1979). *Abnormal psychology: A community mental health perspective.* Chicago: Science Research Associates.

Gottesman, I. I. (1991). *Schizophrenia genesis.* New York: W. H. Freeman & Company.

Gottesman, I. I., & Bertelsen, A. (1989a). Confirming unexpressed genotypes for schizophrenia. *Archives of General Psychiatry, 46*, 867-872.

Gottesman, I. I., & Bertelsen, A. (1989b). Dual mating studies in psychiatry — Offspring of inpatients with examples from reactive (psychogenic) psychoses. *International Review of Psychiatry, 1*, 287-296.

Gottesman, I. I., & Bertelsen, A. (1996). Legacy of German psychiatric genetics: Hindsight is always 20/20. *American Journal of Medical Genetics (Neuropsychiatric Genetics), 67*, 317-322.

Gottesman, I. I., & Erlenmeyer-Kimling, L. (1971). Prologue: A foundation for informed eugenics. In I. Gottesman & L. Erlenmeyer-Kimling (Eds.), Differential reproduction in individuals with physical and mental disorders. *Social Biology (Suppl.)* pp. S1-S8.

Gottesman, I. I., & Gould, T. D. (2003). The endophenotype concept in psychiatry: Etymology and strategic intentions. *American Journal of Psychiatry, 160*, 636-645.

Gottesman, I. I., & Hanson, D. R. (2005). Human development: Biological and genetic processes. *Annual Review of Psychology, 56*, 263-286.

Gottesman, I. I., McGuffin, P., & Farmer, A. E. (1987). Clinical genetics as clues to the "real" genetics of schizophrenia: A decade of modest gains while playing for time. *Schizophrenia Bulletin, 13*, 23-47.

Gottesman, I. I., & Shields, J. (1966a). Contributions of twin studies to perspectives on schizophrenia. In B. Maher (Ed.), *Progress in experimental personality research* (Vol. 3, pp. 1-84). New York: Academic Press.

Gottesman, I. I., & Shields, J. (1966b). Schizophrenia in twins: 16 years' consecutive admissions to a psychiatric clinic. *British Journal of Psychiatry, 112*, 809-818.

Gottesman, I. I., & Shields, J. (1972). *Schizophrenia and genetics: A twin study vantage point.* New York: Academic Press.

Gottesman, I. I., & Shields, J. (1976). A critical review of recent adoption, twin, and family studies of schizophrenia: Behavioral genetics perspectives. *Schizophrenia Bulletin, 2,* 360-401.

Gottesman, I. I., & Shields, J. (1982). *Schizophrenia: The epigenetic puzzle.* New York: Cambridge University Press.

Gralnick, A. (1942). Folie à deux — The psychosis of association. *Psychiatric Quarterly, 16,* 230-263.

Greenberg, D. A., Hodge, S. E., Sowinski, J., & Nicoll, D. (2001). Excess of twins among affected sibling pairs with autism: Implications for the etiology of autism. *American Journal of Human Genetics, 71,* 941-946.

Gütt, A., Rüdin, E., & Ruttke, F. (1934). *Gesetz zur Verhütung erbkranken Nachwuchses* [Law for the prevention of genetically diseased offspring]. Munich: J. F. Lehmanns.

Guze, S., Cloninger, C. R., Martin, R. L., & Clayton, P. J. (1983). A follow-up and family study of schizophrenia. *Archives of General Psychiatry, 40,* 1273-1276.

Haier, R. J., Rosenthal, D., & Wender, P. H. (1978). MMPI assessment of psychopathology in the adopted-away offspring of schizophrenics. *Archives of General Psychiatry, 35,* 171-175.

Hales, R. E., Yudofsky, S. C., & Talbott, J. A. (Eds.). (1999). *Textbook of psychiatry* (3rd ed.). Washington, DC: American Psychiatric Press.

Hallmayer, J., Glasson, E., Bower, C., Petterson, B., Croen, L., Grether, J., & Risch, N. (2002). On the twin risk in autism. *American Journal of Human Genetics, 71,* 941-946.

Halverson, C. F. (1988). Remembering your parents: Reflections on the retrospective method. *Journal of Personality, 56,* 435-449.

Hansen, B. S. (1996). Something rotten in the state of Denmark: Eugenics and the ascent of the welfare state. In G. Broberg & N. Roll-Hansen (Eds.), *Eugenics and the welfare state: Sterilization policy in Denmark, Sweden, Norway, and Finland* (pp. 9-76). East Lansing, MI: Michigan State University Press.

Hanson, D. R. (2004). Getting the bugs into our genetic theories of schizophrenia. In L. DiLalla (Ed.), *Behavior genetics principles* (pp. 205-216). Washington, DC: American Psychological Association Press.

Hanson, D. R., & Gottesman, I. I. (1976). The genetics, if any, of infantile autism and childhood schizophrenia. *Journal of Autism and Childhood Schizophrenia, 6,* 209-234.

Hardt, J., & Rutter, M. (2004). Validity of adult retrospective reports of adverse childhood experiences: Review of the evidence. *Journal of Child Psychology and Psychiatry, 45,* 260-273.

Harrison, P. J., & Owen, M. J. (2003). Genes for schizophrenia: Recent findings and their pathophysiological implications. *Lancet, 361,* 417-419.

Harrison, P. J., & Weinberger, D. R. (2005). Schizophrenia genes, gene expression, and neuropathology: On the matter of their convergence. *Molecular Psychiatry, 10,* 40-68.

Harvald, B., & Haugue, M. (1965). Hereditary factors elucidated by twin studies. In J. Neel, M. Shaw, & W. Schull (Eds.), *Genetics and the epidemiology of chronic diseases* (pp. 61-76). Washington, DC: Public Health Service Publication No. 1163.

Hay, D. A., & Levy, F. (2001). Implications of genetic studies of attentional problems for education and intervention. In F. Levy & D. Hay (Eds.), *Attention, genes, and ADHD* (pp. 214-224). East Sussex, UK: Brunner-Routledge.

Hay, D. A., McStephen, M., & Levy, F. (2001). Introduction to the genetic analysis of attentional disorders. In F. Levy & D. Hay (Eds.), *Attention, genes, and ADHD* (pp. 7-34). East Sussex, UK: Brunner-Routledge.

Healy, D. (1998). *The psychopharmacologists II*. London: Arnold.

Heath, A. C., Jardine, R., & Martin, N. G. (1989). Interactive effects of genotype and social environment on alcohol consumption in female twins. *Journal of Studies on Alcohol, 50,* 38-48.

Hedgecoe, A. (2001). Schizophrenia and the narrative of enlightened geneticization. *Social Studies of Science, 31,* 875-911.

Heinrichs, R. W. (2005). The primacy of cognition in schizophrenia. *American Psychologist, 60,* 229-242.

Henig, R. M. (2004, October 10th). The genome in black and white (and gray). *New York Times Magazine,* pp. 47-51.

Herndon, C. N., & Jennings, R. G. (1951). A twin-family study of susceptibility to poliomyelitis. *American Journal of Human Genetics, 3,* 17-46.

Herrnstein, R. J., & Murray, C. (1994). *The bell curve.* New York: The Free Press.

Heston, L. L. (1966). Psychiatric disorders in foster home reared children of schizophrenic mothers. *British Journal of Psychiatry, 112,* 819-825.

Heston, L. L., & Denny, D. D. (1968). Interactions between early life experience and biological factors in schizophrenia. In D. Rosenthal & S. Kety (Eds.), *The transmission of schizophrenia* (pp. 363-376). New York: Pergamon Press.

Hettema, J. M., Neale, M. C., & Kendler, K. S. (1995). Physical similarity and the equal-environment assumption in twin studies of psychiatric disorders. *Behavior Genetics, 25,* 327-335.

Hill, D. (1983). *The politics of schizophrenia.* Lanham, MD: University Press of America.

Hill, P., Murray, R., & Thorley, A. (Eds.). (1986) *Essentials of postgraduate psychiatry.* London: Grune & Stratton.

Hirsch, J. (1997). Some history of heredity-vs-environment, genetic inferiority at Harvard (?), and The (incredible) Bell Curve. *Genetica, 99,* 207-224.

Hirsch, J. (2004). Uniqueness, diversity, similarity, repeatability, and heritability. In C. Coll, E. Bearer, & R. Lerner (Eds.), *Nature and nurture: The complex interplay of genetic and environmental influences on human behavior and development* (pp. 127-138). Mahwah, NJ: Erlbaum.

Hirsch, S. R., & Weinberger, D. (Eds.). (2003). *Schizophrenia* (2nd ed.). Malden, MA: Blackwell.

Ho, A., Todd, R. D., & Constantino, J. N. (2005). Brief report: Autistic traits in twins vs. non-twins — a preliminary study. *Journal of Autism and Developmental Disorders, 35,* 129-133.

Ho, B. C., Wassink, T. H., O'Leary, D. S., Sheffield, V. C., & Andreasen, N. C. (2005). Catechol-O-methyl transferase Val Met gene polymorphism in schizophrenia: Working memory, frontal lobe MRI morphology and frontal cerebral blood flow. *Molecular Psychiatry, 10,* 287-298.

Hobson, K. (2003, March 17). The vaccine conundrum. *US News and World Report*, pp. 44-45.

Holmberg, D., & Holmes, J. G. (1994). Reconstruction of relationship memories: A mental models approach. In N. Schwarz & S. Sudman (Eds.), *Autobiographical memory and the validity of retrospective reports* (pp. 267-288). New York: Springer Verlag.

Holmes, A. S., Blaxill, M. F., & Haley, B. E. (2003). Reduced levels of mercury in first baby haircuts of autistic children. *International Journal of Toxicology, 22*, 1-9.

Holmes, J., Hever, T., Hewitt, L., Ball, C., Taylor, E., Rubia, K., & Thapar, A. (2002). A pilot twin study of psychological measures of attention deficit hyperactivity disorder. *Behavior Genetics, 32*, 389-395.

Horwitz, A. V., Videon, T. M., Schmitz, M. E., & Davis, D. (2003). Rethinking twins and environments: Possible social sources for assumed genetic influences in twin research. *Journal of Health and Social Behavior, 44*, 111-129.

Hubbard, R., & Armstrong, J. C. (1997). Publication bias against null results. *Psychological Reports, 80*, 337-338.

Hubbard, R., & Wald, E. (1993). *Exploding the gene myth*. Boston: Beacon Press.

Hudziak, J. J., Copeland, W., Rudiger, L. P., Achenbach, T. M., Heath, A. C., & Todd, R. D. (2003). Genetic influences on childhood competencies: A twin study. *Journal of the American Academy of Child and Adolescent Psychiatry, 42*, 357-363.

Hudziak, J. J., Heath, A. C., Madden, P. F., Reich, W., Bucholz, K. K., Slutske, W., Bierut, L. J., Neuman, R. J., & Todd, R. D. (1998). Latent class and factor analysis of DSM-IV ADHD: A twin study of female adolescents. *Journal of the American Academy of Child and Adolescent Psychiatry, 37*, 848-857.

Hultman, C., & Sparen, P. (2004). Autism: Prenatal insults or an epiphenomenon of a strongly genetic disorder? *Lancet, 364*, 485-487.

Hyman, S. E. (1999). Introduction to the complex genetics of mental disorders. *Biological Psychiatry, 45*, 518-521.

Hyman, S. E. (2000). Goals for research on bipolar disorder: The view from NIMH. *Biological Psychiatry, 48*, 436-441.

Hyman, S. E. (2003). Diagnosing disorders. *Scientific American, 289*, 97-102.

Ingraham, L. J., & Kety, S. S. (1988). Schizophrenia spectrum disorders. In M. Tsuang & J. Simpson (Eds.), *Handbook of schizophrenia, Vol. 3: Nosology, epidemiology and genetics* (pp. 117-137). New York: Elsevier Science Publishers.

Inouye, E. (1961). Similarity and dissimilarity of schizophrenia in twins. *Proceedings of the Third World Congress of Psychiatry* (Vol. 1, pp. 524-530). Montreal: University of Toronto Press.

Insel, T. R., & Collins, F. S. (2003). Psychiatry in the genomics era. *American Journal of Psychiatry, 160*, 616-620.

Jackson, D. D. (1960). A critique of the literature on the genetics of schizophrenia. In D. Jackson (Ed.), *The etiology of schizophrenia* (pp. 37-87). New York: Basic Books.

Jackson, G. E. (2003). Rethinking the Finnish adoption studies of schizophrenia: A challenge to genetic determinism. *Journal of Critical Psychology, Counselling and Psychotherapy, 3*, 129-138.

Jang, K. L. (2005). The behavioral genetics of psychopathology: A clinical guide. Mahwah, NJ: Lawrence Erlbaum.

Jones, I., Kent, L., & Craddock, N. (2003). Genetics of affective disorders. In P. McGuffin, M. Owen, & I. Gottesman (Eds.), *Psychiatric genetics and genomics* (pp. 211-245). Oxford: Oxford University Press.

Joseph, J. (1998). The equal environment assumption of the classical twin method: A critical analysis. *Journal of Mind and Behavior, 19,* 325-358.

Joseph, J. (1999a). A critique of the Finnish Adoptive Family Study of Schizophrenia. *Journal of Mind and Behavior, 20,* 133-154.

Joseph, J. (1999b). The genetic theory of schizophrenia: A critical overview. *Ethical Human Sciences and Services, 1,* 119-145.

Joseph, J. (2000a). A critique of the spectrum concept as used in the Danish-American schizophrenia adoption studies. *Ethical Human Sciences and Services, 2,* 135-160.

Joseph, J. (2000b). Inaccuracy and bias in textbooks reporting psychiatric research: The case of the schizophrenia adoption studies. *Politics and the Life Sciences, 19,* 89-99.

Joseph, J. (2000c). Not in their genes: A critical view of the genetics of attention-deficit hyperactivity disorder. *Developmental Review, 20,* 539-567.

Joseph, J. (2000d). Potential confounds in psychiatric genetic research: The case of pellagra. *New Ideas in Psychology, 18,* 83-91.

Joseph, J. (2000e). Problems in psychiatric genetic research: A reply to Faraone and Biederman. *Developmental Review, 20,* 582-593.

Joseph, J. (2001a). The Danish-American Adoptees' Family studies of Kety and associates: Do they provide evidence in support of the genetic basis of schizophrenia? *Genetic, Social, and General Psychology Monographs, 127,* 241-278.

Joseph, J. (2001b). Don Jackson's "A Critique of the Literature on the Genetics of Schizophrenia" — A reappraisal after 40 years. *Genetic, Social, and General Psychology Monographs, 127,* 27-57.

Joseph, J. (2001c). Separated twins and the genetics of personality differences: A critique. *American Journal of Psychology, 114,* 1-30.

Joseph, J. (2002a). Adoption study of ADHD [Letter to the editor]. *Journal of the American Academy of Child and Adolescent Psychiatry, 41,* 1389-1391.

Joseph, J. (2002b). Twin studies in psychiatry and psychology: Science or pseudoscience? *Psychiatric Quarterly, 73,* 71-82.

Joseph, J. (2003). *The gene illusion: Genetic research in psychiatry and psychology under the microscope.* Ross-on-Wye, UK: PCCS Books.

Joseph, J. (2004a). The fruitless search for schizophrenia genes. *Ethical Human Psychology and Psychiatry, 6,* 167-181.

Joseph, J. (2004b). *The gene illusion: Genetic research in psychiatry and psychology under the microscope* [North American Edition]. New York: Algora.

Joseph, J. (2004c). Schizophrenia and heredity: Why the emperor has no genes. In J. Read, L. Mosher, & R. Bentall (Eds.), *Models of madness: Psychological, social and biological approaches to schizophrenia* (pp. 67-83). Andover, UK: Taylor & Francis.

Joseph, J. (2005a). The 1942 "euthanasia" debate in the American Journal of Psychiatry. *History of Psychiatry, 16,* 171-179.

Joseph, J. (2005b). Comments about Frank Miele's article "The Revival of Human Nature? the Denial of Human Nature" [Letter to the editor]. *Skeptic, 11 (3),* 24-26.

Joseph, J. (2005c). Research paradigms of psychiatric genetics [Letter to the editor]. *American Journal of Psychiatry*, *162*, 1985.

Joseph, J., & Baldwin, S. (2000). Four editorial proposals to improve social sciences research and publication. *International Journal of Risk and Safety in Medicine*, *13*, 117-127.

Judd, L. L., & Groves, P. M. (Eds.). (1986). *Psychiatry: Psychobiological foundations of clinical psychiatry* (Vol. 4). New York: Basic Books.

Kalidindi, S., & McGuffin, P. (2003). The genetics of affective disorders: Present and future. In R. Plomin, J. DeFries, I. Craig, & P. McGuffin (Eds.), *Behavioral genetics in the postgenomic era* (pp. 481-501). Washington, DC: American Psychological Association Press.

Kallmann, F. J. (1938a). *The genetics of schizophrenia: A study of heredity and reproduction in the families of 1,087 schizophrenics*. New York: J. J. Augustin.

Kallmann, F. J. (1938b). Heredity, reproduction, and eugenic procedure in the field of schizophrenia. *Eugenical News*, *13*, 105-113.

Kallmann, F. J. (1946). The genetic theory of schizophrenia: An analysis of 691 schizophrenic twin index families. *American Journal of Psychiatry*, *103*, 309-322.

Kamin, L. J. (1974). *The science and politics of I.Q.* Potomac, MD: Lawrence Erlbaum Associates.

Kamin, L. J., & Goldberger, A. S. (2002). Twin studies in behavioral research: A skeptical view. *Theoretical Population Biology*, *61*, 83-95.

Kandel, E. R. (2000). Disorders of thought and volition: Schizophrenia. In E. Kandel, J. Schwartz, & T. Jessell (Eds.), *Principles of neural science* (4th ed., pp. 1188-1208). New York: McGraw-Hill.

Kandel, E. R., Schwartz, J. H., & Jessell, T. M. (2000). *Principles of neural science* (4th ed.). New York: McGraw-Hill.

Kanner, L. (1942). Exoneration of the feebleminded. *American Journal of Psychiatry*, *99*, 17-22.

Kanner, L. (1943). Autistic disturbances of affective contact. *Nervous Child*, *2*, 217-250.

Kaplan, H. I., & Sadock, B. J. (Eds.). (1995). *Comprehensive textbook of psychiatry* (6th ed., Vol. 1). Baltimore: Williams & Wilkins.

Kaprio, J., Koskenvuo, M., & Rose, R. J. (1990). Change in cohabitation and intrapair similarity of monozygotic (MZ) cotwins for alcohol use, extraversion, and neuroticism. *Behavior Genetics*, *20*, 265-276.

Kazdin, A. E., Bellack, A. S., & Hersen, M. (1980). *New perspectives in abnormal psychology*. New York: Oxford University Press.

Kealey, C., Roche, S., Claffey, E., & McKeon, P. (2005). Linkage and candidate gene analysis of 14q22-24 in bipolar disorder: Support for GCHI as a novel susceptibility gene. *American Journal of Medical Genetics Part B (Neuropsychiatric Genetics)*, *136B (1)*, 75-80.

Keefe, R. S. E., & Harvey, P. D. (1994). *Understanding schizophrenia*. New York: The Free Press.

Kelsoe, J. R. (2004). Genomics and the Human Genome Project: Implications for psychiatry. *International Review of Psychiatry*, *16*, 294-300.

Kelsoe, J. R., Ginns, E. I., Egeland, J. A., Gerhard, D. S., Goldstein, A. M., Bale, S. J., Pauls, D. L., Long, R. T., Kidd, K. K., Conte, G., Housman, D. E., & Paul, S. M. (1989). Re-evaluation of the linkage relationship between chromosome 11p loci and the gene for bipolar affective disorder in the Old Order Amish. *Nature*, *342*, 238-243.

Kendler, K. S. (1983). Overview: A current perspective on twin studies of schizophrenia. *American Journal of Psychiatry, 140*, 1413-1425.

Kendler, K. S. (1987). The genetics of schizophrenia: A current perspective. In H. Meltzer (Ed.), *Psychopharmacology: The third generation of progress* (pp. 705-713). New York: Raven Press.

Kendler, K. S. (1993). Twin studies of psychiatric illness: Current status and future directions. *Archives of General Psychiatry, 50*, 905-915.

Kendler, K. S. (2001). Twin studies of psychiatric illness: An update. *Archives of General Psychiatry, 58*, 1005-1014.

Kendler, K. S. (2003a). The genetics of schizophrenia: Chromosomal deletions, attentional disturbances, and spectrum boundaries. *American Journal of Psychiatry, 160*, 1549-1553.

Kendler, K. S. (2003b). Of genes and twins [editorial]. *Psychological Medicine, 33*, 763-768.

Kendler, K. S. (2004). Schizophrenia genetics and dysbindin: A corner turned? *American Journal of Psychiatry, 161*, 1533-1536.

Kendler, K. S. (2005a). "A gene for ...": The nature of gene actions in psychiatric disorders. *American Journal of Psychiatry, 162*, 1243-1252.

Kendler, K. S. (2005b). Psychiatric genetics: A methodologic critique. *American Journal of Psychiatry, 162*, 3-11.

Kendler, K. S. (2005c). Towards a philosophical structure for psychiatry. *American Journal of Psychiatry, 162*, 433-440.

Kendler, K. S., & Diehl, S. R. (1993). The genetics of schizophrenia: A current, genetic-epidemiologic perspective. *Schizophrenia Bulletin, 19*, 261-285.

Kendler, K. S., & Diehl, S. R. (1995). Schizophrenia: Genetics. In H. Kaplan & B. Sadock (Eds.), *Comprehensive textbook of psychiatry* (6th ed., Vol. 1, pp. 942-957). Baltimore: Williams & Wilkins.

Kendler, K. S., & Eaves, L. J. (1986). Models for the joint effect of genotype and environment on liability to psychiatric illness. *American Journal of Psychiatry, 143*, 279-289.

Kendler, K. S., & Gardner, C. O. (1998). Twin studies of adult psychiatric and substance dependent disorders: Are they biased by differences in the environmental experiences of monozygotic and dizygotic twins in childhood and adolescence? *Psychological Medicine, 28*, 625-633.

Kendler, K. S., & Gruenberg, A. M. (1984). An independent analysis of the Danish adoption study of schizophrenia. *Archives of General Psychiatry, 41*, 555-564.

Kendler, K. S., Gruenberg, A. M., & Tsuang, M. T. (1985). Psychiatric illness in first-degree relatives of schizophrenic and surgical control patients. *Archives of General Psychiatry, 42*, 770-779.

Kendler, K. S., Heath, A. C., Martin, N. G., & Eaves, L. J. (1986). Symptoms of anxiety and depression in a volunteer twin population. *Archives of General Psychiatry, 43*, 213-221.

Kendler, K. S., Heath, A. C., Neale, M. C., Kessler, R. C., & Eaves, L. J. (1992). A population-based study of alcoholism in women. *Journal of the American Medical Association, 268*, 1877-1882.

Kendler, K. S., Neale, M. C., Kessler, R. C., Heath, A. C., & Eaves, L. J. (1992a). The genetic epidemiology of phobias in women. *Archives of General Psychiatry, 49*, 273-281.

Kendler, K. S., Neale, M. C., Kessler, R. C., Heath, A. C., & Eaves, L. J. (1992b). A population-based twin study of major depression in women. *Archives of General Psychiatry, 49,* 257-266.

Kendler, K. S., Neale, M. C., Kessler, R. C., Heath, A. C., & Eaves, L. J. (1992c). Generalized anxiety disorder in women. *Archives of General Psychiatry, 49,* 267-272.

Kendler, K. S., Neale, M. C., Kessler, R. C., Heath, A. C., & Eaves, L. J. (1993). A test of the equal-environment assumption in twin studies of psychiatric illness. *Behavior Genetics, 23,* 21-27.

Kendler, K. S., Neale, M. C., Kessler, R. C., Heath, A. C., & Eaves, L. J. (1994). Parental treatment and the equal environment assumption in twin studies of psychiatric illness. *Psychological Medicine, 24,* 579-590.

Kendler, K. S., Pedersen, N. L., Johnson, L., Neale, M. C., & Mathe, A. A. (1993). A pilot Swedish twin study of affective illness, including hospital- and population-ascertained subsamples. *Archives of General Psychiatry, 50,* 669-706.

Kendler, K. S., & Robinette, C. D. (1983). Schizophrenia in the national academy of sciences-national research council twin registry: A 16-year update. *American Journal of Psychiatry, 140,* 1551-1563.

Kennedy, F. (1942). The problem of social control of the congenital defective: Education, sterilization, euthanasia. *American Journal of Psychiatry, 99,* 13-16.

Kennedy, J. L., Farrer, L. A., Andreasen, N. C., Mayeux, R., & St. George-Hyslop, P. (2003). The genetics of adult-onset neuropsychiatric disease: Complexities and conundra? *Science, 302,* 822-826.

Keski-Rahkonen, A., Viken, R. J., Kaprio, J., Rissanen, A., & Rose, R. J. (2004). Genetic and environmental factors in breakfast eating patterns. *Behavior Genetics, 34,* 503-514.

Kessler, R. C., Berglund, P., Demler, O., Jin, R., Merikangas, K. R., & Walters, E. E. (2005). Lifetime prevalence and age-of-onset distributions of DSM-IV disorders in the National Comorbidity Survey replication. *Archives of General Psychiatry, 62,* 593-602.

Kety, S. S. (1959). Biochemical theories of schizophrenia, part II. *Science, 129,* 1590-1596.

Kety, S. S. (1970). Genetic-environmental interactions in the schizophrenic syndrome. In R. Cancro (Ed.), *The schizophrenic reactions* (pp. 233-244). New York: Brunner/Mazel.

Kety, S. S. (1975). Mental illness in the biological and adoptive families of adopted individuals who have become schizophrenic. In H. van Praag (Ed.), *On the origin of schizophrenic psychosis* (pp. 19-26). Amsterdam: De Ervin Bohn BV.

Kety, S. S. (1980). The syndrome of schizophrenia: unresolved questions and opportunities for research. *British Journal of Psychiatry, 136,* 421-436.

Kety, S. S. (1983a). Mental illness in the biological and adoptive relatives of schizophrenia adoptees: Findings relevant to genetic and environmental factors in etiology. *American Journal of Psychiatry, 140,* 720-727.

Kety, S. S. (1983b). Dr. Kety responds [Letter to the editor]. *American Journal of Psychiatry, 140,* 964.

Kety, S. S. (1985a). The concept of schizophrenia. In M. Alpert (Ed.), Controversies in schizophrenia: Changes and constancies — Proceedings of the 74th annual meeting of the psychopathological association, New York, March 1-3, 1984 (pp. 3-11). New York: Guilford Press.

Kety, S. S. (1985b). Schizotypal personality disorder: An operational definition of Bleuler's latent schizophrenia? *Schizophrenia Bulletin, 11,* 590-594.

Kety, S. S. (1987). The significance of genetic factors in the etiology of schizophrenia: Results from the national study of adoptees in Denmark. *Journal of Psychiatric Research*, 21, 423-429.

Kety, S. S. (1988). Schizophrenic illness in the families of schizophrenic adoptees: Findings from the Danish national sample. *Schizophrenia Bulletin*, 14, 217-222.

Kety, S. S., & Ingraham, L. J. (1992). Genetic transmission and improved diagnosis of schizophrenia from pedigrees of adoptees. *Journal of Psychiatric Research*, 26, 247-255.

Kety, S. S., & Matthysse, S. (1988). Genetic and biological aspects of schizophrenia. In A. Nicholi (Ed.), *The new Harvard guide to psychiatry* (pp. 139-151). Cambridge, MA: Harvard University Press.

Kety, S. S., Rosenthal, D., & Wender, P. W. (1978). Genetic relationships within the schizophrenia spectrum: Evidence from adoption studies. In R. Spitzer & D. Klein (Eds.), *Critical issues in psychiatric diagnosis* (pp. 213-223). New York: Raven Press.

Kety, S. S., Rosenthal, D., Wender, P. H., & Schulsinger, F. (1968). The types and prevalence of mental illness in the biological and adoptive families of adopted schizophrenics. In D. Rosenthal & S. Kety (Eds.), *The transmission of schizophrenia* (pp. 345-362). New York: Pergamon Press.

Kety, S. S., Rosenthal, D., Wender, P. H., & Schulsinger, F. (1976). Studies based on a total sample of adopted individuals and their relatives: Why they were necessary, what they demonstrated and failed to demonstrate. *Schizophrenia Bulletin*, 2, 413-427.

Kety, S. S., Rosenthal, D., Wender, P. H., Schulsinger, F., & Jacobsen, B. (1975). Mental illness in the biological and adoptive families of adopted individuals who have become schizophrenic: A preliminary report based on psychiatric interviews. In R. Fieve, D. Rosenthal, & H. Brill (Eds.), *Genetic research in psychiatry* (pp. 147-165). Baltimore: The Johns Hopkins Press.

Kety, S. S., Rosenthal, D., Wender, P. H., Schulsinger, F., & Jacobsen, B. (1978). The biologic and adoptive families of adopted individuals who became schizophrenic: Prevalence of mental illness and other characteristics. In L. Wynne, R. Cromwell, & S. Matthysse (Eds.), *The nature of schizophrenia: New approaches to research and treatment* (pp. 25-37). New York: John Wiley & Sons.

Kety, S. S., Wender, P. H., Jacobsen, B., Ingraham, L. J., Jansson, L., Faber, B., & Kinney, D. K. (1994). Mental illness in the biological and adoptive relatives of schizophrenic adoptees: Replication of the Copenhagen study to the rest of Denmark. *Archives of General Psychiatry*, 51, 442-455.

Kieseppä, T., Partonen, T., Haukka, J., Kaprio, J., & Lönnqvist, J. (2004). High concordance of bipolar I in a nationwide sample of twins. *American Journal of Psychiatry*, 161, 1814-1821.

King, R. A., Rotter, J. I., & Motulsky, O. G. (2002a). Approach to genetic basis of common diseases. In R. King, J. Rotter, & A. Motulsky (Eds.), *The genetic basis of common diseases* (2nd ed., pp. 3-38). Oxford: Oxford University Press.

King, R. A., Rotter, J. I., & Motulsky, O. G. (2002b). *The genetic basis of common diseases*. Oxford: Oxford University Press.

Kirby, D. (2005). *Evidence of harm. Mercury in vaccines and the autism epidemic: A medical controversy*. New York: St. Martin's.

Kirk, S. A., & Kutchins, H. (1992). *The selling of DSM: The rhetoric of science in psychiatry*. New York: Aldine De Gruyter.

Kirov, G., O'Donovan, M. C., & Owen, M. J. (2005). Finding schizophrenia genes. *Journal of Clinical Investigation, 115*, 1440-1448.

Klinger, L. G., Dawson, G., & Renner, P. (2003). Autistic disorder. In E. Mash & R. Barkley (Eds.), *Child psychopathology* (2nd ed., pp. 409-454). New York: Guilford.

Klump, K. L., Holley, A., Iacono, W. G., McGue, M., & Willson, L. E. (2000). Physical similarity and twin resemblance for eating attitudes and behaviors: A test of the equal environments assumption. *Behavior Genetics, 30*, 51-58.

Koenig, R. (2000). Reopening the darkest chapter in German science. *Science, 288*, 1576-1577.

Kolb, L., & Brodie, H. K. (1982). *Modern clinical psychiatry* (10th ed.). Philadelphia: W. B. Saunders Company.

Koskenvuo, M., Langinvainio, H., Kaprio, J., Lönnqvist, J., & Tienari, P. (1984). Psychiatric hospitalization in twins. *Acta Geneticae Medicae et Gemellologiae, 33*, 321-332.

Kraepelin, E. (1976). *Manic-depressive insanity and paranoia.* New York: Arno Press. (Originally published in English in 1921)

Kringlen, E. (1967). *Heredity and environment in the functional psychoses: An epidemiological-clinical study.* Oslo: Universitetsforlaget.

Kuhn, T. S. (1996). *The structure of scientific revolutions* (3rd ed.). Chicago: The University of Chicago Press. (Originally published in 1962)

Kutchins, H., & Kirk, S. A. (1997). *Making us crazy: DSM, the psychiatric bible and the creation of mental disorders.* New York: The Free Press.

LaBuda, M. C., Gottesman, I. I., & Pauls, D. L. (1993). Usefulness of twin studies for exploring the etiology of childhood and adolescent psychiatric disorders. *American Journal of Medical Genetics (Neuropsychiatric Genetics), 48*, 47-59.

LaBuda, M. C., Svikis, D. S., & Pickens, R. V. (1997). Twin closeness and co-twin risk for substance use disorders: Assessing the impact of the equal environment assumption. *Psychiatry Research, 70*, 155-164.

Lacasse, J. R., & Gomory, T. (2003). Is graduate social work education promoting a critical approach to mental health practice? *Journal of Social Work Education, 39*, 383-408.

Lamb, J. A., Moore, J., Bailey, A., & Monaco, A. P. (2000). Autism: Recent molecular genetic advances. *Human Molecular Genetics, 9*, 861-868.

Langley, K., Marshall, L., van den Bree, M., Thomas, H., Owen, M., O'Donovan, M., & Thapar, A. (2004). Association of the dopamine D4 receptor gene 7-repeat allele with neuropsychological test performance of children with ADHD. *American Journal of Psychiatry, 161*, 133-138.

Lenox, R. H., Gould, T. D., & Manji, H. K. (2002). Endophenotypes in bipolar disorder. *American Journal of Medical Genetics (Neuropsychiatric Genetics), 11*, 391-406.

Leo, J. (2002, January/February). American preschoolers on Ritalin. *Society, 39, (2)*, 52-60.

Leo, J. (2003). The fallacy of the 50% concordance rate for schizophrenia in identical twins. [Review of the book *The Gene Illusion: Genetic Research in Psychiatry and Psychology Under the Microscope*]. *Human Nature Review, 3*, 406-415.

Leo, J., & Cohen, D. (2003). Broken brains or flawed studies? A critical review of ADHD neuroimaging research. *Journal of Mind and Behavior, 24*, 29-55.

Leo, J., & Cohen, D. (2004). An update on ADHD neuroimaging research. *Journal of Mind and Behavior, 25*, 161-166.

Leo, J., & Joseph, J. (2002). Schizophrenia: Medical students are taught it's all in the genes, but are they hearing the whole story? *Ethical Human Sciences and Services, 4,* 17-30.

Lerer, B., & Segman, R. H. (1997). Correspondence regarding German psychiatric genetics and Ernst Rüdin [Letter to the editor]. *American Journal of Human Genetics (Neuropsychiatric Genetics), 74,* 459-460.

Lerner, R. M. (1995). The limits of biological influence: Behavioral genetics as the emperor's new clothes [Review of the book *The Limits of Family Influence*]. *Psychological Inquiry, 6,* 145-156.

Leverich, G. S., McElroy, S. L., Suppes, T., Keck, P. E., Jr., Denicoff, K. D., Nolen, W. A., Altshuler, L. L., Rush, A. J., Frye, M. A., Autio, K. A., & Post, R. M. (2002). Early physical and sexual abuse associated with an adverse course of bipolar illness. *Biological Psychiatry, 51,* 288-297.

Levinson, D. F., & Mowry, B. J. (2000). Genetics of schizophrenia. In D. Pfaff, W. Berrettini, T. Joh, & S. Maxson (Eds.), *Genetic influences on neural and behavioral functions* (pp. 47-82). Boca Raton, FL: CRC Press.

Levy, F., Hay, D. A., McStephen, M., Wood, C., & Waldman, I. (1997). Attention-deficit hyperactivity disorder: A category or a continuum? Genetic analysis of a large-scale twin study. *Journal of the American Academy of Child and Adolescent Psychiatry, 36,* 737-744.

Lewis, C. M., Levinson, D. F., Wise, L. H., DeLisi, L. E., Straub, R. E., Hovatta, I. et al. (2003). Genome scan meta-analysis of schizophrenia and bipolar disorder, part II: Schizophrenia. *American Journal of Human Genetics, 73,* 34-48.

Lewontin, R. C. (1974). The analysis of variance and the analysis of causes. *American Journal of Human Genetics, 26,* 400-411.

Lewontin, R. C., Rose, S., & Kamin, L. J. (1984). *Not in our genes.* New York: Pantheon.

Li, J., Tabor, H. K., Nguyen, L., Gleason, C., Lotspeich, L. J., Spiker, D., Risch, N., & Meyers, R. M. (2002). Lack of association between HOXA1 and HOXB1 gene variants and autism in 110 multiplex families. *American Journal of Medical Genetics (Neuropsychiatric Genetics), 114,* 24-30.

Lidz, T. (1976). Commentary on a critical review of recent adoption, twin, and family studies of schizophrenia: Behavioral genetics perspectives. *Schizophrenia Bulletin, 2,* 402-412.

Lidz, T., & Blatt, S. (1983). Critique of the Danish-American studies of the biological and adoptive relatives of adoptees who became schizophrenic. *American Journal of Psychiatry, 140,* 426-435.

Lidz, T., Blatt, S., & Cook, B. (1981). Critique of the Danish-American studies of the adopted-away offspring of schizophrenic parents. *American Journal of Psychiatry, 138,* 1063-1068.

Lifton, R. J. (1986). *The Nazi doctors.* New York: Basic Books.

Lilienfeld, S. O., Lynn, S. J., & Lohr, J. M. (2003). Science and pseudoscience in clinical psychology: Initial thoughts, reflections, and considerations. In S. Lilienfeld, S. Lynn, & J. Lohr (Eds.), *Science and pseudoscience in clinical psychology* (pp. 1-14). New York: Guilford.

Lippman, A. (1992). Led (astray) by genetic maps: The cartography of the human genome and heath care. *Social Science and Medicine, 35,* 1469-1476.

Lippman, A. (1998). The politics of heath: Geneticization versus heath promotion. In S. Sherwin (Ed.), *The politics of women's health* (pp. 64-82). Philadelphia: Temple University Press.

Loehlin, J. C., & Nichols, R. C. (1976). *Heredity, environment, and personality*. Austin: University of Texas Press.

Lopez, R. E. (1965). Hyperactivity in twins. *Canadian Psychiatric Association Journal, 10*, 421-426.

Lord, C., & Bailey, A. (2002). Autism spectrum disorders. In M. Rutter & E. Taylor (Eds.), *Child and adolescent psychiatry* (4th ed., pp. 636-663). Oxford, UK: Blackwell.

Lowing, P. A., Mirsky, A. F., & Pereira, R. (1983). The inheritance of schizophrenia spectrum disorders: A reanalysis of the Danish adoptee study data. *American Journal of Psychiatry, 140*, 1167-1171.

Lush, J. L. (1949). Heritability of quantitative characteristics in farm animals. *Hereditas* (Suppl.). G. Bonnier & R. Larsson (Eds.), 356-375.

Luxenburger, H. (1928). Vorläufiger Bericht über psychiatrische Serienuntersuchungen an Zwillingen [Provisional report on a series of psychiatric investigations of twins]. *Zeitschrift fur die Gesamte Neurologie und Psychiatrie, 116*, 297-347.

Luxenburger, H. (1931). Psychiatrische Erbprognose und Eugenik. [Psychiatric genetic prognosis and eugenics]. *Eugenik, 1*, 117-124.

Luxenburger, H. (1934). Rassenhygienisch wichtige Probleme und Ergebnisse der Zwillingspathologie [Racial hygienic important problems and results of twin pathology]. In E. Rüdin (Ed.), *Erblehre und rassenhygiene im völkischen staat* [Genetics and racial hygiene in the völkish state] (pp. 303-316). Munich: J. F. Lehmanns.

Lykken, D. T., McGue, M., Bouchard, T. J., Jr., & Tellegen, A. (1990). Does contact lead to similarity or similarity to contact? *Behavior Genetics, 20*, 547-561.

Lykken, D. T., McGue, M., & Tellegen, A. (1987). Recruitment bias in twin research: The rule of two-thirds reconsidered. *Behavior Genetics, 17*, 343-362.

Lykken, D. T., Tellegen, A., & DeRubels, R. (1978). Volunteer bias in twin research: The rule of two-thirds. *Social Biology, 25*, 1-9.

Lynch, T. (2004). *Beyond Prozac: Healing mental distress*. Ross-on-Wye, UK: PCCS Books.

Lyons, M. J., Kendler, K. S., Provet, A., & Tsuang, M. T. (1991). The genetics of schizophrenia. In M. Tsuang, K. Kendler, & M. Lyons (Eds.), *Genetic issues in psychosocial epidemiology* (pp. 119-152). New Brunswick, NJ: Rutgers University Press.

Lytton, H. (1973). Three approaches to the study of parent-child interaction: Ethological, interview and experimental. *Journal of Child Psychology and Psychiatry, 14*, 1-17.

Lytton, H. (1977). Do parents create, or respond to, differences in twins? *Developmental Psychology, 13*, 456-459.

Lytton, H., & Zwirner, W. (1975). Compliance and its controlling stimuli observed in a natural setting. *Developmental Psychology, 11*, 769-779.

Macgregor, S., Visscher, P. M., Knott, S. A., Porteous, D. J., Millar, J. K., Devon, R. S., Blackwood, D., & Muir, W. J. (2004). A genome scan and follow-up study identify a bipolar disorder susceptibility locus on chromosome 1q42. *Molecular Psychiatry, 9*, 1083-1090.

Maier, W. (2002). Genetics of mood disorders: Current status of knowledge and prospects. In W. Kaschka (Ed.), *Perspectives in affective disorders* (pp. 35-62). New York: Karger.

Marteau, T. M. (2001). Genotype or genohype? *The Psychologist, 14*, 148-149.

Martin, B. (1981). *Abnormal psychology: Clinical and scientific perspectives* (2nd ed.). New York: Holt, Rinehart, and Winston.

Martin, N. G., Eaves, L. J., Heath, A. C., Jardine, R., Feingold, L. M., & Eysenck, H. J. (1986). Transmission of social attitudes. *Proceedings of the National Academy of Science, 83*, 4364-4368.

Matheny, A. P. (1979). Appraisal of parental bias in twin studies. *Acta Geneticae Medicae et Gemellologiae, 28*, 155-160.

Matheny, A. P., Wilson, R. S., & Brown, A. B. (1976). Relations between identical twins' similarity of appearance and behavioral similarity: Testing an assumption. *Behavior Genetics, 6*, 343-351.

Maxmen, J. S., & Ward, N. G. (1995). *Essential psychopathology and its treatment* (2nd ed., revised for DSM-IV). New York: W. W. Norton.

Maziade, M., Roy, M. A., Chagnon, Y. C., Cliché, D., Fournier, J. P., Montgrain, N., Dion, C., Lavallée, J. C., Garneau, Y., Gingras, N., Nicole, L., Pirès, A., Ponton, A. M., Potvin, A., Wallot, H., & Mérette, C. (2005). Shared and specific susceptibility loci for schizophrenia and bipolar disorder: A dense genome scan in Eastern Quebec families. *Molecular Psychiatry, 10*, 486-499.

McGuffin, P. (2002). Genetic research in psychiatry. In J. López-Ibor, W. Gaebel, M. Maj, & N. Sartorius (Eds.), *Psychiatry as a neuroscience* (pp. 1-27). New York: John Wiley & Sons.

McGuffin, P. (2004). Behavioral genomics: Where molecular genetics is taking psychiatry and psychology. In L. DiLalla (Ed.), *Behavior genetics principles* (pp. 191-204). Washington, DC: American Psychological Association Press.

McGuffin, P., Owen, M. J., O'Donovan, M. C., Thapar, A., & Gottesman, I. I. (1994). *Seminars in psychiatric genetics.* London: Gaskell Press.

McGuffin, P., Rijsdijk, F., Andrew, M., Sham, P., Katz, R., Cardno, A. (2003). The heritability of bipolar affective disorder and genetic relationship to unipolar depression. *Archives of General Psychiatry, 60*, 497-502.

McGuire, T. R., & Hirsch, J. (1977). General intelligence (g) and heritability (H2, h2). In I. Uzgiris & F. Weitzmann (Eds.), *The structuring of experience* (pp. 25-72). New York: Plenum Press.

McInnis, M. G., & Potash, J. B. (2004). Psychiatric genetics: Into the 21st century. *International Review of Psychiatry, 16*, 301-310.

McMahon, R. C. (1980). Genetic etiology in the hyperactive child syndrome: A critical review. *American Journal of Orthopsychiatry, 50*, 145-150.

Mednick, S. A., & Hutchings, B. (1977). Some considerations in the interpretation of the Danish adoption studies in relation to asocial behavior. In S. Mednick & K. Christiansen (Eds.), *Biosocial bases of criminal behavior* (pp. 159-164). New York: Gardner Press.

Melnick, M., Myrianthopoulos, N. C., & Christian, J. C. (1978). The effects of chorion type on variation in IQ in the NCPP twin population. *American Journal of Human Genetics, 30*, 425-433.

Menand, L. (2002, November 25th). What comes naturally [Review of the book *The Blank Slate*]. *The New Yorker*, 96-101.

Mendlewicz, J. (1988). Genetics of depression and mania. In A. Georgotas & R. Cancro (Eds.), *Depression and mania* (pp. 197-212). New York: Elsevier.

Mendlewicz, J., & Rainer, J. D. (1977). Adoption study supporting genetic transmission in manic-depressive illness. *Nature, 268*, 327-329.

Merikangas, K. R., Chakravarti, A., Moldin, S. O., Araj, H., Blangero, J., Burmeister, M., Crabbe, J. C., Jr., Depaulo, J. R., Jr., Foulks, E., Freimer, N. B., Koretz, D. S., Lichtenstein, W., Mignot, E., Reiss, A. L., Risch, N. J., & Takahashi, J. S. (2002). Future of genetics of mood disorder research. *Biological Psychiatry, 52*, 457-477.

Merikangas, K. R., & Risch, N. (2003a). Genomic priorities and public health. *Science, 302*, 599-601.

Merikangas, K. R., & Risch, N. (2003b). Will the genomics revolution revolutionize psychiatry? *American Journal of Psychiatry, 160*, 625-635.

Miele, F. (2005). Nature and nurture: Putting all the pieces together — Miele replies to Joseph and Schlinger. *Skeptic, 11 (3)*, 26-30.

Mill, J., Xiaohui, X., Ronald, A., Curran, S., Price, T., Knight, J., Sham, P., Plomin, R., & Asherson, P. (2005). Quantitative trait locus analysis of candidate gene alleles associated with attention deficit hyperactivity disorder (ADHD) in five genes: DRD4, DAT1, DRD5, SNAP-25, and 5HT1B. *American Journal of Medical Genetics (Series B, Neuropsychiatric Genetics), 133B*, 68-73.

Miller, A. (1984). *Thou shalt not be aware: Society's betrayal of the child.* New York: Meridian.

Miller, A. (1997). *Banished knowledge: Facing childhood injuries.* New York: Anchor.

Moises, H., Zoega, T., Li, L., & Hood, L. (2004). Genes and neurodevelopment in schizophrenia. In L. DiLalla (Ed.), *Behavior genetics principles* (pp. 145-157). Washington, DC: American Psychological Association Press.

Morris, D. W., McGhee, K. A., Schwaiger, S., Scully, P., Quinn, J., Meagher, D., Waddington, J. L., Gill, M., & Corvin, A. P. (2003). No evidence for association of the dysbindin gene [DTNBP1] with schizophrenia in an Irish population-based study. *Schizophrenia Research, 60*, 167-172.

Morris-Yates, A., Andrews, G., Howie, P., & Henderson, S. (1990). Twins: A test of the equal environments assumption. *Acta Psychiatrica Scandinavica, 81*, 322-326.

Morrison, J. R., & Stewart, M. A. (1971). A family study of the hyperactive child syndrome. *Biological Psychiatry, 3*, 189-195.

Morrison, J. R., & Stewart, M. A. (1973). The psychiatric status of the legal families of adopted hyperactive children. *Archives of General Psychiatry, 28*, 888-891.

Mountain, J. L., & Risch, N. (2004). Assessing genetic contributions to phenotypic differences among "racial" and "ethnic" groups. *Nature Genetics 36, (Suppl.)*, S48-S53.

Muhle, R., Trentacoste, S. V., & Rapin, I. (2004). The genetics of autism. *Pediatrics, 113*, e472-e486.

Müller-Hill, B. (1998a). *Murderous science.* Plainview, NY: Cold Spring Harbor Laboratory Press. (Original English version published in 1988)

Müller-Hill, B. (1998b). Psychiatric genetics in Germany 1939-1945: Can anything be learned from history? [Abstract]. *American Journal of Medical Genetics (Neuropsychiatric Genetics), 81*, 473.

Munsinger, H., & Douglass, A. (1976). The syntactic abilities of identical twins, fraternal twins, and their siblings. *Child Development, 47*, 40-50.

Murray, R. M. (1986). Schizophrenia. In P. Hill, R. Murray, & A. Thorley (Eds.), *Essentials of postgraduate psychiatry* (2nd ed., pp. 339-379). London: Grune & Stratton.

Murray, R. M., & Castle, D. J. (2000). Genetic and environmental risk factors for schizophrenia. In M. Gelder, J. Lopez-Ibor, & N. Andreasen (Eds.), *New Oxford textbook of psychiatry* (pp. 599-605). New York: Oxford University Press.

Murray, R. M., Jones, P. B., Susser, E., van Os, J., & Cannon, M. (Eds.). (2003). *The epidemiology of schizophrenia.* Cambridge, UK: Cambridge University Press.

Nadder, T. S., Silberg, J. L., Eaves, L. J., Maes, H. H., & Meyer, J. M. (1998). Genetic effects on ADHD symptomatology in 7- 13- year-old twins: Results from a telephone survey. *Behavior Genetics, 28,* 83-99.

National Institutes of Health. (2003). *New program will pursue schizophrenia gene leads.* Retrieved 10/1/03, from http://www.nih.gov/news/pr/sep2003/nimh-12.htm

Neale, J. M., & Oltmanns, T. F. (1980). *Schizophrenia.* New York: John Wiley & Sons.

Newman, H. H., Freeman, F. N., & Holzinger, K. J. (1937). *Twins: A study of heredity and environment.* Chicago: The University of Chicago Press.

Nicholi, A. M. (Ed.). (1988). *The new Harvard guide to psychiatry.* Cambridge, MA: Harvard University Press.

Nichols, P. L., & Chen, T. C. (1981). *Minimal brain dysfunction.* Hillsdale, NJ: Lawrence Erlbaum Associates.

Nolen-Hoeksema, S. (1998). *Abnormal psychology.* Boston: McGraw-Hill.

Ogdie, M. N., Fisher, S. E., Yang, M., Ishii, J., Francks, C., Loo, S., Cantor, R. M., McCracken, J. T., McGough, J. J., Smalley, S. L., & Nelson, S. F. (2004). Attention deficit hyperactivity disorder: Fine mapping supports linkage to 5p13, 6q12, 16p13, and 17p11. *American Journal of Human Genetics, 75,* 661-668.

Ogdie, M. N., Macphie, I. L., Minassian, S. L., Yang, M., Fisher, S. E., Francks, C., Cantor, R. M., McCracken, J. T., McGough, J. J., Nelson, S. F., Monaco, A. P., & Smalley, S. L. (2003). A genomewide scan for attention-deficit/hyperactivity disorder in an extended sample: Suggestive linkage on 17p11. *American Journal of Human Genetics, 72,* 1268-1279.

Oldham, J., & Riba, M. (Eds.). (1995). *Review of psychiatry* (Vol. 14). Washington, DC: American Psychiatric Press.

Oltmanns, T. F., & Emery, R. E. (1998). *Abnormal psychology* (2nd ed.). Upper Saddle River, NJ: Prentice Hall.

Owen, M. J. (1992). Will schizophrenia become a graveyard for molecular geneticists? *Psychological Medicine, 22,* 289-293.

Owen, M. J., Cardno, A. G., & O'Donovan, M. C. (2000). Psychiatric genetics: Back to the future. *Molecular Psychiatry, 5,* 22-31.

Page, G. P., Varghese, G., Go, R. C., Page, P. Z., & Allison, D. B. (2003). "Are we there yet?": Deciding when one has demonstrated specific genetic causation in complex diseases and quantitative traits. *American Journal of Human Genetics, 73,* 711-719.

Pam, A. (1995). Biological psychiatry: Science or pseudoscience? In C. Ross & A. Pam (Eds.), *Pseudoscience in biological psychiatry: Blaming the body* (pp. 7-84). New York: John Wiley & Sons.

Pam, A., Kemker, S. S., Ross, C. A., & Golden, R. (1996). The "equal environment assumption" in MZ-DZ comparisons: An untenable premise of psychiatric genetics? *Acta Geneticae Medicae et Gemellologiae, 45,* 349-360.

Parens, E. (2004). Genetic differences and human identities: On why talking about behavioral genetics is important and difficult. *Hastings Center Report, Special Supplement 34, (1),* S1-S36.

Paris, J. (1999). *Nature and nurture in psychiatry: A predisposition-stress model of mental disorders.* Washington, DC: American Psychiatric Press.

Pastore, N. (1949). The genetics of schizophrenia: A special review. *Psychological Bulletin, 46*, 285-302.

Paul, D. B. (1985). Textbook treatments of the genetics of intelligence. *Quarterly Review of Biology, 60*, 317-326.

Paul, D. B. (1998). *The politics of heredity: Essays on eugenics, biomedicine, and the nature-nurture debate.* Albany, NY: State University of New York Press.

Pericak-Vance, M. A. (2003). The genetics of autistic disorder. In R. Plomin, J. DeFries, I. Craig, & P. McGuffin (Eds.), *Behavioral genetics in the postgenomic era* (pp. 267-288). Washington, DC: American Psychological Association Press.

Peters, U. H. (1999). German psychiatry. In H. Freeman (Ed.), *A century of psychiatry* (pp. 86-93). London: Mosby.

Pfaff, D., Berrettini, W., Joh, T., & Maxson, S. (Eds.). (2000). *Genetic influences on neural and behavioral functions.* Boca Raton, FL: CRC Press.

Phillips, D. I. W. (1993). Twin studies in medical research: Can they tell us whether diseases are genetically determined? *Lancet, 341*, 1008-1009.

Pinker, S. (2002). *The blank slate.* New York: Viking.

Pittelli, S. J. (2002). Meta-analysis and psychiatric genetics [Letter to the editor]. *American Journal of Psychiatry, 159*, 496.

Pittelli, S. J. (2003). Genetic link in schizophrenia [Letter to the editor]. *American Journal of Psychiatry, 160*, 597.

Pittelli, S. J. (2004). Genetic linkage for schizophrenia? [Letter to the editor]. *American Journal of Psychiatry, 161*, 1134.

Pittelli, S. J. (2005). Nonsignificant but suggestive results? [Letter to the editor]. *American Journal of Psychiatry, 162*, 633.

Piven, J. (2002). Genetics of personality: The example of the broad autism phenotype. In J. Benjamin, R. Ebstein, & R. Belmaker (Eds.), *Molecular genetics and the human personality* (pp. 43-62). Washington, DC: American Psychiatric Press.

Plomin, R. (2004). Genetics and developmental psychology. *Merrill-Palmer Quarterly, 50*, 341-352.

Plomin, R. (2005). Finding genes in child psychology and psychiatry: When are we going to be there? *Journal of Child Psychology and Psychiatry, 46*, 1030-1038.

Plomin, R., Corley, R., Caspi, A., Fulker, D. W., & DeFries, J. C. (1998). Adoption results for self-reported personality: Evidence for nonadditive genetic effects? *Journal of Personality and Social Psychology, 75*, 211-218.

Plomin, R., DeFries, J. C., & McClearn, G. E. (1990). *Behavioral genetics: A primer* (2nd ed.). New York: W. H. Freeman and Company.

Plomin, R., DeFries, J. C., McClearn, G. E., & Rutter, M. (1997). *Behavioral genetics* (3rd ed.). New York: W. H. Freeman and Company.

Plomin, R., & McGuffin, P. (2003). Psychopathology in the postgenomic era. *Annual Review of Psychology, 54*, 205-228.

Plomin, R., Willerman, L., & Loehlin, J. C. (1976). Resemblance in appearance and the equal environments assumption in twin studies of personality traits. *Behavior Genetics, 6*, 43-52.

Pollin, W., Allen, M. G., Hoffer, A., Stabenau, J. R., & Hrubec, Z. (1969). Psychopathology in 15,909 pairs of veteran twins: Evidence for a genetic factor in the pathogenesis of schizophrenia and its relative absence in psychoneurosis. *American Journal of Psychiatry, 126,* 597-610.

Pope, H. G., Jonas, J. M., Cohen, B. M., & Lipinski, J. F. (1982). Failure to find evidence of schizophrenia in first-degree relatives of schizophrenic probands. *American Journal of Psychiatry, 139,* 826-828.

Potash, J. B., & DePaulo, J. R., Jr. (2000). Searching high and low: A review of the genetics of bipolar disorder. *Bipolar Disorders, 2,* 8-26.

Price, R. H., & Lynn, S. J. (1986). *Abnormal psychology* (2nd ed.). Chicago: The Dorsey Press.

Proctor, R. N. (1988). *Racial hygiene: Medicine under the Nazis.* Cambridge, MA: Harvard University Press.

Propping, P. (2005). The biography of psychiatric genetics: From early achievements to historical burden, from an anxious society to critical geneticists. *American Journal of Medical Genetics Part B (Neuropsychiatric Genetics), 136B (1),* 2-7.

Pulver, A. E., Pearlson, G. D., McGrath, J. A., Lasseter, V. K., Swartz, K. L., & Papadimitriou, G. (2002). Schizophrenia. In R. King, J. Rotter, & A. Motulsky (Eds.), *The genetic basis of common diseases* (2nd ed., pp. 850-875). Oxford: Oxford University Press.

Ratner, C. (2004). Genes and psychology in the news. *New Ideas in Psychology, 22,* 29-47.

Raybould, R., Green, E. K., MacGregor, S., Gordon-Smith, K., Heron, J., Hyde, S., Caesar, S., Nikolov, I., Williams, N., Jones, L., O'Donovan, M. C., Owen, M. J., Jones, I., Kirov, G., & Craddock, N. (2005). Bipolar disorder and polymorphisms in the dysbindin gene (DTNBP1). *Biological Psychiatry, 57,* 696-701.

Read, J., Goodman, L., Morrison, A. P., Ross, C. A., & Aderhold, V. (2004). Childhood trauma, loss and stress. In J. Read, L. Mosher, & R. Bentall (Eds.), *Models of madness: Psychological, social and biological approaches to schizophrenia* (pp. 223-252). Andover, UK: Taylor & Francis.

Reber, A. S. (1985). *The Penguin dictionary of psychology.* London: Penguin Books.

Rehm, L. P., Mehta, P., & Dodrill, C. L. (2001). Depression. In M. Herson & V. Van Hasselt (Eds.), *Advanced abnormal psychology* (2nd ed., pp. 307-324). New York: Kluwer.

Rende, R. (2004). Beyond heritability: Biological process in social context. In C. Coll, E. Bearer, & R. Lerner (Eds.), *Nature and Nurture: The complex interplay of genetic and environmental influences on human behavior and development* (pp. 107-126). Mahwah, NJ: Erlbaum.

Reuband, K. (1994). Reconstructing social change through retrospective questions: Methodological problems and prospects. In N. Schwarz & S. Sudman (Eds.), *Autobiographical memory and the validity of retrospective reports* (pp. 305-311). New York: Springer Verlag.

Richardson, K. & Norgate, S. (in press) The equal environment assumption of classical twin studies may not hold. *British Journal of Educational Psychology.*

Ridley, M. (2003). *The agile gene: How nature turns on nurture* [Originally published as *Nature via nurture*]. New York: Perennial.

Rieder, R. O. (1979). Children at risk. In L. Bellack (Ed.), *Disorders of the schizophrenic syndrome* (pp. 232-263). New York: Basic Books.

Rietveld, M. J. H., Hudziak, J. J., Bartels, M., van Beijsterveldt, C. E. M., & Boomsma, D. I. (2004). Heritability of attention problems in children: Longitudinal results from a study of twins, age 3 to 12. *Journal of Child Psychology and Psychiatry, 45*, 577-588.

Riley, B., Asherson, P. J., & McGuffin, P. (2003). Genetics and schizophrenia. In S. Hirsch & D. Weinberger (Eds.), *Schizophrenia* (2nd ed., pp. 251-276). Malden, MA: Blackwell.

Risch, N. J. (2000). Searching for genetic determinants in the new millennium. *Nature, 405*, 847-856.

Risch, N., & Botstein, D. (1996). A manic depressive history. *Nature Genetics, 12*, 351-353.

Ritter, M. (2004, October 21). How humans compare to worms. *The San Francisco Chronicle*, p. A12.

Ritvo, E. R., Freeman, B. J., Mason-Brothers, A., Mo, A., & Ritvo, A. M. (1985a). Concordance for the syndrome of autism on 40 pairs of affected twins. *American Journal of Psychiatry, 142*, 74-77.

Ritvo, E. R., Freeman, B. J., Mason-Brothers, A., Mo, A., & Ritvo, A. M. (1985b). Dr. Ritvo and associates reply [Letter to the editor]. *American Journal of Psychiatry, 142*, 1521.

Robbins, L. C. (1963). The accuracy of parental recall of aspects of child development and of child rearing practices. *Journal of Abnormal and Social Psychology, 66*, 261-270.

Roe, D. A. (1973). *The plague of corn: The social history of pellagra*. Ithaca, NY: Cornell University Press.

Roelcke, V., Hohendorf, G., & Rotzoll, M. (1998). Genetic research in the context of Nazi "euthanasia": New documents and aspects on Ernst Rüdin and the Deutsche Forschungsanstalt für Psychiatrie [Abstract]. *American Journal of Medical Genetics (Neuropsychiatric Genetics), 81*, 474.

Rosanoff, A. J., Handy, L. M., Plesset, I. R., & Brush, S. (1934). The etiology of so-called schizophrenic psychoses. *American Journal of Psychiatry, 91*, 247-286.

Rose, R. J. (1980). Genetic factors. In A. Kazdin, A. Bellack, & M. Hersen (Eds.), *New perspectives in abnormal psychology* (pp. 85-109). New York: Oxford University Press.

Rosenthal, D. (1960). Confusion of identity and the frequency of schizophrenia in twins. *Archives of General Psychiatry, 3*, 297-304.

Rosenthal, D. (1961). Sex distribution and the severity of illness among samples of schizophrenic twins. *Journal of Psychiatric Research, 1*, 26-36.

Rosenthal, D. (1962a). Familial concordance by sex with respect to schizophrenia. *Psychological Bulletin, 59*, 401-421.

Rosenthal, D. (1962b). Problems of sampling and diagnosis in the major twin studies of schizophrenia. *Journal of Psychiatric Research, 1*, 116-134.

Rosenthal, D. (1970). *Genetic theory and abnormal behavior*. New York: McGraw-Hill.

Rosenthal, D. (1971a). *Genetics of psychopathology*. New York: McGraw-Hill.

Rosenthal, D. (1971b). A program of research on heredity in schizophrenia. *Behavioral Science, 16*, 191-201.

Rosenthal, D. (1972). Three adoption studies of heredity in the schizophrenic disorders. *International Journal of Mental Health, 1*, 63-75.

Rosenthal, D. (1974). The genetics of schizophrenia. In S. Arieti & E. Brody (Eds.), *American handbook of psychiatry* (2nd ed., pp. 588-600). New York: Basic Books.

Rosenthal, D. (1975). The spectrum concept in schizophrenic and manic-depressive disorders. In D. Freedman (Ed.), *Biology of the major psychoses* (pp. 19-25). New York: Raven Press.

Rosenthal, D. (1979). Genetic factors in behavioural disorders. In M. Roth & V. Cowie (Eds.), *Psychiatry, genetics and pathography: A tribute to Eliot Slater* (pp. 22-33). London: Oxford University Press.

Rosenthal, D. (1980). Genetic aspects of schizophrenia. In H. van Praag, M. Lader, O. Rafaelsen, & E. Sachar (Eds.), *Handbook of biological psychiatry* (Part 3, pp. 3-34). New York: Marcel Dekker.

Rosenthal, D., Wender, P. H., Kety, S. S., Schulsinger, F., Welner, J., & Østergaard, L. (1968). Schizophrenics' offspring reared in adoptive homes. In D. Rosenthal & S. Kety (Eds.), *The transmission of schizophrenia* (pp. 377-391). New York: Pergamon Press.

Rosenthal, D., Wender, P. H., Kety, S. S., Welner, J., & Schulsinger, F. (1971). The adopted-away offspring of schizophrenics. *American Journal of Psychiatry, 128*, 307-311.

Rüdin, E. (1916). *Zur Vererbung und Neuentstehung der Dementia praecox* [On the heredity and new development of dementia praecox]. Berlin: Springer Verlag OHG.

Rüdin, E. (1933). Eugenic sterilization: An urgent need. *Birth Control Review, 17*, 102-104.

Rüdin, E. (1939). Bedeutung der Forschung und Mitarbeit von Neurologen und Psychiatrie im nationalsozialistischen Staat [The meaning of research and cooperation of neurologists and psychiatry in the National Socialist state]. *Zeitschrift für die Gesamte Neurologie und Psychiatrie, 165*, 7-17.

Rüdin, E. (1942). Zehn Jahre nationalsozialistischer Staat [Ten years of the National Socialist state]. *Archiv für Rassen- und Gesellschaftsbiologie, 36*, 321-322.

Ruscio, J. (2001). *Clear thinking with psychology: Separating sense from nonsense.* Pacific Grove, CA: Wadsworth.

Rutter, M. (2001). Child psychiatry in the era following sequencing the genome. In F. Levy & D. Hay (Eds.), *Attention, genes, and ADHD* (pp. 225-248). East Sussex, UK: Brunner-Routledge.

Rutter, M. (2003). Commentary: Nature-nurture interplay in emotional disorders. *Journal of Child Psychology and Psychiatry, 44*, 934-944.

Rutter, M. (2005). Aetiology of autism: Findings and questions. *Journal of Intellectual Disability Research, 49*, 231-238.

Rutter, M., Bolton, P., Harrington, R., Le Couteur, A., Macdonald, H., & Simonoff, E. (1990). Genetic factors in child psychiatric disorders — I. A review of research strategies. *Journal of Child Psychology and Psychiatry, 31*, 3-37.

Rutter, M., Pickles, A., Murray, R., & Eaves, L. (2001). Testing hypotheses on specific environmental causal effects on behavior. *Psychological Bulletin, 127*, 291-324.

Sadock, B. J., & Sadock, V. A. (2003). *Kaplan and Sadock's synopsis of psychiatry: Behavioral sciences/clinical psychiatry* (9th ed.). Philadelphia: Lippincott, Williams, & Williams.

Safe Minds. (2003). Analysis and critique of the CDC's handling of the Thimerosal exposure assessment based on vaccine safety datalink (VSD) information. *Unpublished Document.*

Safer, D. J. (1973). A familial factor in minimal brain dysfunction. *Behavior Genetics, 3*, 175-186.

Sarbin, T. R., & Mancuso, J. C. (1980). *Schizophrenia: Medical diagnosis or moral verdict?* New York: Pergamon Press.

Sarason, I. G., & Sarason, B. R. (1984). *Abnormal psychology: The problem of maladaptive behavior* (4th ed.). Englewood Cliffs, NJ: Prentice-Hall.

Saudino, K. J., Ronald, A., & Plomin, R. (2005). The etiology of behavior problems in 7-year-old twins: Substantial genetic influence and negligible shared environmental influence for parent ratings and ratings by same and different teachers. *Journal of Abnormal Child Psychology, 33,* 113-130.

Scarr, S. (1968). Environmental bias in twin studies. *Eugenics Quarterly, 15,* 34-40.

Scarr, S., & Carter-Saltzman, L. (1979). Twin method: Defense of a critical assumption. *Behavior Genetics, 9,* 527-542.

Schachar, R., & Tannock, R. (2002). Syndromes of hyperactivity and attention deficit. In M. Rutter & E. Taylor (Eds.), *Child and adolescent psychiatry* (4th ed., pp. 399-418). Malden, MA: Blackwell Science.

Scharfetter, C., & Nüsperli, M. (1980). The group of schizophrenias, schizoaffective psychoses, and affective disorders. *Schizophrenia Bulletin, 6,* 586-591.

Schuckit, M. A. (1986). Trait (and state) markers of a predisposition to psychopathology. In L. Judd & P. Groves (Eds.), *Psychiatry: Psychobiological foundations of clinical psychiatry* (Vol. 4, pp. 145-163). New York: Basic Books.

Schultz, S. C. (1991). Genetics of schizophrenia: A status report. In A. Tasman & S. Goldfinger (Eds.), *Annual review of psychiatry* (Vol. 10, pp. 79-97). Washington, DC: American Psychiatric Press.

Schulz, B. (1934). Rassenhygienische Eheberatung [Racial hygienic marriage counseling]. *Volk und Rasse, 9,* 138-143.

Schulze, T. G., Fangerau, H., & Propping, P. (2004). From degeneration to genetic susceptibility, from eugenics to genethics, from Bezugsziffer to LOD score: The history of psychiatric genetics. *International Review of Psychiatry, 16,* 260-283.

Segurado, R., Detera-Wadleigh, S. D., et al. (2003). Genome scan meta-analysis of schizophrenia and bipolar disorder, Part III: Bipolar disorder. *American Journal of Human Genetics, 73,* 49-62.

Seligman, K. (2005, February 4). Scientists baffled as autism cases soar in state, with no relief in sight. *The San Francisco Chronicle,* p. A1, A11.

Shastry, B. S. (2004). Molecular genetics of attention-deficit hyperactivity disorder (ADHD): An update. *Neurochemistry International, 44,* 469-474.

Shastry, B. S. (2005). Bipolar disorder: An update. *Neurochemistry International, 46,* 273-279.

Sherman, D. K., Iacono, W. G., & McGue, M. K. (1997). Attention-deficit hyperactivity disorder dimensions: A twin study of inattention and impulsivity-hyperactivity. *Journal of the American Academy of Child and Adolescent Psychiatry, 36,* 745-753.

Sherrington, R., Brynjolfsson, J., Petursson, H., Potter, M., Duddleston, K., Barraclough, B., Wasmuth, J., Dobbs, M., & Gurling, H. (1988). Localization of a susceptibility locus for schizophrenia on chromosome 5. *Nature, 336,* 164-167.

Shields, J., Gottesman, I. I., & Slater, E. (1967). Kallmann's 1946 schizophrenia twin study in the light of new information. *Acta Psychiatrica Scandinavica, 43, Fasc. 4,* 385-396.

Shih, R. A., Belmonte, P. L., & Zandi, P. P. (2004). A review of the evidence from family, twin and adoption studies for a genetic contribution to adult psychiatric disorders. *International Review of Psychiatry, 16,* 260-283.

Shink, E., Morissette, J., & Barden, N. (2005). A genome-wide scan points to a susceptibility locus for bipolar disorder on chromosome 12. *Molecular Psychiatry, 10,* 545-552.

Siever, L. J., & Gunderson, J. G. (1979). Genetic determinants of borderline conditions. *Schizophrenia Bulletin, 5,* 59-86.

Silberg, J. L., Rutter, M., Meyer, J. M., Maes, H. H., Hewitt, J., Simonoff, E., Pickles, A., & Loeber, R. (1996). Genetic and environmental influences on the covariation between hyperactivity and conduct disturbance in juvenile twins. *Journal of Child Psychology and Psychiatry, 37,* 803-816.

Silverman, C., & Herbert, M. (2003). Autism and genetics. *Gene Watch, 18.* Available online: http://www.gene-watch.org/genewatch/articles/16-2herbert_silverman.html.

Sklar, P. (2002). Linkage analysis in psychiatric disorders. The emerging picture. *Annual Review of Genomics and Human Genetics, 3,* 371-413.

Skuse, D. H. (2001). Endophenotypes and child psychiatry. *British Journal of Psychiatry, 178,* 395-396.

Slaats-Willemse, D., Swaab-Barneveld, H., de Sonneville, L., van der Meulen, E., & Buitelaar, J. (2003). Deficient response inhibition as a cognitive endophenotype of ADHD. *Journal of the American Academy of Child and Adolescent Psychiatry, 42,* 1242-1248.

Slater, E. (1953). Psychotic and neurotic illnesses in twins. *Medical Research Council Special Report Series No. 278.* London: Her Majesty's Stationary Office.

Slater, E., & Cowie, V. (1971). *The genetics of mental disorders.* London: Oxford University Press.

Smalley, S. L., Asarnow, R. F., & Spence, M. A. (1988). Autism and genetics. *Archives of General Psychiatry, 45,* 953-961.

Smalley, S. L., Kustanovich, V., Minassian, S. L., Stone, J. L., Ogdie, M. N., McGough, J. J., McCracken, J. T., MacPhie, I. L., Franks, C., Fisher, S. E., Cantor, R. M., Monaco, A. P., & Nelsen, S. F. (2002). Genetic linkage of attention-deficit/hyperactivity disorder on chromosome 16p13, in a region implicated in autism. *American Journal of Human Genetics, 71,* 959-963.

Smith, R. T. (1965). A comparison of socioenvironmental factors in monozygotic and dizygotic twins, testing an assumption. In S. Vandenberg (Ed.), *Methods and goals in human behavior genetics* (pp. 45-61). New York: Academic Press.

Smoller, J. W., & Finn, C. T. (2003). Family, twin, and adoption studies of bipolar disorder. *American Journal of Medical Genetics (Series C, Semin. Med. Genet.), 123C,* 48-58.

Spitzer, R. L., & Endicott, J. (1979). Justification for separating schizotypal and borderline personality disorders. *Schizophrenia Bulletin, 6,* 95-104.

Sprich, S., Biederman, J., Crawford, M. H., Mundy, E., & Faraone, S. V. (2000). Adoptive and biological families of children and adolescents with ADHD. *Journal of the American Academy of Child and Adolescent Psychiatry, 39,* 1432-1437.

Stefansson, H., Sigurdsson, E., et al. (2002). Neuregulin 1 and susceptibility to schizophrenia. *American Journal of Human Genetics, 71,* 877-892.

Steffenburg, S., Gillberg, C., Hellgren, L., Andersson, L., Gillberg, I. C., Jakobsson, G., & Bohman, M. (1989). A twin study of autism in Denmark, Finland, Iceland, Norway and Sweden. *Journal of Child Psychology and Psychiatry, 30,* 405-416.

Steffensson, B., Larsson, J., Fried, I., El-Sayed, E., Rydelius, P., & Lichtenstein, P. (1999). Genetic disposition for global maturity: An explanation for genetic effects on parental report of ADHD. *International Journal of Behavioral Development, 23*, 357-374.

Stein, D. B. (1998). *Ritalin is not the answer.* San Francisco: Jossey-Bass.

Stevenson, J. (1992). Evidence for a genetic etiology in hyperactivity in children. *Behavior Genetics, 22*, 337-344.

Straub, R. E., Jiang, Y., MacLean, C. J., Ma, Y., Webb, B. T., Myakishev, M. V., Harris-Kerr, C., Wormley, B., Sadek, H., Kadambi, B., Cesare, A. J., Gibberman, A., Wang, X., O'Neill, F. A., Walsh, D., & Kendler, K. S. (2002). Genetic variation in the 6p22.3 gene DTNBP1, the human ortholog of the mouse dysbindin gene, is associated with schizophrenia. *American Journal of Human Genetics, 71*, 337-348.

Sturt, E. (1985). Is autism the expression of a recessive gene? [Letter to the editor]. *American Journal of Psychiatry, 142*, 1520-1521.

Sullivan, P. F., & Kendler, K. S. (2003). Schizophrenia as a complex trait: Evidence from a meta-analysis of twin studies. *Archives of General Psychiatry, 60*, 1187-1192.

Surgeon General. (1999). *Mental Health: A report of the Surgeon General.* Chapter 4, p. 256. Retrieved online 10/22/2004 from http://www.surgeongeneral.gov/library/mentalhealth/pdfs/c4.pdf

Sutker, P. B., & Adams, H. E. (Eds.). (1993). *Comprehensive textbook of psychopathology* (2nd ed.). New York: Plenum Press.

Szasz, T. S. (1964). The moral dilemma of psychiatry: Autonomy or heteronomy? *American Journal of Psychiatry, 121*, 521-528.

Szasz, T. S. (1976). *Schizophrenia: The sacred symbol of psychiatry.* New York: Basic Books.

Tannock, R. (1998). Attention deficit hyperactivity disorder: Advances in cognitive, neurobiological, and genetic research. *Journal of Child Psychology and Psychiatry, 39*, 65-99.

Tasman, A., & Goldfinger, S. (1991). *Review of psychiatry* (Vol. 10). Washington, DC: American Psychiatric Association Press.

Tasman, A., Kay, J., & Lieberman, J. A. (Eds.). (1997). *Psychiatry* (Vol. 2). Philadelphia: W.B. Saunders Company.

Taylor, H. F. (1980). *The IQ game: A methodological inquiry into the heredity-environment controversy.* New Brunswick, NJ: Rutgers University Press.

Thapar, A. (2003). Attention deficit hyperactivity disorder: New genetic findings, new directions. In R. Plomin, J. DeFries, I. Craig, & P. McGuffin (Eds.), *Behavioral genetics in the postgenomic era* (pp. 445-462). Washington, DC: American Psychological Association Press.

Thapar, A., Hervas, A., & McGuffin, P. (1995). Childhood hyperactivity scores are highly heritable and show sibling competition effects: Twin study evidence. *Behavior Genetics, 25*, 537-544.

Thapar, A., Holmes, J., Poulton, K., & Harrington, R. (1999). Genetic basis of attention-deficit and hyperactivity. *British Journal of Psychiatry, 174*, 105-111.

Thapar, A., & Scourfield, J. (2003). Childhood disorders. In P. McGuffin, M. Owen, & I. Gottesman (Eds.), *Psychiatric genetics and genomics* (pp. 147-180). Oxford: Oxford University Press.

Tienari, P. (1963). *Psychiatric illnesses in identical twins.* Copenhagen: Munksgaard.

Tienari, P. (1971). Schizophrenia and monozygotic twins. *Psychiatria Fennica, 1971*, 97-104.

Tienari, P. (1975). Schizophrenia in Finnish male twins. *British Journal of Psychiatry Special Publication, No. 10*. M. Lader (Ed.). pp. 29-35.

Tienari, P., Sorri, A., Lahti, I., Naarala, M., Wahlberg, K., Moring, J., Pohjola, J., & Wynne, L. C. (1987). Genetic and psychosocial factors in schizophrenia: The Finnish Adoptive Family Study. *Schizophrenia Bulletin, 13*, 477-484.

Tienari, P., Wynne, L. C., Läksy, K., Moring, J., Nieminen, P., Sorri, A., Lahti, I., & Wahlberg, K. E. (2003). Genetic boundaries of the schizophrenia spectrum: Evidence from the Finnish Adoptive Family Study. *American Journal of Psychiatry, 160*, 1587-1594.

Tienari, P., Wynne, L. C., Sorri, A., Lahti, I., Läksy, K., Moring, J., Naarala, M., Nieminen, P., & Wahlberg, K. E. (2004). Genotype-environment interaction in schizophrenia-spectrum disorders. *British Journal of Psychiatry, 184*, 216-222.

Timimi, S., & 33 Coendorsers. (2004). A critique of the international consensus statement on ADHD. *Clinical Child and Family Psychology Review, 7*, 59-63.

Todd, R. D., Rasmussen, E. R., Neuman, R. J., Reich, W., Hudziak, J. J., Bucholz, K. K., Madden, P. A. F., & Heath, A. (2001). Familiality and heritability of subtypes of attention deficit hyperactivity disorder in a population sample of adolescent female twins. *American Journal of Psychiatry, 158*, 1891-1898.

Tohen, M., & Angst, J. (2002). Epidemiology of bipolar disorder. In M. Tsuang & M. Tohen (Eds.), *Textbook in psychiatric epidemiology* (2nd ed., pp. 427-444). New York: Wiley-Liss.

Torgersen, S. (1986). Genetic factors in moderately severe and mild affective disorders. *Archives of General Psychiatry, 43*, 222-226.

Torrey, E. F. (1992). Are we overestimating the genetic contribution to schizophrenia? *Schizophrenia Bulletin, 18*, 159-170.

Torrey, E. F., Bowler, A. E., Taylor, E. H., & Gottesman, I. I. (1994). *Schizophrenia and manic-depressive disorder: The biological roots of mental illness as revealed by the landmark study of identical twins.* New York: Basic Books.

Tozzi, F., Aggen, S. H., Neale, B. N., Anderson, C. B., Mazzeo, S. E., Neale, M. C., & Bulik, C. M. (2004). The structure of perfectionism: A twin study. *Behavior Genetics, 34*, 483-494.

Trimble, M. R. (1988). *Biological psychiatry.* New York: John Wiley & Sons.

Tsai, L. Y. (2004). Autistic disorder. In J. Wiener & M. Dulcan (Eds.), *Textbook of child and adolescent psychiatry* (3rd ed., pp. 261-315). Washington, DC: American Psychiatric Association Press.

Tsuang, M. T., & Faraone, S. V. (1990). *The genetics of mood disorders.* Baltimore: Johns Hopkins University Press.

Tsuang, M. T., Faraone, S. V., & Green, R. R. (1994). Genetic epidemiology of mood disorders. In D. Papolos & H. Lachman (Eds.), *Genetic studies of affective disorders* (pp. 3-27). New York: John Wiley and Sons.

Tsuang, M. T., Nossova, N., Yager, T., Tsuang, M., Guo, S., Shyu, K. G., Glatt, S. J., & Liew, C. C. (2005). Assessing the validity of blood-based gene expression profiles for the classification of schizophrenia and bipolar disorder: A preliminary report. *American Journal of Medical Genetics (Series B, Neuropsychiatric Genetics), 133B*, 1-5.

Tsuang, M. T., Winokur, G., & Crowe, R. R. (1980). Morbidity risks of schizophrenia and affective disorders among first degree relatives of patients with schizophrenia, mania, depression, and surgical conditions. *American Journal of Psychiatry, 137*, 497-504.

Van den Bogaert, A., Schumacher, J., et al. (2003). The DTNBP1 (Dysbindin) gene contributes to schizophrenia, depending on family history of the disease. *American Journal of Human Genetics, 73,* 1438-1443.

Van den Oord, E. J. C. G., Boomsma, D. I., & Verhulst, F. C. (1994). A study of problem behaviors in 10- to 15-year-old biologically related and unrelated international adoptees. *Behavior Genetics, 24,* 193-205.

Van den Oord, E. J. C. G., Verhulst, F. C., & Boomsma, D. I. (1996). A genetic study of maternal and paternal ratings of problem behaviors in 3-year-old twins. *Journal of Abnormal Psychology, 105,* 349-357.

Van Hasselt, V. B., & Hersen, M. (1994). *Advanced abnormal psychology.* New York: Plenum Press.

Van Praag, H. M., Lader, M. H., Rafaelsen, O. J., & Sachar, E. J. (1980). *Handbook of Biological Psychiatry* (Part III). New York: Marcel Dekker.

Veenstra-VanderWeele, J., Christian, S. L., & Cook, E. H., Jr. (2004). Autism as a paradigmatic complex genetic disorder. *Annual Review of Genomics and Human Genetics, 5,* 379-405.

Veenstra-VanderWeele, J., & Cook, E. H., Jr. (2004). Molecular genetics of autism spectrum disorder. *Molecular Psychiatry, 9,* 819-832.

Venken, T., Claes, S., Sluijs, S., Paterson, A. D., van Duijn, C., Adolfsson, R., Del-Favero, J., & Van Broeckhoven, C. (2005). Genomewide scan for affective disorder susceptibility loci in families of a northern Swedish isolated population. *American Journal of Human Genetics, 76,* 237-248.

Verhulst, F. C., Althaus, M., & Versluis-den Bieman, H. J. M. (1990). Problem behavior in international adoptees: II. Age at placement. *Journal of the American Academy of Child and Adolescent Psychiatry, 29,* 104-111.

Verhulst, F. C., Althaus, M., & Versluis-den Bieman, H. J. M. (1990). Problem behavior in international adoptees: I. An epidemiological study. *Journal of the American Academy of Child and Adolescent Psychiatry, 29,* 94-103.

Verstraeten, T., Davis, R. L., DeStefano, F., Lieu, T. A., Rhodes, P. H., Black, S. B., Shinefield, H., & Chen, R. T. (2003). Safety of Thimerosal-containing vaccines: A two-phased study of computerized heath maintenance organization databases. *Pediatrics, 112,* 1039-1048.

Volkmar, F. R., & Pauls, D. (2003). Autism. *Lancet, 362,* 1133-1141.

Von Knorring, A., Cloninger, C. R., Bohman, M., & Sigvardsson, S. (1983). An adoption study of depressive disorders and substance abuse. *Archives of General Psychiatry, 40,* 943-950.

Walker, E., Downey, G., & Caspi, A. (1991). Twin studies of psychopathology: Why do concordance rates vary? *Schizophrenia Research, 5,* 211-221.

Walster, G., & Cleary, T. (1970). A proposal for a new editorial policy in the social sciences. *The American Statistician, 26,* 16-19.

Waslick, B., & Greenhill, L. L. (2004). Attention-deficit/hyperactivity disorder. In J. Wiener & M. Dulcan (Eds.), *Textbook of child and adolescent psychiatry* (3rd ed., pp. 485-507). Washington, DC: American Psychiatric Association Press.

Wassink, T. H., Brzustowicz, L. M., Bartlett, C. W., & Szatmari, P. (2004). The search for autism genes. *Mental Retardation and Developmental Disabilities Research Reviews, 10,* 272-283.

Watson, J. (2003). The double helix 50 years later: Implications for psychiatry. *American Journal of Psychiatry, 160,* 614.

Weber, M. M. (1996). Ernst Rüdin, 1874-1952. *American Journal of Medical Genetics (Neuropsychiatric Genetics), 67,* 323-331.

Weber, M. M. (2000). Psychiatric research and science policy in German: The history of the *Deutsche Forschungsanstalt für Psychiatrie* (German Institute for Psychiatric Research) in Munich from 1917 to 1945. *History of Psychiatry, 11,* 235-258.

Weindling, P. (1989). *Health, race, and German politics between national unification and Nazism, 1870-1945.* Cambridge: Cambridge University Press.

Weingart, P. (1989). German eugenics between science and politics. *Osiris, 5 (second series),* 260-282.

Weinreich, M. (1946). *Hitler's professors.* New York: Yiddish Scientific Institute — Yivo.

Weller, E. A., Weller, R. A., & Danielyan, A. K. (2004). Mood disorders in adolescents. In J. Wiener & M. Dulcan (Eds.), *Textbook of child and adolescent psychiatry* (3rd ed., pp. 437-481). Washington, DC: American Psychiatric Association Press.

Welner, Z., Welner, A., Stewart, M., Palkes, H., & Wish, E. (1977). A controlled study of siblings of hyperactive children. *Journal of Nervous and Mental Disease, 165,* 110-117.

Wender, P. H. (1995). *Attention-deficit hyperactivity disorder in adults.* New York: Oxford University Press.

Wender, P. H., Kety, S. S., Rosenthal, D., Schulsinger, F., Ortmann, J., & Lunde, I. (1986). Psychiatric disorders in the biological and adoptive families of adopted individuals with affective disorders. *Archives of General Psychiatry, 43,* 923-929.

Wender, P. H., & Klein, D. F. (1981). *Mind, mood, and medicine.* New York: Farrar, Straus, & Giroux.

Wender, P. H., Rosenthal, D., & Kety, S. S. (1968). A psychiatric assessment of the adoptive parents of schizophrenics. In D. Rosenthal & S. Kety (Eds.), *The transmission of schizophrenia* (pp. 235-250). New York: Pergamon Press.

Wender, P. H., Rosenthal, D., Kety, S. S., Schulsinger, F., & Welner, J. (1974). Crossfostering: A research strategy for clarifying the role of genetic and experiential factors in the etiology of schizophrenia. *Archives of General Psychiatry, 30,* 121-128.

Willcutt, E. G., Pennington, B. F., & DeFries, J. C. (2000). Etiology of inattention and hyperactivity/impulsivity in a community sample of twins with learning disabilities. *Journal of Abnormal Child Psychology, 28,* 149-159.

Wilens, T. E., Biederman, J., & Spencer, T. J. (2002). Attention-deficit/ hyperactivity disorder across the lifespan. *Annual Review of Medicine, 53,* 113-131.

Willerman, L. (1973). Activity level and hyperactivity in twins. *Child Development, 44,* 288-293.

Williams, N. M., Norton, N., et al. (2003). A systematic genomewide linkage study in 353 sib pairs with schizophrenia. *American Journal of Human Genetics, 73,* 1355-1367.

Wilson, P. T. (1934). A study of twins with special reference to heredity as a factor determining differences in environment. *Human Biology, 6,* 324-354.

Wright, W. (1998). *Born that way.* New York: Alfred A. Knopf.

Wynne, L. C., Singer, M. T., & Toohey, M. L. (1976). Communication of the adoptive parents of schizophrenics. In J. Jorstad & E. Ugelstad (Eds.), *Schizophrenia 75 — psychotherapy family studies, research* (pp. 413-451). Oslo: University of Oslo Press.

Yarrow, M. R., Campbell, J. D., & Burton, R. V. (1970). Recollections of childhood: A study of the retrospective method. *Monographs of the Society for Research in Child Development, 35,* (5, Serial No. 138).

Zametkin, A., Ernst, M., & Cohen, R. (2001). Single gene studies of ADHD. In F. Levy & D. Hay (Eds.), *Attention, genes, and ADHD* (pp. 157-172). East Sussex, UK: Brunner-Routledge.

Zammit, S., O'Donovan, M., & Owen, M. J. (2002). Neurogenetics of schizophrenia. In H. D'haenen, J. den Boer, & P. Willner (Eds.), *Biological psychiatry* (Vol. 1, pp. 663-671). New York: John Wiley and Sons.

Zerbin-Rüdin, E., & Kendler, K. S. (1996). Ernst Rüdin (1874-1952) and his genealogic-demographic department in Munich (1917-1986): An introduction to their family studies of schizophrenia. *American Journal of Medical Genetics (Neuropsychiatric Genetics), 67,* 332-337.

Zwaigenbaum, L., Szatmari, P., Jones, M. B., Bryson, S. E., MacLean, J. E., Mahoney, W. J., Bartolucci, G., & Tuff, L. (2002). Pregnancy and birth complications in autism and liability to the broader autism phenotype. *Journal of the American Academy of Child and Adolescent Psychiatry, 41,* 572-579.

Acknowledgments

Writing a book of this type requires a lot of time, energy, and support. First, I would like to thank my family for the support they provided over the past three years. I would also like to thank David Cohen of Florida International University, and Jonathan Leo and Laurence Simon, co-Editors of *Ethical Human Psychology and Psychiatry*, for reading the entire manuscript and providing helpful feedback. Jonathan Leo also contributed to portions of Chapter 6. Jonathan Beckwith provided critical feedback to an early version of Chapter 11, and Garland E. Allen helped improve the glossary. I also thank Mark Blaxill, Marcia Campos, and Martha Herbert for helping me improve my understanding of autism. The views expressed in this book are my own, and do not necessarily reflect the opinions of people who have assisted me. I take full responsibility for any errors found in the text. Finally, I thank Algora Publishing for making the publication of this book possible.

INDEX OF NAMES

INDEX OF SUBJECTS

Printed in the United States
By Bookmasters